FUNDAMENTALS OF MATHEMATICS

FUNDAMENTALS OF MATHEMATICS

MICHAEL SENTLOWITZ
ROCKLAND COMMUNITY COLLEGE

PATRICIA HIRSCHY
DELAWARE TECHNICAL COMMUNITY COLLEGE

HBJ

HARCOURT BRACE JOVANOVICH, PUBLISHERS
AND ITS SUBSIDIARY, ACADEMIC PRESS
San Diego New York Chicago Austin Washington, D.C.
London Sydney Tokyo Toronto

ISBN: 0-15-529466-0
Library of Congress Catalog Card Number: 90-82258
Printed in the United States of America

TO THE STUDENT

Fundamentals of Mathematics is designed to help you learn or relearn basic topics in mathematics. Whether you are registered for a lecture section or independent study, we suggest that you use all the learning features contained in this textbook. The major features are as follows:

1. **Preview Test:** Each section begins with a preview test. If you are working through the material as part of an independent study program, you can use these tests to identify your level of mastery of the topics in the section. If you are in a lecture course, the tests will provide an overview of the kinds of problems your instructor will address in class.
2. **Practice Problems:** When you are studying the material in the book, you will find that the examples illustrate the skills and methods to be mastered. To give you an immediate opportunity to try a similar problem, we have included a practice problem for each example. Answers to the practice problems are found before the exercise sets for a given section.
3. **Exercise Sets:** Each section contains two exercise sets, A and B, constructed in parallel format. Answers to the A exercises are at the end of the book. Your instructor has the answers for the B exercises. The following types of problems are included in the exercise sets:
   - *Skill Development:* These problems allow you to practice the skills and methods illustrated in the section.
   - *Applications:* These problems provide further practice of the skills and methods offered through word problems.
   - *For Some Extra Thought:* These problems allow you to go beyond the level of difficulty found in the skill development and applications problems. Working these problems will provide a better understanding of the math concepts that underlie the skills you have mastered.
   - *Mental Estimation:* These problems appear in some sections of Chapters 1, 3, and 4. Mental estimation is an important skill for many jobs.
   - *Review Problems for Maintaining Your Skills:* Review problems are found in every section after Section 1.1. Working the review problems section by section will help you retain your skills throughout the course.
4. **Practice Tests:** There are two practice tests at the end of each section. They will help you prepare for chapter quizzes and chapter examinations.
5. **Cumulative Review Exercises:** In addition to the review exercises in each section, there are cumulative reviews at the ends of Chapters 2 through 10.

Working through each of these elements—Practice Tests, Practice Problems, Exercise Sets, Practice Tests, and Cumulative Review Exercises—should help you master the basic concepts, reduce your math anxiety, and shorten the time you spend studying for examinations.

Michael Sentlowitz
Patricia Hirschy

CONTENTS

TO THE STUDENT v

CHAPTER 1 **WHOLE NUMBERS** 1

1.1 Place Value 2
1.2 Rounding 15
1.3 Addition 23
1.4 Subtraction 35
1.5 Multiplication 47
1.6 Division 61
1.7 Mixed Operations 75
Practice Test 1A 81
Practice Test 1B 83

CHAPTER 2 **FRACTIONS AND MIXED NUMBERS** 85

2.1 Equivalent Fractions 86
2.2 Mixed Numbers and Improper Fractions 95
2.3 Multiplying Fractions and Mixed Numbers 101
2.4 Dividing Fractions and Mixed Numbers 109
2.5 Adding Fractions and Mixed Numbers 117
2.6 Subtracting Fractions and Mixed Numbers 125
2.7 Mixed Operations 131
Practice Test 2A 137
Practice Test 2B 139
Cumulative Review Exercises For Chapters 1 and 2 141

CHAPTER 3 **DECIMALS** 143

3.1 Place Value 144
3.2 Rounding 151
3.3 Addition and Subtraction 157
3.4 Multiplication 169
3.5 Division 183
3.6 Mixed Operations 197
Practice Test 3A 203
Practice Test 3B 205
Cumulative Review Exercises For Chapters 1 Through 3 207

CHAPTER 4 **PERCENTS** 209

4.1 Percent/Decimal and Percent/Fraction Conversions 210
4.2 Finding the Amount 219
4.3 Finding the Rate 227
4.4 Finding the Base 235
4.5 Finding the Amount, Rate, or Base 241

4.6 Percent Increase and Decrease 249
Practice Test 4A 255
Practice Test 4B 257
Cumulative Review Exercises For Chapters 1 Through 4 259

CHAPTER 5 **RATIO AND PROPORTION 261**

5.1 Writing Ratios 262
5.2 Checking Proportions 269
5.3 Solving Proportions 275
Practice Test 5A 287
Practice Test 5B 289
Cumulative Review Exercises For Chapters 1 Through 5 291

CHAPTER 6 **SIGNED NUMBERS 293**

6.1 An Introduction to Signed Numbers 294
6.2 Adding Signed Numbers 301
6.3 Subtracting Signed Numbers 311
6.4 Multiplying Signed Numbers 317
6.5 Dividing Signed Numbers 325
6.6 Mixed Operations 331
Practice Test 6A 337
Practice Test 6B 339
Cumulative Review Exercises For Chapters 1 Through 6 341

CHAPTER 7 **ALGEBRAIC EXPRESSIONS AND EQUATIONS 343**

7.1 Algebraic Expressions 344
7.2 Equivalent Equations 355
7.3 Solving Equations by Addition 363
7.4 Solving Equations by Multiplication or Division 371
7.5 Solving Equations by Multiple Operations 379
Practice Test 7A 389
Practice Test 7B 391
Cumulative Review Exercises For Chapters 1 Through 7 393

CHAPTER 8 **POWERS AND ROOTS 395**

8.1 Finding Powers 396
8.2 Properties of Exponents 403
8.3 Scientific Notation 413
8.4 Finding Square Roots 421
Practice Test 8A 429
Practice Test 8B 431
Cumulative Review Exercises For Chapters 1 Through 8 433

CHAPTER 9 **MEASUREMENTS 435**

9.1 United States Customary and Metric Length Measurements 436
9.2 Converting Lengths Into Different Systems 447

9.3 Liquid Volume Measurements 455
9.4 Weight Measurements 465
Practice Test 9A 475
Practice Test 9B 477
Cumulative Review Exercises For Chapters 1 Through 9 479

CHAPTER 10 GEOMETRY 481

10.1 Some Common Geometric Forms and Perimeter 482
10.2 Area 501
10.3 Circles 517
10.4 Volume 527
Practice Test 10A 535
Practice Test 10B 537
Cumulative Review Exercises For Chapters 1 Through 10 539

USING A CALCULATOR 543

ANSWER KEY 547

INDEX 562

C H A P T E R 1

WHOLE NUMBERS

1.1 PLACE VALUE

1.2 ROUNDING

1.3 ADDITION

1.4 SUBTRACTION

1.5 MULTIPLICATION

1.6 DIVISION

1.7 MIXED OPERATIONS

The numbers 0, 1, 2, 3, and so on, are called **whole numbers**. There are many situations where a whole number is used.

- Babe Ruth's retired New York Yankee uniform number: 3
- A grade on a history test: 87
- The first year of the twenty-first century: 2001
- The price for a box of computer disks: $16

You may have to write a whole number in words when you are writing a check.

- $97 is written as "ninety-seven dollars"
- $4,568 is written as "four thousand five hundred sixty-eight dollars"

You may have to round a whole number to a place value position.

- Your automobile dealer's service department asks you for the approximate mileage on your car. Your odometer reading is 65,897, approximately 66,000 miles to the nearest thousand miles.
- You are applying for a loan at a bank and are asked for your approximate monthly salary. Your monthly salary of $2,841 is approximately $2,800 to the nearest hundred dollars.

You may have to add, subtract, multiply, or divide whole numbers to answer questions such as:

- What is your total SAT score if your math score is 540 and your verbal score is 586?
- How many miles did you drive if your car odometer changes from 5,267 to 5,384?
- At $28 for one share of AT&T stock, what is the cost for 50 shares?
- If 14 tickets to a concert cost $490, what is the price for one ticket?

1.1 PLACE VALUE

Preview Test

1. In the number 539,820,001,346
   the 5 is in the (a) ____________ place,
   the 2 is in the (b) ____________ place,
   the 1 is in the (c) ____________ place,
   the 6 is in the (d) ____________ place,
   the 9 is in the (e) ____________ place,
   the 8 is in the (f) ____________ place, and
   the 4 is in the (g) ____________ place.
2. Write 2,405 in expanded form.

   ____

3. The name for the number 56,092,000,107 in words is

____.

4. The number whose name is one hundred twenty-three billion, forty million, nine is written in digits as

______________________.

The system that you use for writing numbers originated in India over 2,000 years ago and is called the **Hindu-Arabic system of numeration**. Since this system is probably the only one you have used to solve arithmetic problems, it is easy to fail to recognize its superiority over other systems.

Consider the Roman system of numeration in which the numbers eighteen and nine are written as XVIII and IX, respectively. How can you multiply XVIII by IX on paper using Roman notation? The answer is, "With great difficulty." The Romans did not use pencil and paper (neither item was readily available in those days) but instead used a device called an *abacus.*

The pencil and paper methods you have learned for addition, subtraction, multiplication, and division evolved out of the Hindu-Arabic system, whose significant feature is **place value**. That is, all numbers are expressed using the set of ten digits (0, 1, 2, 3, 4, 5, 6, 7, 8, 9), and the value of each digit depends on its location.

Place Value Names

In Figure 1.1.1 the names of the place values from the units place on the right through the hundred billions place on the left are shown. It is important that you memorize the names and their respective positions in this chart.

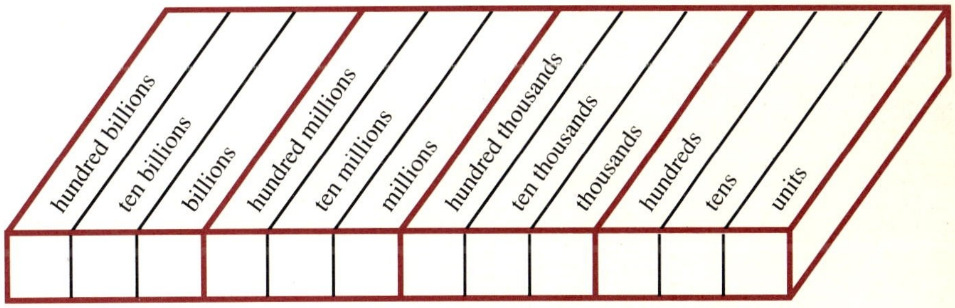

Figure 1.1.1

ANSWERS FOR PREVIEW TEST

1. **a)** hundred billions **b)** ten millions **c)** thousands **d)** units **e)** billions **f)** hundred millions **g)** tens
2. 2,000 + 400 + 5
3. fifty-six billion, ninety-two million, one hundred seven
4. 123,040,000,009

EXAMPLE 1: In the number 632,051,470,809 name the place value position for the indicated digit.

a) 3 **b)** 1 **c)** 9 **d)** 6 **e)** 8

Solution:

a) 3 is in the ten billions place.

b) 1 is in the millions place.

c) 9 is in the units place.

d) 6 is in the hundred billions place.

e) 8 is in the hundreds place.

PRACTICE PROBLEM 1: In the number 47,608,195,023 name the place value position for the indicated digit.

a) 5 **b)** 6 **c)** 2 **d)** 4 **e)** 9

Each place value name shown in Figure 1.1.1 is worth ten times the value of the place on its right. In Figure 1.1.2 the information from Figure 1.1.1 is repeated, along with the numerical values for each place.

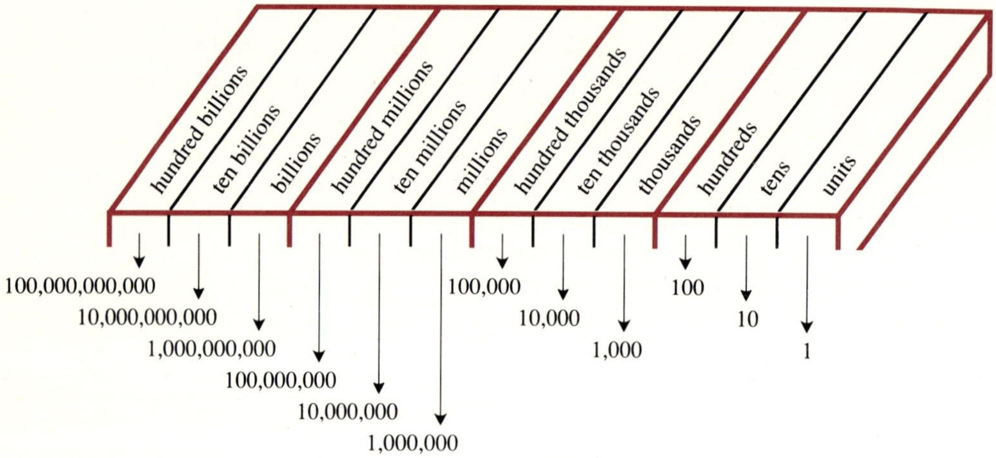

Figure 1.1.2

Expanded Form

Using the numerical place values you can see that

$$2{,}614 \text{ means } 2 \times 1{,}000 + 6 \times 100 + 1 \times 10 + 4 \times 1$$
$$\text{or} \quad 2{,}000 + 600 + 10 + 4$$

Writing 2,614 as 2,000 + 600 + 10 + 4 is called writing the number in **expanded form**.

EXAMPLE 2: Write each number in expanded form.

a) 647 **b)** 4,509 **c)** 12,800

Solution:

a) $647 = 6 \times 100 + 4 \times 10 + 7 \times 1$
$= 600 + 40 + 7$

b) $4{,}509 = 4 \times 1{,}000 + 5 \times 100 + 0 \times 10 + 9 \times 1$
$= 4{,}000 + 500 + 0 + 9$
$= 4{,}000 + 500 + 9$

Notice that the zero in the tens place means that there are no multiples of ten.

c) $12{,}800 = 1 \times 10{,}000 + 2 \times 1{,}000 + 8 \times 100 + 0 \times 10 + 0 \times 1$
$= 10{,}000 + 2{,}000 + 800$

PRACTICE PROBLEM 2: Write each number in expanded form.

a) 43 **b)** 56,804 **c)** 2,003

Names for Numbers

Without names for numbers, we could write them down but we could never talk about them because there would be no words to describe them. Suppose your friend tells you that last week she drove 362 miles to a small town with a population of 4,687. Without names for 362 and 4,687 the sentence would sound like this:

> Last week I drove three six two miles to a small town with a population of four six eight seven.

The word names for the numbers 0 through 19 are listed below.

0	zero	5	five	10	ten	15	fifteen
1	one	6	six	11	eleven	16	sixteen
2	two	7	seven	12	twelve	17	seventeen
3	three	8	eight	13	thirteen	18	eighteen
4	four	9	nine	14	fourteen	19	nineteen

For two-digit numbers larger than nineteen the word names of the tens place digits from 20 through 90 are used along with the names for 1 through 9. Notice the hyphen (-) between the tens place word name and the units place word name.

twenty	twenty-one	twenty-two	...	twenty-nine
thirty	thirty-one	thirty-two	...	thirty-nine
forty	forty-one	forty-two	...	forty-nine
fifty	fifty-one	fifty-two	...	fifty-nine
sixty	sixty-one	sixty-two	...	sixty-nine
seventy	seventy-one	seventy-two	...	seventy-nine
eighty	eighty-one	eighty-two	...	eighty-nine
ninety	ninety-one	ninety-two	...	ninety-nine

For numbers between 100 and 999 the word name for the number of hundreds is followed by the word name of the two-digit number in the tens and units position. Some examples are

128	one hundred twenty-eight	374	three hundred seventy-four
603	six hundred three	817	eight hundred seventeen

Be Careful About This

Notice that the word names for three-digit numbers such as the four shown here do not include the word "and." That is, the name for 349 is "three hundred forty-nine," not "three hundred and forty-nine." The word "and" will be used in Chapter 3 for naming decimal numbers.

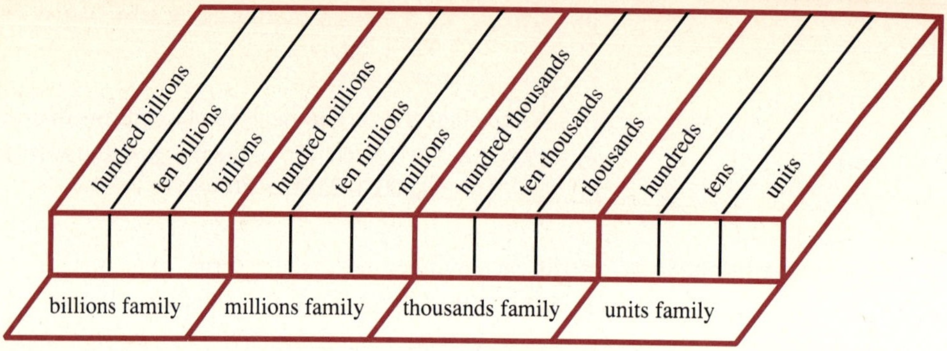

Figure 1.1.3

For numbers larger than 999 it is necessary to know the names for each group of three digits. Notice that in Figure 1.1.3 the name of each group of three digits is taken from the place value name of the rightmost digit in the group.

When a number larger than 999 is written, commas are used to separate the groups. That is, starting your count from the right, place a comma after every three digits. So for 37209145 the commas would be placed as shown here.

37,209,145

To write 37,209,145 in words, the names for 37 and 209 are followed by their group names. However, the name "units" does not follow the name for 145.

thirty-seven **million**, two hundred nine **thousand**, one hundred forty-five

EXAMPLE 3: Write each number in words.

a) 52,578 **b)** 406,712,084 **c)** 708,000,003,286

Solution:

a) fifty-two thousand, five hundred seventy-eight

b) four hundred six million, seven hundred twelve thousand, eighty-four

c) seven hundred eight billion, three thousand, two hundred eighty-six

PRACTICE PROBLEM 3: Write each number in words.

a) 214,917 **b)** 6,402 **c)** 2,504,718,000

EXAMPLE 4: Write each number using digits.

a) forty-two thousand, six hundred thirty-eight

b) five hundred million, twenty-six thousand, eight hundred sixty-five

c) seven billion, six thousand.

Solution:

a) 42,638 **b)** 500,026,865 **c)** 7,000,006,000

PRACTICE PROBLEM 4: Write each number using digits.

a) five hundred six million, seven hundred

b) seventy-eight thousand

c) two billion, seven hundred eighty million, nine hundred twenty-five

Be Careful About This

It is very easy to lose zeros when writing the digits for a number whose word name is given. Make sure that you account for all groups in order—billions, millions, thousands, units—from the largest group named, and that you write three digits for each group, except perhaps for the leftmost group.

ANSWERS FOR PRACTICE PROBLEMS

1. **a)** thousands **b)** hundred millions **c)** tens **d)** ten billions
e) ten thousands

2. **a)** 40 + 3 **b)** 50,000 + 6,000 + 800 + 4 **c)** 2,000 + 3

3. **a)** two hundred fourteen thousand, nine hundred seventeen
b) six thousand, four hundred two
c) two billion, five hundred four million, seven hundred eighteen thousand

4. **a)** 506,000,700 **b)** 78,000 **c)** 2,780,000,925

EXERCISE SET 1.1A

Skill Development

1. In the number 483,762
   a) 4 is in the ____________________ place.
   b) 7 is in the ____________________ place.
   c) 6 is in the ____________________ place.
2. In the number 73,401,862,955
   a) 1 is in the ____________________ place.
   b) 7 is in the ____________________ place.
   c) 0 is in the ____________________ place.
3. In the number 485,762,493,122
   a) 8 is in the ____________________ place.
   b) 1 is in the ____________________ place.
   c) 9 is in the ____________________ place.

In 4 through 13, write each number in words.

4. 67,039

5. 849,000

6. 1,328

7. 31,600,521

8. 26,000,000,426

9. 897,276,400,000

10. 9,000,000

11. 43,651

12. 463,000,600,005

13. 95,650,000,000

In 14 through 23, write each number using digits.

14. two thousand, eighty-three

15. five million, seven hundred forty-three thousand, nine

16. eighteen thousand, five hundred

17. seventy-three million, six hundred thousand, four hundred

18. twenty-three billion, fifty thousand, three hundred twenty-eight

19. six hundred ninety-two thousand, one hundred seventeen

20. thirty-seven billion

21. nine million, eight

22. four hundred billion, six million, twenty-three thousand, eight hundred sixty-two

23. eight hundred forty-two million, seventy thousand

Applications

24. In 1980 the population of the United States was about two hundred twenty-seven million. Write this population total using digits.

25. The size of Texas is 262,017 square miles. Write the size of Texas in words.

26. In 1984 the number of Christians in South America was 196,599,780. In this number the 5 represents five ______________________

For Some Extra Thought

27. To the left of the billions group is the trillions group. Write 409,736,000,000,000 in words.

28. Use digits to write seventy-two trillion, six billion, five hundred million.

29. Which number is larger, seven hundred million or two billion?

EXERCISE SET 1.1B

Skill Development

1. In the number 56,970

a) 6 is in the ____________ place.

b) 0 is in the ____________ place.

c) 9 is in the ____________ place.

2. In the number 426,712,098

a) 4 is in the ____________ place.

b) 1 is in the ____________ place.

c) 0 is in the ____________ place.

3. In the number 735,000,002,000

a) 3 is in the ____________ place.

b) 2 is in the ____________ place.

c) 7 is in the ____________ place.

In 4 through 13, write each number in words.

4. 6,800

5. 942,760

6. 12,500

7. 9,000,000,000

8. 27,006,800

9. 62,400,000,000

10. 5,600,003,045

11. 125,971

12. 10,000,000

13. 305,008,050

In 14 through 23, write each number using digits.

14. fifty-six thousand, seven hundred eighty

15. ninety million

16. twenty-three thousand, eight hundred two

17. six billion, four hundred thousand

18. one hundred thirty-two thousand, five hundred

19. eighty-nine million, five hundred seventy-two thousand, six hundred

20. sixty-two billion, two million, six

21. fourteen thousand, three hundred twenty-one

22. seven million, thirty thousand, two hundred seventy-one

23. six hundred thousand, five hundred eight

Applications

24. In 1986 the magazine with the largest circulation was TV Guide, with 16,800,441 copies. Write the circulation in words.

25. In 1981, the United States exported 43,909,000 metric tons of wheat. In this number the four represents four ____________________

26. The public library of Tulsa, Oklahoma has approximately five hundred twenty-five thousand bound books. Write the number of bound books using digits.

For Some Extra Thought

27. If the largest single digit number were 4, then you would group a number into units, fives, twenty-fives, and so on. For example, if eight were grouped into fives and units it would be written as 13 (one five and three units). If you use the fives and units grouping, how would you write the digits for

a) nine **b)** thirteen **c)** fifteen?

28. Refer to question 27. If you use a fours and units grouping, how would you write the digits for

a) six **b)** ten **c)** fifteen?

29. Which number is larger, sixty billion or eight hundred forty-five million?

1.2 ROUNDING

Preview Test

Use the number below to answer questions 1 through 4.

35,196,045

1. This number rounded to the nearest ten is

________________________.

2. This number rounded to the nearest million is

________________________.

3. This number rounded to the nearest ten thousand is

________________________.

4. This number rounded to the nearest hundred is

________________________.

There are times when you do not need an exact answer but can just use an estimate. For example, when you are grocery shopping you may estimate your bill as you are placing items in the cart. Rounding a number is a way of estimating.

ROUNDING USING THE METHOD "WHICH NUMBER IS CLOSER?"

Consider the numbers shown in Figure 1.2.1.

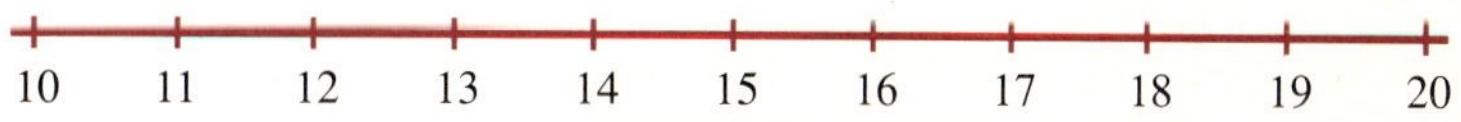

Figure 1.2.1

To round 14 to the nearest tens place you have to find out if 14 is closer to 10 or 20. Since 14 is 4 units from 10 and 6 units from 20, it is closer to 10. Therefore, 14 rounded to the nearest tens place is 10.

To round 17 to the nearest tens place notice that 17 is 7 units from 10 and 3 units from 20. Since 17 is closer to 20, 17 rounded to the nearest tens place is 20.

Now what about 15? It's 5 units from 10 and 5 units from 20. We will follow the practice of rounding **up** numbers that are exactly halfway between two endpoints (in this case 10 and 20). So 15 rounded to the nearest ten is 20.

This same "Which number is closer?" idea works when rounding larger numbers. Suppose you want to round 328 to the nearest hundred. First notice that 328 is between

ANSWERS FOR PREVIEW TEST

1. 35,196,050 **2.** 35,000,000 **3.** 35,200,000 **4.** 35,196,000

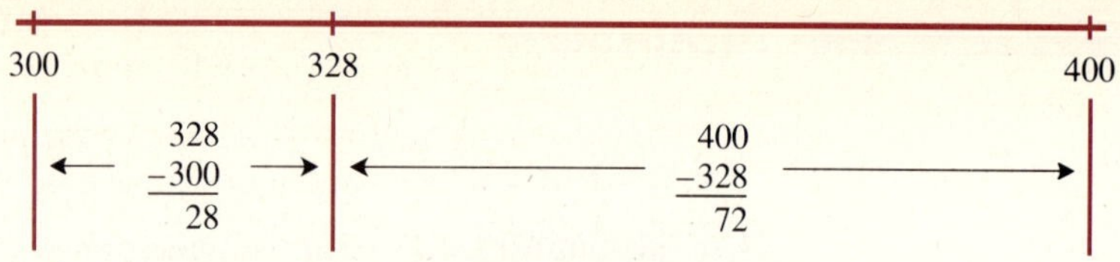

Figure 1.2.2

the two "hundreds" numbers 300 and 400. Which number is it closer to? Refer to Figure 1.2.2 and notice that 328 is 28 units from 300 and 72 units from 400. Since it is closer to 300, 328 rounded to the nearest hundred is 300.

EXAMPLE 1: Use the "Which number is closer?" method to round each number.

a) 1,475 to the nearest hundred.

b) 1,475 to the nearest thousand.

c) 1,475 to the nearest ten.

Solution:

a) 1,475 is between the two hundreds numbers 1,400 and 1,500. Since 1,475 is closer to 1,500 than to 1,400, 1,475 rounded to the nearest hundred is 1,500.

b) Although this is the same number as in part (a), you are rounding to the thousands place. Therefore, you have to determine if 1,475 is closer to 1,000 or 2,000. Since 1,475 is 475 units from 1,000 and 525 units from 2,000, it is closer to 1,000. So, 1,475 rounded to the nearest thousand is 1,000.

c) 1,475 is halfway between the two tens numbers 1,470 and 1,480. Since numbers halfway between two endpoints are rounded up, 1,475 rounded to the nearest ten is 1,480.

PRACTICE PROBLEM 1: Use the "Which number is closer?" method to round each number.

a) 48,350 to the nearest thousand.

b) 48,350 to the nearest ten thousand.

c) 48,350 to the nearest hundred.

ROUNDING WHOLE NUMBERS BY THE "DIGIT TO THE RIGHT" METHOD

Although the "Which number is closer?" method can be used to round a whole number to any place value position, there are other methods that may be faster. One method is outlined here.

> To round a whole number to a place value position
>
> **Step 1:** Look at the digit to the right of the place value rounding position.
>
> **Step 2:** If the digit in Step 1 is 0, 1, 2, 3, or 4, then the digit in the rounding position is not changed. If the digit in Step 1 is 5, 6, 7, 8, or 9, then the digit in the rounding position is increased by 1.
>
> **Step 3:** Change all digits to the right of the place value rounding position to zero.

Let's use this rule to round 4,823 and 4,863 to the nearest hundred.

Step 1: For 4,823 the digit that lies to the right of the hundreds position is 2.

Step 2: Do not change the 8 in the hundreds position.

Step 3: Change the 2 and 3 to 0.
Therefore, 4,823 rounded to the nearest hundred equals 4,800.

Step 1: For 4,863 the digit that lies to the right of the hundreds position is 6.
Step 2: Add 1 to the 8 in the hundreds position.
Step 3: Change 2 and 3 to 0.
Therefore, 4,863 rounded to the nearest hundred equals 4,900.

Notice that the three-step method gives you the same result as the "Which number is closer?" method. Both 4,823 and 4,863 lie between the two hundreds numbers, 4,800 and 4,900. As you can see in Figure 1.2.3, 4,823 is closer to 4,800 and 4,863 is closer to 4,900.

Figure 1.2.3

EXAMPLE 2: Round each number using the digit to the right method.

a) 76,395 to the nearest thousand.

b) 405,734,123 to the nearest ten million.

c) 7,396 to the nearest ten.

Solution:

a) **Step 1:** The digit to the right of the thousands position is 3.
Step 2: Do not change the 6 in the thousands position.
Step 3: Change the 3, 9, and 5 to 0.
Therefore, 76,395 rounded to the nearest thousand equals 76,000.

b) **Step 1:** The digit to the right of the ten millions position is 5.
Step 2: Add 1 to the 0 in the ten millions position.
Step 3: Change 5, 7, 3, 4, 1, 2, and 3 to 0.
Therefore, 405,734,123 rounded to the nearest ten million equals 410,000,000.

c) **Step 1:** The digit to the right of the tens position is 6.
Step 2: Add 1 to the 9 in the tens position.
Step 3: Change the 6 to 0.
Therefore, 7,396 rounded to the nearest ten equals 7,400.

PRACTICE PROBLEM 2: Round each number using the digit to the right method.

a) 530,245,721 to the nearest ten thousand.

b) 475,020,749 to the nearest million.

c) 259,900 to the nearest thousand.

Be Careful About This

Since you may picture rounding as what happens to the digit in the rounding position after completing Step 2, it is very easy to forget Step 3. Make sure that after completing Step 2 you change all digits to the right of the place value rounding position to zero.

ANSWERS FOR PRACTICE PROBLEMS

1. **a)** 48,000 **b)** 50,000 **c)** 48,400

2. **a)** 530,250,000 **b)** 475,000,000 **c)** 260,000

EXERCISE SET 1.2A

Skill Development

In 1 through 4, use the "Which number is closer?" method to round each number.

1. 5,670 to the nearest hundred.

2. 749,862 to the nearest ten thousand.

3. 498,694 to the nearest ten.

4. 12,634,850 to the nearest million.

In 5 through 12, use the digit to the right method to round each number.

5. 23,987,300 to the nearest hundred thousand.

6. 56,450 to the nearest hundred.

7. 5,723,569 to the nearest ten thousand.

8. 44,700 to the nearest thousand

9. 1,230,560 to the nearest hundred thousand.

10. 56 to the nearest ten.

11. 6,780,478 to the nearest million.

12. 23,879,452,000 to the nearest billion.

Applications

13. The circulation for the Sunday edition of a large city newspaper is 1,593,107. Round the circulation to the nearest hundred thousand.

14. The length of the Grand Coulee Dam on the Columbia River is 5,673 feet. Round this length to the nearest hundred feet.

15. The area of Wilmington, Delaware is 396 square miles. Round this area to the nearest ten square miles.

For Some Extra Thought

16. Is 4,444,444,444,444 closer to 4,000,000,000,000 or 5,000,000,000,000?

17. Can a rounded result ever be the same as the unrounded result? If yes, give an example.

Review Problems for Maintaining Your Skills

18. In the number 8,907,154

a) 7 is in the ______________________ place.

b) 9 is in the ______________________ place.

c) 5 is in the ______________________ place.

19. Write each number in words.

a) 4,587

b) 62,008,400

c) 3,845,000,200

20. Write each number using digits.

a) fifty-two thousand, seven hundred nineteen.

b) nine million, sixty thousand, five hundred eight.

c) two hundred thirty billion, one hundred thousand.

EXERCISE SET 1.2B

Skill Development

In 1 through 4, use the "Which number is closer?" method to round each number.

1. 52,890 to the nearest thousand.
2. 2,678,512 to the nearest hundred thousand.
3. 87 to the nearest ten.
4. 23,500 to the nearest thousand.

In 5 through 12, use the digit to the right method to round each number.

5. 5,643,972,500 to the nearest ten million.
6. 623 to the nearest hundred.
7. 879,600 to the nearest thousand.
8. 4,600,000 to the nearest hundred thousand.
9. 52,210,590 to the nearest million.
10. 6,948,200 to the nearest ten thousand.
11. 41,856 to the nearest ten.
12. 649,700,000,000 to the nearest ten billion.

Applications

13. In 1981 the number of Canadian citizens whose native language was French was 6,249,095. Round this total to the nearest million.

14. In 1982 there were 1,193,000 cars stolen in the United States. Round this statistic to the nearest hundred thousand.

15. The height of the Sears Tower in Chicago is 1,454 feet. Round this height to the nearest hundred feet.

For Some Extra Thought

16. Is 999,999,999,999 closer to 999,999,999,990 or 1,000,000,000,000?

17. If 1 is not added to the digit in the place value rounding position does this mean that the rounded answer is the same as the original number?

Review Problems for Maintaining Your Skills

18. In the number 45,307,921
    - a) 4 is in the ______________________ place.
    - b) 7 is in the ______________________ place.
    - c) 3 is in the ______________________ place.

19. Write each number in words.

a) 12,600

b) 1,387,632

c) 13,000,000,000

20. Write each number using the digits 0 through 9.

a) four hundred seventy-two thousand, six hundred

b) seventy-two billion, nine thousand

c) eight hundred forty-two million, seventy-three

1.3 ADDITION

Preview Test

In 1 through 3, find each sum.

1.
$$\begin{array}{r}7\\6\\9\\\underline{5}\end{array}$$

2.
$$\begin{array}{r}34\\67\\136\\\underline{198}\end{array}$$

3.
$$\begin{array}{r}846\\1{,}756\\651\\\underline{2{,}581}\end{array}$$

In 4 through 7, add each set of numbers.

4. 8, 4, 8, 5, 3, 1, and 6 **5.** 4, 3, 1, 9, 7, 7, and 8

6. 49, 36, 78, and 64 **7.** 949, 87, 1,268, and 379

Adding Three or More One-Digit Numbers

Once you have memorized the basic addition facts from $0 + 0 = 0$ through $9 + 9 = 18$, you are ready to master the skill of adding any set of whole numbers. Let's first consider how you add three one-digit whole numbers, such as $3 + 5 + 4$. You add two numbers at a time and carry the sum to the next number. This can be done by adding from top to bottom or from bottom to top.

Adding from Top to Bottom

$$\begin{array}{l}\left.\begin{array}{r}3\\5\end{array}\right\}\ 3+5=8\\ \left.\begin{array}{r}8\\4\end{array}\right\}\ 8+4=12\\ \underline{4}\end{array}$$

Adding from Bottom to Top

$$\begin{array}{l}3\\ \left.\begin{array}{r}5\\\underline{4}\end{array}\right\}\ 4+5=9\\ \left.\begin{array}{r}3\\9\end{array}\right\}\ 9+3=12\end{array}$$

ANSWERS FOR PREVIEW TEST

1. 27 **2.** 435 **3.** 5,834 **4.** 35 **5.** 39 **6.** 227
7. 2,683

EXAMPLE 1: Add the numbers 4, 7, 8, 9, and 8 in a column from top to bottom with 4 at the top.

Solution:

4
7 } 4 + 7 = 11
8 } 11 + 8 = 19
9 } 19 + 9 = 28
8 } 28 + 8 = 36

So, 4 + 7 + 8 + 9 + 8 = 36

PRACTICE PROBLEM 1: Add the numbers 6, 4, 1, 9, 5, 3, and 7 in a vertical column from bottom to top with 7 on the bottom.

Adding with Carrying

How do we add sets of whole numbers with more than one digit? Consider 43 + 25. We can carry out this addition by lining up the numbers in place value positions and adding the digits in each column.

43 ⟶ 4 tens + 3 units
25 ⟶ 2 tens + 5 units
6 tens + 8 units = 68

Of course we can abbreviate the above without place value names and write the addition as

$$\begin{array}{r} 43 \\ \underline{25} \\ 68 \end{array}$$

Let's now try 29 + 54 with place value names.

29 ⟶ 2 tens + 9 units
54 ⟶ 5 tens + 4 units
7 tens + 13 units

We have a two-digit total in the units column. Some shifting is required, because the largest number we can have in any place value position is 9. Of the 13 units in the units column, 10 have to be shifted to the tens column as 1 ten.

7 tens + 13 units
= 7 tens + (10 units + 3 units)
= (7 tens + 1 ten) + 3 units
= 8 tens + 3 units
= 83

This process of shifting the 10 units to the tens column as 1 ten can be carried out by simply carrying the number 1 to the tens column as shown here.

$$\begin{array}{rr} ① & \\ 2 & 9 \\ 5 & 4 \\ \hline 8 & 3 \end{array}$$

← 9 + 4 = 13. Write the 3 in the units column and carry 1 (which is circled) to the tens column.

The "carry" procedure applies to shifts from the tens column to the hundreds column, and in general from any column to the column on its immediate left.

EXAMPLE 2: Add the numbers 489 and 376.

Solution:

$$\begin{array}{rrr} ① & ① & \\ 4 & 8 & 9 \\ 3 & 7 & 6 \\ \hline 8 & 6 & 5 \end{array}$$

9 + 6 = 15. Write 5 in the units column and carry 1 to the tens column.

1 + 8 + 7 = 16. Write 6 in the tens column and carry 1 to the hundreds column.

1 + 4 + 3 = 8

PRACTICE PROBLEM 2: Add the numbers 739 and 174.

Adding more than two whole numbers, each with more than one digit, is done the same way as in Example 2. That is, line up the place values and carry digits to the left as necessary.

EXAMPLE 3: Add the numbers 376, 894, 38, and 500.

Solution:

$$\begin{array}{rrrr} & ② & ① & \\ & 3 & 7 & 6 \\ & 8 & 9 & 4 \\ & & 3 & 8 \\ & 5 & 0 & 0 \\ \hline 1 & 8 & 0 & 8 \end{array}$$

6 + 4 + 8 = 18. Write the 8 in the units column and carry 1 to the tens column.

1 + 7 + 9 + 3 + 0 = 20. Write the 0 in the tens column and carry 2 to the hundreds column.

2 + 3 + 8 + 5 = 18. Since there are no place value columns to the left of the hundreds column, we do not have to carry 1 from 18 to the top of the next column.

PRACTICE PROBLEM 3: Add the numbers 89, 436, and 987.

Mental Estimation of Sums

Sometime you may find yourself in a situation where you have to estimate an answer mentally. Consider the following situations:

1. You are a new department manager attending your first monthly meeting at corporate headquarters. In the meeting, numbers relating to current and projected sales and expenses are being discussed. You notice that all of the managers are responding to questions with numerical results which they have approximated mentally.
2. In your job, most of the arithmetic calculations are done on a computer. You are responsible for judging whether or not the computer-generated results are reasonably accurate.

To estimate a sum of whole numbers it is helpful to round each number to a place value position to the left of the units position.

Suppose you purchase four textbooks in the college bookstore. The prices are \$22, \$17, \$38, and \$29. You can estimate the total cost before going to the cash register by first rounding each number to the nearest ten dollars.

Actual Cost	Rounded to the Nearest Ten Dollars
\$22	\$ 20
17	20
38	40
29	30
	Estimated Sum = \$110

The \$110 estimate is reasonably close to the exact sum, which is \$106.

EXAMPLE 4: Mentally estimate the sum for 1,252, 690, and 2,335 by rounding each number to the nearest thousand.

Solution:

Original Number	Rounded to the Nearest Thousand
1,252	1,000
690	1,000
2,335	2,000
	Estimated Sum = 4,000

PRACTICE PROBLEM 4: Mentally estimate the sum for \$785, \$942, \$521, and \$168 by rounding each amount to the nearest hundred dollars.

ANSWERS FOR PRACTICE PROBLEMS

1. 35 **2.** 913 **3.** 1,512 **4.** \$2,400

EXERCISE SET 1.3A

Skill Development

In 1 through 12, find each sum.

1.
```
 6
 7
 4
 3
 8
——
```

2.
```
 8
 3
 1
 9
 4
——
```

3.
```
 6
 5
 3
 9
 8
——
```

4.
```
 23
 86
 50
 41
 57
———
```

5.
```
 56
  8
 89
 44
———
```

6.
```
 256
  72
 325
 892
————
```

7.
```
 923
 844
 264
  51
————
```

8.
```
 1,260
 2,651
 7,630
   835
——————
```

9.
```
 6,500
 7,000
 8,423
   530
 3,410
——————
```

10.
```
 13,560
 43,103
  1,230
 57,971
  7,000
———————
```

11.
```
 54,900
  9,700
 62,743
 12,210
  6,744
———————
```

12.
```
 126,872
  23,560
 560,000
 885,220
 788,200
————————
```

13. Add the numbers 2, 8, 9, 6, 5, 9, 0, and 6.

14. Add the numbers 6, 9, 8, 5, 4, 6, 2, 1, 9, and 7.

15. Add the numbers 23, 56, 45, 17, 8, 26, and 52.

16. Add the numbers 65, 41, 20, 78, 63, and 48.

17. Add the numbers 291, 156, 87, 414, 23, 505, and 419.

18. Add the numbers 126, 450, 671, 890, 153, and 226.

19. Add the numbers 12,600, 2,800, 1,600, 45,981, and 23,600.

20. Add the numbers 300,000, 450,636, 23,544, 454,870, and 65,776.

Applications

21. In the 1984 presidential election the numbers of votes for Ronald Reagan in the counties of Kent, New Castle, and Sussex, Delaware were respectively, 21,531, 102,322, and 28,337. Find the total vote for Reagan.

22. The numbers of sexual harassment charges filed with the Equal Opportunities Commission from 1981 through 1984 were, respectively, 4,272, 5,100, 5,566, and 6,442. Find the total number of charges filed during the four-year period.

23. Henry Ropper deposited checks for $29, $184, $261, $58, and $80 in his checking account. Find the total amount deposited.

24. On a test with 5 20-point questions, Kathy Collins scored 19, 16, 17, 20, and 18 points. Find her grade on the test.

25. During the 7 days after winning $2,000,000 in the state lottery, Mike Grogan received the following number of congratulatory phone calls: 54, 71, 39, 41, 19, 17, and 8. Find the number of calls for the week.

For Some Extra Thought

26. When adding a long column of numbers, the total of the digits in the units column is 126. What two-digit number would you carry to the tens column?

27. If 2 is carried from the hundreds column to the thousands column, how many hundreds does the 2 represent?

28. What is the most number of two-digit whole numbers you can add if the sum is under 100?

Mental Estimation

In 29 through 32, estimate each sum mentally by rounding each number to the nearest hundred.

29. 230 + 765 + 327 =

30. 977 + 845 + 210 + 360 =

31. 1,361 + 2,549 =

32. 2,978 + 5,400 + 1,231 =

In 33 through 36, estimate each sum mentally by rounding each amount to the nearest ten dollars.

33. $265 + $741 + $833 + $126 =

34. $75 + $84 + $23 + $9 + $17 =

35. $7 + $13 + $39 + $2 + $18 =

36. $415 + $688 + $741 =

Review Problems for Maintaining Your Skills

37. In the number 348,720 the 4 is in the ________________________ place.

38. Write 7,489,603 in words.

39. Write nine billion, six hundred seventy thousand using digits.

In 40 through 42, round each number.

40. 46,863 to the nearest thousand.

41. 1,387,944 to the nearest hundred thousand.

42. 684 to the nearest ten.

24. Robert Thelen scored 513 on the math part of the SAT exam and 467 on the verbal part of the exam. Find his combined math and verbal score.

25. Between July 18 and July 23, Ruth Simmons received the following number of birthday cards each day: 8, 12, 17, 9, 10, and 7. Find the number of cards she received during the six days.

For Some Extra Thought

26. If you add eight whole numbers, what is the largest possible number that you would have to carry to the tens column?

27. If 3 is carried from the thousands column to the ten thousands column, how many thousands does the 3 represent?

28. Five two-digit numbers are added. The units column is totaled and 4 is carried to the tens column. What conclusion can we make about the units digit of the fifth number if the units digits of the first four numbers are 7, 8, 8, and 8?

Mental Estimation

In 29 through 32, estimate each sum mentally by rounding each number to the nearest hundred.

29. 123 + 650 + 45 =

30. 875 + 452 + 792 + 231 =

31. 2,368 + 4,533 =

32. 6,470 + 4,312 + 3,756 =

In 33 through 36, estimate each sum mentally by rounding each amount to the nearest ten dollars.

33. \$561 + \$422 + \$684 =

34. \$8 + \$15 + \$27 + \$13 + \$3 =

35. \$234 + \$125 + \$47 + \$72 =

36. \$145 + \$368 + \$227 + \$192 =

Review Problems for Maintaining Your Skills

37. In the number 23,874,000 2 is in the ______________________ place.

38. Write 56,871,307,000 in words.

39. Write thirty-two million, eight hundred using digits.

In 40 through 42, round each number.

40. 1,357 to the nearest hundred.

41. 36,741,988 to the nearest ten thousand.

42. 24,378,400 to the nearest million.

1.4 SUBTRACTION

Preview Test

In 1 through 3, find each difference.

1. $\begin{array}{r} 89 \\ -23 \\ \hline \end{array}$ **2.** $\begin{array}{r} 714 \\ -81 \\ \hline \end{array}$ **3.** $\begin{array}{r} 832 \\ -679 \\ \hline \end{array}$

4. Subtract 38 from 97 **5.** Subtract 405 from 722

6. Subtract 1,490 from 2,000 **7.** Subtract 324 from 1,132

Subtraction Without Borrowing

Knowing your addition facts helps you in other areas of arithmetic. Consider, for example, subtraction. Subtraction is the inverse of addition. For example, to subtract 9 from 17 answer the question, "9 plus what number equals 17?" The result is 8. That is

$$17 - 9 = 8, \text{ because } 9 + 8 = 17$$

Similarly,

$$5 - 3 = 2, \text{ because } 3 + 2 = 5$$
$$10 - 3 = 7, \text{ because } 3 + 7 = 10$$
and $16 - 8 = 8$, because $8 + 8 = 16$

Subtracting one whole number from another can be done using the same place value alignment we used for addition in Section 1.3. Consider $87 - 31$.

$$\begin{array}{rcl} 87 & \longrightarrow & 8 \text{ tens} + 7 \text{ units} \\ -31 & \longrightarrow & 3 \text{ tens} + 1 \text{ unit} \\ \hline & & 5 \text{ tens} + 6 \text{ units} = 56 \end{array}$$

As in addition of whole numbers, we can carry out the subtraction without place value names.

$$\begin{array}{r} 87 \\ -31 \\ \hline 56 \end{array}$$

ANSWERS FOR PREVIEW TEST

1. 66 **2.** 633 **3.** 153 **4.** 59 **5.** 317 **6.** 510
7. 808

When we are subtracting one whole number from another we have names for each whole number and the result. If the numbers are aligned vertically, the top number is called the **minuend**, the number being subtracted is called the **subtrahend**, and the result is called the **difference**.

$$\begin{array}{r l}
87 & \longleftarrow \text{Minuend} \\
-31 & \longleftarrow \text{Subtrahend} \\
\hline
56 & \longleftarrow \text{Difference}
\end{array}$$

We can check the subtraction by adding the difference (56) to the subtrahend (31). Since the sum is the minuend (87), the difference is correct.

EXAMPLE 1: Subtract 258 from 879

Solution:

$$\begin{array}{r}
879 \\
-258 \\
\hline
621
\end{array}$$

PRACTICE PROBLEM 1: Subtract 473 from 885

Subtraction with Borrowing

Consider the subtraction 62 − 25 with place value names.

$$\begin{array}{r l}
62 & \longrightarrow 6 \text{ tens} + 2 \text{ units} \\
-25 & \longrightarrow 2 \text{ tens} + 5 \text{ units} \\
\hline
\end{array}$$

We have a problem in the units column because we cannot subtract 5 units from 2 units. To deal with this we can shift 1 ten from the 6 tens into the units column as 10 units. That is, we can express 62 as 5 tens and 12 units. We can now proceed with the subtraction.

$$\begin{array}{l l}
6 \text{ tens} + 2 \text{ units} \longrightarrow & 5 \text{ tens} + 12 \text{ units} \\
 & 2 \text{ tens} + 5 \text{ units} \\
\hline
 & 3 \text{ tens} + 7 \text{ units} = 37
\end{array}$$

How do we show what we have done without place value names? In the illustration that follows notice that the 6 and 2 are crossed out and replaced with 5 and 12. This shows the shift of 1 ten from 6 tens to the 10 units that are added to the 2 units. We refer to this procedure as the **borrow method**, since we are "borrowing" from the tens position and shifting what is borrowed to the units position.

$$\begin{array}{r r}
5 & 12 \\
\not{6} & \not{2} \\
-2 & 5 \\
\hline
3 & 7
\end{array}$$

EXAMPLE 2: Subtract 29 from 74.

Solution:
Since we cannot subtract 9 units from 4 units, we borrow 1 ten from 7 tens and add 10 units to 4 units.

$$\begin{array}{rr} 6 & 14 \\ \not 7 & \not 4 \\ -2 & 9 \\ \hline 4 & 5 \end{array}$$

PRACTICE PROBLEM 2: Subtract 35 from 91.

EXAMPLE 3: Subtract 245 from 812.

Solution:
This subtraction requires two steps.

Step 1: Borrow 1 ten from 1 ten and add 10 units to 2 units.
Step 2: Borrow 1 hundred from 8 hundreds and add 10 tens to 0 tens.

Step 1

$$\begin{array}{rrr} & 0 & 12 \\ 8 & \not 1 & \not 2 \\ -2 & 4 & 5 \\ \hline \end{array}$$

Step 2

$$\begin{array}{rrr} & 10 & \\ 7 & \not 0 & 12 \\ \not 8 & \not 1 & \not 2 \\ -2 & 4 & 5 \\ \hline 5 & 6 & 7 \end{array}$$

PRACTICE PROBLEM 3: Subtract 764 from 913.

EXAMPLE 4: Subtract 73 from 4,000

Solution:
This subtraction requires three steps.

Step 1: Since we cannot borrow from zero, we must borrow 1 thousand from 4 thousands and add 10 hundreds to 0 hundreds.
Step 2: Borrow 1 hundred from 10 hundreds and add 10 tens to 0 tens.
Step 3: Borrow 1 ten from 10 tens and add 10 units to 0 units.

Step 1

$$\begin{array}{rrrr} 3 & 10 & & \\ \not 4 & \not 0 & 0 & 0 \\ - & & 7 & 3 \\ \hline \end{array}$$

Step 2

$$\begin{array}{rrrr} & 9 & & \\ 3 & \not{10} & 10 & \\ \not 4 & \not 0 & \not 0 & 0 \\ - & & 7 & 3 \\ \hline \end{array}$$

Step 3

$$\begin{array}{rrrr} & 9 & 9 & \\ 3 & \not{10} & \not{10} & 10 \\ \not 4 & \not 0 & \not 0 & \not 0 \\ - & & 7 & 3 \\ \hline 3 & 9 & 2 & 7 \end{array}$$

PRACTICE PROBLEM 4: Subtract 23 from 5,000

Mental Estimation of Differences

To mentally estimate the difference between two whole numbers, such as 720 and 190, it is helpful to avoid the borrow method. The following approach is suggested. Think in terms of what you must add to 190 to reach 720.

First move from 190 to the next multiple of 100.

$$190 + 10 = 200$$

Then go from 200 to the largest multiple of 100 that is less than 720.

$$200 + 500 = 700$$

Finally, go from 700 to 720.

$$700 + 20 = 720$$

The sum of the numbers which are added to 190, 200, and 700 is the difference between 720 and 190. Therefore,

$$720 - 190 = 10 + 500 + 20 = 530$$

If you've ever noticed the way a person who operates a cash register counts your change, the method suggested above may look familiar.

EXAMPLE 5: Mentally estimate the difference between \$3,392 and \$867 by rounding each amount to the nearest hundred dollars.

Solution:

The rounded amounts are \$3,400 and \$900. Using the addition method we have

$$\$900 + \$100 = \$1{,}000$$
$$\$1{,}000 + \$2{,}000 = \$3{,}000$$
$$\$3{,}000 + \$400 = \$3{,}400$$

So, $\$3{,}400 - \$900 = \$100 + \$2{,}000 + \$400 = \$2{,}500$

PRACTICE PROBLEM 5: Mentally estimate the difference between 791 and 236 by rounding each number to the nearest multiple of ten.

If another approach for mental estimation works better for you, use it. For example, you may feel that the solution for \$3,400 − \$900 could have been found by subtracting \$1,000 and adding \$100. That is,

$$\begin{aligned} \$3{,}400 - \$900 &= (\$3{,}400 - \$1{,}000) + \$100 \\ &= \$2{,}400 + \$100 \\ &= \$2{,}500 \end{aligned}$$

If this works better for you, great! If not, try the method illustrated in this section, or some other method that you may discover.

ANSWERS FOR PRACTICE PROBLEMS

1. 412 **2.** 56 **3.** 149 **4.** 4,977 **5.** 550

EXERCISE SET 1.4A

Skill Development

In 1 through 12, find each difference.

1. $\begin{array}{r} 78 \\ -45 \\ \hline \end{array}$ 2. $\begin{array}{r} 350 \\ -267 \\ \hline \end{array}$ 3. $\begin{array}{r} 641 \\ -589 \\ \hline \end{array}$ 4. $\begin{array}{r} 80 \\ -29 \\ \hline \end{array}$

5. $\begin{array}{r} 250 \\ -120 \\ \hline \end{array}$ 6. $\begin{array}{r} 1{,}230 \\ -\ 453 \\ \hline \end{array}$ 7. $\begin{array}{r} 3{,}000 \\ -1{,}587 \\ \hline \end{array}$ 8. $\begin{array}{r} 672 \\ -55 \\ \hline \end{array}$

9. $\begin{array}{r} 2{,}600 \\ -1{,}532 \\ \hline \end{array}$ 10. $\begin{array}{r} 5{,}780 \\ -2{,}387 \\ \hline \end{array}$ 11. $\begin{array}{r} 12{,}000 \\ -\ 651 \end{array}$ 12. $\begin{array}{r} 360{,}000 \\ -275{,}981 \\ \hline \end{array}$

13. Subtract 36 from 80.

14. Subtract 48 from 92.

15. Subtract 367 from 891.

16. Subtract 25 from 350.

17. Subtract 540 from 895.

18. Subtract 1,563 from 6,871.

19. Subtract 2,541 from 20,000.

20. Subtract 30,211 from 97,890.

Applications

21. John Howard's birthday is October 23. If today is October 8, how many days remain until his birthday?

22. Mary Brett purchases a video recorder that has a list price of \$375. The store offers her a discount of \$57 off the list price. What price does she pay for the recorder?

23. Before writing checks for \$56, \$73, and \$121, the balance in Robert Cowan's checking account is \$678. Find his balance after the checks are deducted.

24. On his trip home from college Michael Burns drove 140 miles in three hours. If he covered 56 miles during the first hour and 41 miles during the second hour, what distance did he drive during the third hour?

25. The enrollment at Preston University is 12,651. If there are 6,826 females enrolled, what is the male enrollment?

For Some Extra Thought

26. If the minuend equals 87 and the difference equals 40, find the value of the subtrahend.

27. What number must we subtract from one million for the difference to be nine hundred ninety-nine thousand, nine hundred ninety-nine?

28. In a subtraction problem, if you borrow 1 ten million from 7 ten millions how much is being added to the millions position?

Mental Estimation

In 29 through 32, estimate each difference mentally by rounding each amount to the nearest ten dollars.

29. \$644 − \$126 =

30. \$1,232 − \$980 =

31. \$475 − \$92 =

32. \$855 − \$439 =

33. The attendance at a country and western concert was 3,542 the first night and 5,891 the second night. Round each number to the nearest hundred and mentally estimate how many more people attended the concert the second night.

Review Problems for Maintaining Your Skills

34. Write 56,970,200 in words.

35. Write sixty-seven billion, nine thousand, forty using digits.

In 36 through 38, round each number.

36. 2,578 to the nearest hundred.

37. 6,189,784 to the nearest ten thousand.

38. 45,475,382 to the nearest million.

39. Add the numbers 569, 47, 964, 235, and 90.

40. Add the numbers 12,980, 8,500, and 18,631.

EXERCISE SET 1.4B

Skill Development

In 1 through 12, find each difference.

1. $\begin{array}{r} 92 \\ -21 \\ \hline \end{array}$ 2. $\begin{array}{r} 87 \\ -59 \\ \hline \end{array}$ 3. $\begin{array}{r} 421 \\ -278 \\ \hline \end{array}$ 4. $\begin{array}{r} 600 \\ -79 \\ \hline \end{array}$

5. $\begin{array}{r} 642 \\ -233 \\ \hline \end{array}$ 6. $\begin{array}{r} 230 \\ -93 \\ \hline \end{array}$ 7. $\begin{array}{r} 1{,}635 \\ -1{,}268 \\ \hline \end{array}$ 8. $\begin{array}{r} 6{,}490 \\ -326 \\ \hline \end{array}$

9. $\begin{array}{r} 6{,}544 \\ -5{,}322 \\ \hline \end{array}$ 10. $\begin{array}{r} 30{,}511 \\ -8{,}974 \\ \hline \end{array}$ 11. $\begin{array}{r} 12{,}000 \\ -10{,}672 \\ \hline \end{array}$ 12. $\begin{array}{r} 600{,}000 \\ -456{,}999 \\ \hline \end{array}$

13. Subtract 23 from 89.

14. Subtract 51 from 100.

15. Subtract 236 from 871.

16. Subtract 1,654 from 3,914.

17. Subtract 631 from 9,780.

18. Subtract 4,533 from 11,652.

19. Subtract 5,647 from 9,589.

20. Subtract 36,400 from 51,260.

Applications

21. Spring vacation begins on March 31. If today's date is March 9, how many days remain until the beginning of spring vacation?

22. The price of a car before the sales tax is added is $11,300 and the price after the sales tax is added is $11,978. Find the amount of the sales tax.

23. Mary Graf has $300 credit at the college bookstore. After charging four textbooks that cost $23, $19, $15, and $30 against her account, how much credit remains?

24. A concert hall has 1,523 seats. During the first two days that tickets are on sale for a modern dance performance, 650 and 721 tickets are sold. How many tickets are still available after the two days?

25. The height of Mt. Everest is 29,028 feet. When you climb to the 15,300 foot level, how much higher is it to the top of the mountain?

For Some Extra Thought

26. If the subtrahend equals 52 and the difference equals 35, find the value of the minuend.

27. What number must we add to forty-five thousand, seven hundred sixty-two for the sum to be forty-six thousand?

28. In a subtraction problem, if you borrow 1 billion from the billions position, how much will you add to the hundred millions position?

Mental Estimation

In 29 through 32, estimate each difference mentally by rounding each amount to the nearest hundred dollars.

29. $931 – $226 =

30. $2,460 – $1,355 =

31. $1,674 – $812 =

32. $563 – $422 =

33. A discount of $284 is deducted from the $926 list price of a stereo. Round each amount to the nearest ten dollars and mentally estimate the net amount paid for the stereo.

Review Problems for Maintaining Your Skills

34. Write 952,300 in words.

35. Write eight million, four hundred thousand using digits.

In 36 through 38, round each number.

36. 65,874 to the nearest thousand.

37. 4,629 to the nearest hundred.

38. 6,874,322 to the nearest hundred thousand.

39. Add the numbers 45, 87, 214, 487, and 300.

40. Add the numbers 54,600, 45,841, and 102,350.

1.5 MULTIPLICATION

Preview Test

Find each product.

1. $\begin{array}{r} 27 \\ \times\ 6 \\ \hline \end{array}$ **2.** $\begin{array}{r} 841 \\ \times\ 74 \\ \hline \end{array}$ **3.** $\begin{array}{r} 6{,}201 \\ \times\ 894 \\ \hline \end{array}$

4. $\begin{array}{r} 421 \\ \times\ 9 \\ \hline \end{array}$ **5.** $\begin{array}{r} 2{,}087 \\ \times\ 603 \\ \hline \end{array}$ **6.** $\begin{array}{r} 192 \\ \times 100 \\ \hline \end{array}$

7. $4 \times 28 =$ **8.** $10 \times 690 =$

9. $834 \times 62 =$ **10.** $10{,}000 \times 73 =$

One-Digit Multipliers

Once you've memorized the table of basic multiplication facts ($0 \times 0 = 0$ through $9 \times 9 = 81$), you are ready to master the skill of multiplying any two whole numbers. Let's first consider how we multiply a whole number with only one nonzero digit (for example, 40, 700, or 60,000) by a one-digit number.

Recall that multiplication can be carried out by repeated addition. Using repeated addition for 2×30, 4×700, and $3 \times 4{,}000$, we get the following results.

$$2 \times 30 = 30 + 30 = 60$$
$$4 \times 700 = 700 + 700 + 700 + 700 = 2{,}800$$
$$3 \times 4{,}000 = 4{,}000 + 4{,}000 + 4{,}000 = 12{,}000$$

Instead of using repeated addition, we can find the product by multiplying the nonzero digits and annexing (adding) the same number of zeros on the right as the number of zeros in the number being multiplied. So,

$$2 \times 30 = 60$$

(6: $2 \times 3 = 6$; 0: Annex 1 zero.)

ANSWERS FOR PREVIEW TEST

1. 162 **2.** 62,234 **3.** 5,543,694 **4.** 3,789 **5.** 1,258,461
6. 19,200 **7.** 112 **8.** 6,900 **9.** 51,708 **10.** 730,000

$$4 \times 700 = 2{,}800$$

Annex 2 zeros.

$4 \times 7 = 28$

$$3 \times 4{,}000 = 12{,}000$$

Annex 3 zeros.

$3 \times 4 = 12$

EXAMPLE 1: Find the product for 8×600.

Solution:

$$8 \times 600 = 4{,}800$$

PRACTICE PROBLEM 1: Find the product for 7×90.

Let's now consider multiplying any whole number by a one-digit whole number. Suppose you have \$25 (1 twenty dollar bill and 1 five dollar bill) and you wish to double the amount. You need one additional twenty dollar bill and one additional five dollar bill. So,

			The additional twenty dollar bill ↓				The additional five dollar bill ↓
	\$20	+	\$20	+	\$5	+	\$5
=	(2	×	\$20)	+	(2	×	\$5)
=		\$40		+		\$10	
=				\$50			

The above calculation suggests the following method for multiplying 25 by 2.

Step 1: Express 25 as $20 + 5$

Step 2: Multiply 20 and 5 by 2. Then add.

$$\begin{aligned} &2 \times (20 + 5) \\ &= (2 \times 20) + (2 \times 5) \\ &= 40 + 10 \\ &= 50 \end{aligned}$$

The multiplication 2×25 can be shown vertically.

```
 25
× 2
---
 10   ←——   2 × 5 = 10
 40   ←——   2 × 20 = 40
---
 50
```

We can abbreviate this further to a one-line calculation by multiplying each digit (2 and 5) by 2. However, when we multiply 5×2, instead of writing 10, we write 0 for the units position and carry 1 ten (add 1 ten) to the result for 2×2 tens. This is shown below.

```
  25
 × 2
 ---
  50
  ↑↑
  | 2 × 5 = 10
  | Write the 0 and carry 1 ten.
  |
  2 × 2 tens = 4 tens
  4 tens + (the 1 ten we carried) = 5 tens
```

EXAMPLE 2: Find the product for 4 × 28.

Solution:

```
   28
  × 4
  ---
  112
  ↑ ↑
  |  4 × 8 = 32
  |  Write 2 and carry 3.
  |
  4 × 2 = 8
  8 + (the 3 we carried) = 11
```

PRACTICE PROBLEM 2: Find the product for 7 × 36.

EXAMPLE 3: Find the product for 9 × 483.

Solution:

```
    483
  ×   9
  -----
  4,347
  ↑  ↑↑
  |  | 9 × 3 = 27
  |  | Write 7 and carry 2.
  |  |
  |  9 × 8 = 72
  |  72 + (the 2 we carried) = 74
  |  Write 4 and carry 7.
  |
  9 × 4 = 36
  36 + (the 7 we carried) = 43
```

PRACTICE PROBLEM 3: Find the product for 6 × 274

Multipliers of Two or More Digits

If we multiply by a number with more than one digit the procedure is essentially the same as we used for a one-digit multiplier, but is repeated for each digit in the multiplier. Consider 35 × 47.

Step 1: First multiply 47 by 5.

```
   47
 × 35
 ----
  235
```

Step 2: Then multiply 47 by 3.

```
   47
 × 35
  235
 141    ←——   Since the 3 for 3 × 7 represents 3 tens,
              the product, 21, represents 21 tens.
              So, we must place the 1 in the tens
              place under the 3 in the multiplier 35.
```

Step 3: Finally, add the two products.

```
   47
 × 35
  235
 141
1,645
```

For any multiplication problem the numbers being multiplied are called **factors**. If the problem is done vertically, the top number is called the **multiplicand** and the bottom number is called the **multiplier**. When the multiplier has two or more digits, each multiplication gives us a **partial product**. We add the partial products to get our final **product**. The solution for 35 × 47 is shown again with the labels for each part.

```
   47  ←——  Multiplicand or Factor
 × 35  ←——  Multiplier or Factor
  235  ←——  Partial Product
 141   ←——  Partial Product
1,645  ←——  Product
```

EXAMPLE 4: Find the product for 49 × 507.

Solution:

```
   507
  × 49
  4563
 2028
24,843
```

PRACTICE PROBLEM 4: Find the product for 64 × 723.

EXAMPLE 5: Find the product for 426 × 863.

Solution:

```
    863
  × 426
   5178
  1726
 3452    ←——   Since the 4 from 4 × 3 represents
367,638        4 hundreds, the product, 12,
               represents 12 hundreds. So, we
               must place the 2 in the hundreds
               position under the 4 in 426.
```

PRACTICE PROBLEM 5: Find the product for 834 × 197.

EXAMPLE 6: Find the product for 402×671.

Solution:

$$\begin{array}{r} 671 \\ \times\ 402 \\ \hline 1342 \\ 0000 \\ 2684 \\ \hline 269{,}742 \end{array}$$

When we are multiplying by zero we can write one zero under the zero in the multiplier and omit the remaining zeros in the row. The partial product for the next line can be written on the same line to the left of the zero. The solution for 402×671 is repeated below with only one zero for 0×671, and the partial product for 4×671 on the same line.

$$\begin{array}{r} 671 \\ \times\ 402 \\ \hline 1342 \\ 26840 \\ \hline 269{,}742 \end{array}$$

← Write only one 0 under the 0 in 402, and then to its left write the partial product for 4×671.

PRACTICE PROBLEM 6: Find the product for 706×983.

EXAMPLE 7: Find the product for $2{,}300 \times 7{,}925$.

Solution:

$$\begin{array}{r} 7925 \\ \times\ 2300 \\ \hline 2377500 \\ 15850 \\ \hline 18{,}227{,}500 \end{array}$$

← Notice that a zero is placed on this line under each zero in 2,300. Then the partial product for $3 \times 7{,}925$ is written.

PRACTICE PROBLEM 7: Find the product for $7{,}060 \times 4{,}892$

When the multiplier is a power of ten (10, 100, 1,000, and so on), we can find the product by simply annexing a zero or zeros. Consider $10 \times 3{,}891$, $100 \times 3{,}891$, and $1{,}000 \times 3{,}891$ done the usual way.

$$\begin{array}{r} 3891 \\ \times\ 10 \\ \hline 38{,}910 \end{array} \qquad \begin{array}{r} 3891 \\ \times\ 100 \\ \hline 389{,}100 \end{array} \qquad \begin{array}{r} 3891 \\ \times\ 1000 \\ \hline 3{,}891{,}000 \end{array}$$

From the above results you can see that we can find each product by annexing as many zeros as the number of zeros following the 1 in the power of ten (one zero for 10, two zeros for 100, three zeros for 1,000, and so on). Therefore,

$10 \times 47 = 470$ ← annex 1 zero

$100 \times 39 = 3{,}900$ ← annex 2 zeros

$1{,}000 \times 293 = 293{,}000$ ← annex 3 zeros

Be Careful About This

For 100 × 30 we must annex two zeros after 30. The 0 in 30 is *not* one of the two zeros. So

$$100 \times 30 = 3{,}\underbrace{00}_{\text{The two annexed zeros.}}0$$

EXAMPLE 8: Find the product for 10,000 × 705.

Solution:
We must annex 4 zeros after 705. So,

$$10{,}000 \times 705 = 7{,}050{,}000$$

PRACTICE PROBLEM 8: Find the product for 1,000 × 6.

Mental Estimation of Products

We can mentally estimate a product by rounding each number to the place value position occupied by the lefthand digit. For example, to estimate the product for 72 × 875 we can round 72 to the nearest ten and 875 to the nearest hundred.

72 = 70 rounded to the nearest ten

875 = 900 rounded to the nearest hundred

The product of the rounded numbers is

$$70 \times 900 = 63{,}000$$

EXAMPLE 9: Mentally estimate the product for 834 × 4,731 by rounding each number to the lefthand place value position.

Solution:

834 = 800 rounded to the nearest hundred.

4,731 = 5,000 rounded to the nearest thousand.

Estimated Product = 800 × 5,000 = 4,000,000

PRACTICE PROBLEM 9: Mentally estimate the product for 82 × 57 by rounding each number to the lefthand place value position.

ANSWERS FOR PRACTICE PROBLEMS

1. 630 **2.** 252 **3.** 1,644 **4.** 46,272 **5.** 164,298
6. 693,998 **7.** 34,537,520 **8.** 6,000 **9.** 4,800

EXERCISE SET 1.5A

Skill Development

In 1 through 20, find each product.

1. $\begin{array}{r} 64 \\ \times\ 7 \\ \hline \end{array}$ 2. $\begin{array}{r} 42 \\ \times\ 9 \\ \hline \end{array}$ 3. $\begin{array}{r} 121 \\ \times\ 6 \\ \hline \end{array}$ 4. $\begin{array}{r} 452 \\ \times\ 23 \\ \hline \end{array}$

5. $\begin{array}{r} 56 \\ \times 43 \\ \hline \end{array}$ 6. $\begin{array}{r} 89 \\ \times 20 \\ \hline \end{array}$ 7. $\begin{array}{r} 248 \\ \times\ 39 \\ \hline \end{array}$ 8. $\begin{array}{r} 702 \\ \times\ 58 \\ \hline \end{array}$

9. $\begin{array}{r} 150 \\ \times\ 64 \\ \hline \end{array}$ 10. $\begin{array}{r} 1{,}350 \\ \times\ 200 \\ \hline \end{array}$ 11. $\begin{array}{r} 3{,}871 \\ \times\ 108 \\ \hline \end{array}$ 12. $\begin{array}{r} 12{,}781 \\ \times\ 670 \\ \hline \end{array}$

13. $9 \times 78 =$
14. $42 \times 57 =$
15. $78 \times 654 =$
16. $100 \times 431 =$
17. $500 \times 800 =$
18. $78 \times 604 =$
19. $1{,}000 \times 1{,}000 =$
20. $826 \times 437 =$

Applications

21. George Minton is in the lawn care business. He can cut 250 square feet of lawn per minute. How many square feet can he cut in 27 minutes?

22. If the TV advertising cost for a prime time program is $50,000 per minute, what is the cost for 7 minutes of advertising?

23. If Mary Easler saves $35 per week, how much will she save in 52 weeks?

24. How many hours are there in 7 days?

25. Michael Kass purchases a car for a down payment of $2,000 plus payments of $150 per month for 36 months. Find his total cost.

For Some Extra Thought

26. Find a pair of two-digit numbers whose product is 195.

27. If a five-digit whole number is multiplied by 1,000, how many digits will the product have?

28. If we multiply a pair of two-digit whole numbers, can the product be more than 10,000? Explain.

Mental Estimation

In 29 through 35, estimate each product mentally by rounding each number to the lefthand place value position.

29. $31 \times 84 =$

30. $78 \times 43 =$

31. $281 \times 68 =$

32. $56 \times 675 =$

33. $114 \times 187 =$

34. $620 \times 180 =$

35. At $22 per ticket, what are the approximate receipts for the sale of 47,950 concert tickets?

Review Problems for Maintaining Your Skills

36. Write 3,000,873,000 in words.

37. Round 78,497 to the nearest thousand.

38. Round 892 to the nearest ten.

39. Add the numbers: 54, 89, 147, and 62.

40. Add the numbers: 458, 371, 1,400, 966, and 700.

41. Subtract 72 from 158.

42. Subtract 843 from 5,000.

EXERCISE SET 1.5B

Skill Development

In 1 through 20, find each product.

1. 52×6 **2.** 71×7 **3.** 85×23 **4.** 452×9

5. 120×92 **6.** 200×100 **7.** 267×32 **8.** 507×670

9. 789×72 **10.** 412×176 **11.** $2{,}500 \times 408$ **12.** $40{,}000 \times 10{,}000$

13. $8 \times 68 =$

14. $37 \times 50 =$

15. $12 \times 19 =$

16. $34 \times 400 =$

17. $1{,}000 \times 56 =$

18. $47 \times 830 =$

19. $650 \times 890 =$

20. $354 \times 783 =$

Applications

21. In a 1-mile walk you burn 112 calories. How many calories will you burn in a 7-mile walk?

22. In a 10-month budget plan the Keller family pays $350 per month for gas and electric. What do they pay over the 10-month period?

23. How many minutes are there in 24 hours?

24. A bookstore purchases 260 textbooks at a cost of $23 per book. What is the total cost?

25. Betty Lane purchases a computer by making a down payment of $200 and paying the balance in 24 monthly installments of $50. What is her total cost?

For Some Extra Thought

26. Find a pair of two-digit numbers whose product is 192.

27. If a seven-digit number is multiplied by a one-digit number can the product be more than 8 digits? Explain.

28. If we multiply a two-digit number by a three-digit number can the product be less than 1,000? Explain.

Mental Estimation

In 29 through 35, estimate each product mentally by rounding each number to the lefthand place value position.

29. $56 \times 37 =$

30. $21 \times 58 =$

31. $26 \times 480 =$

32. $67 \times 229 =$

33. $320 \times 792 =$

34. $194 \times 215 =$

35. Ellen Sawyer earns $689 per week. What are her approximate yearly earnings?

Review Problems for Maintaining Your Skills

36. Write eight million, six hundred thousand, eighty-four using digits.

37. Round 709,344 to the nearest thousand.

38. Round 45,837 to the nearest hundred.

39. Add the numbers: 48, 77, 106, 72, and 158.

40. Add the numbers: 2,875, 364, 1,080, and 3,097.

41. Subtract 53 from 122.

42. Subtract 542 from 2,500.

1.6 DIVISION

Preview Test

Find each quotient. If the division is not exact, identify the remainder.

1. $3\overline{)36}$ **2.** $7\overline{)483}$ **3.** $13\overline{)650}$ **4.** $27\overline{)89}$

5. $93\overline{)727}$ **6.** $829\overline{)1{,}045}$ **7.** $43\overline{)17{,}946}$ **8.** $71\overline{)156{,}413}$

9. $79 \div 5 =$ **10.** $863 \div 3 =$

11. $479 \div 26 =$ **12.** $1{,}604 \div 71 =$

13. $10{,}440 \div 18 =$ **14.** $22{,}290 \div 55 =$

Another way of thinking about $42 \div 6$ is to ask, "How many multiples of 6 are contained in 42?" If we list the multiples of 6 we find that there are 7 multiples of 6 in 42.

6	12	18	24	30	36	42
↑ The first multiple of 6						↑ The seventh multiple of 6

Suppose we want to divide 45 one-dollar bills among 6 people so that each person receives the same number of bills. We already know that $42 \div 6 = 7$. So we can give each person 7 one-dollar bills, and 3 one-dollar bills will remain undistributed. We can say that \$45 divided by 6 equals \$7 with \$3 remaining. The division can be shown as follows.

ANSWERS FOR PREVIEW TEST

1. 12 **2.** 69 **3.** 50 **4.** 3 R 8 **5.** 7 R 76 **6.** 1 R 216
7. 417 R 15 **8.** 2,203 **9.** 15 R 4 **10.** 287 R 2 **11.** 18 R 11
12. 22 R 42 **13.** 580 **14.** 405 R 15

Multiply 7 × 6 and write the product under 45. ⟶ 42

Subtract 42 from 45 to find that 3 remains. ⟶ 3

$$\begin{array}{r} 7 \\ 6\overline{)45} \\ \underline{42} \\ 3 \end{array}$$

⟵ 45 ÷ 6 = 7, since 7 × 6 = 42 is the largest multiple of 6 that is less than 45

We have names for 45, 6, 7 and 3. The division is repeated below with the labels for each number.

$$\begin{array}{r} 7 \\ 6\overline{)45} \\ \underline{42} \\ 3 \end{array}$$

Divisor ⟶ 6; 7 ⟵ Quotient; 45 ⟵ Dividend; 3 ⟵ Remainder

When we have a remainder we say that the division is not exact, or that the dividend is not an exact multiple of the divisor. When there is a remainder we write the result this way.

$$45 \div 6 = 7 \text{ R } 3$$

That is, the quotient, 7, with the letter R (for remainder) preceding the remainder, 3.

We can check a division problem with a remainder by using the formula

QUOTIENT × DIVISOR + REMAINDER = DIVIDEND

For 45 ÷ 6 = 7 R 3 we have

$$7 \times 6 + 3 = 45$$

EXAMPLE 1: 78 ÷ 9 =

Solution:

$$\begin{array}{r} 8 \\ 9\overline{)78} \\ \underline{72} \\ 6 \end{array}$$

⟵ 78 ÷ 9 = 8, because 8 × 9 = 72 is the largest multiple of 9 that is less than 78.

So, 78 ÷ 9 = 8 R 6

PRACTICE PROBLEM 1: 65 ÷ 7 =

It is important to know if the number we choose in the quotient is too large or too small. Consider 29 ÷ 8. Let's try 4 as the quotient.

$$\begin{array}{r} 4 \\ 8\overline{)29} \\ \underline{32} \end{array}$$

32 ⟵ Since 32 is larger than 29, we know that 4 is too large for the quotient.

Now let's try 2 for the quotient.

$$\begin{array}{r} 2 \\ 8\overline{)29} \\ \underline{16} \\ 13 \end{array}$$

13 ⟵ Since 13 is larger than the divisor, 8, the quotient is too small.

The correct quotient for 29 ÷ 8 is 3, with a remainder of 5.

$$\begin{array}{r} 3 \\ 8\overline{)29} \\ \underline{24} \\ 5 \end{array}$$

Let's now consider 47 ÷ 3. The quotient is 15, since 15 × 3 = 45 is the largest multiple of 3 that is less than 47.

$$\begin{array}{r} 15 \\ 3\overline{)47} \\ \underline{45} \\ 2 \end{array}$$

So, 47 ÷ 3 = 15 R 2.

Instead of finding the quotient, 15, in one step, we can find it one digit at a time.

Step 1: Divide 4 by 3 and write the result above 4.

$$\begin{array}{r} 1 \\ 3\overline{)47} \end{array}$$

Step 2: Multiply 1 × 3 and write the product under 4. Then subtract the product from 4.

$$\begin{array}{r} 1 \\ 3\overline{)47} \\ \underline{3} \\ 1 \end{array}$$

Step 3: Bring down the next digit from the dividend, which is 7, and place it next to the difference you found in step 2.

$$\begin{array}{r} 1 \\ 3\overline{)47} \\ \underline{3} \\ 17 \end{array}$$

Step 4: Divide 17 by 3 and write the result (5 in this case) next to the 1 in the quotient. Then multiply 5 × 3 and write the product under 17.

$$\begin{array}{r} 15 \\ 3\overline{)47} \\ \underline{3} \\ 17 \\ 15 \end{array}$$

Step 5: Subtract 15 from 17 to find the remainder.

$$\begin{array}{r} 15 \\ 3\overline{)47} \\ \underline{3} \\ 17 \\ \underline{15} \\ 2 \end{array}$$

If the number formed by the lefthand digit or digits in the dividend is less than the divisor, then we have to move to the right until we have a number that is larger than the divisor. This is illustrated with 734 ÷ 8.

Since 7 is smaller than 8, we have to divide 73 by 8, which equals 9. Write 9 over the 3 in 73. ⟶

14 ÷ 8 = 1 Write 1 next to 9 in the quotient. ⟶

$$\begin{array}{r} 91 \\ 8\,\overline{)734} \\ \underline{72} \\ 14 \\ \underline{8} \\ 6 \end{array}$$

- 72 ⟵ 9 × 8 = 72
- 14 ⟵ 73 − 72 = 1. Bring down 4.
- 8 ⟵ 1 × 8 = 8
- 6 ⟵ 14 − 8 = 6

So, 734 ÷ 8 = 91 R 6

EXAMPLE 2: 276 ÷ 4 =

Solution:

$$\begin{array}{r} 69 \\ 4\,\overline{)276} \\ \underline{24} \\ 36 \\ \underline{36} \end{array}$$

36 ⟵ Since 36 − 36 = 0 the remainder is 0.

When the remainder is 0 we do not write

$$276 \div 4 = 69 \text{ R } 0$$

but instead write

$$276 \div 4 = 69$$

PRACTICE PROBLEM 2: 763 ÷ 9 =

If the divisor contains 2 or more digits, the division process is essentially the same. We will illustrate this with 674 ÷ 23.

Since 6 is smaller than 23, we have to divide 67 by 23. 67 ÷ 23 = 2. ⟶

$$\begin{array}{r} 29 \\ 23\,\overline{)674} \\ \underline{46} \\ 214 \\ \underline{207} \\ 7 \end{array}$$

- 46 ⟵ 2 × 23 = 46
- 214 ⟵ Subtract 46 from 67 and bring down the 4.
- 9 × 23 = 207 ⟶ 207
- 7 ⟵ Subtract 207 from 214.

So, 674 ÷ 23 = 29 **R** 7.

In the above example, suppose we had chosen 3 for 67 ÷ 23. How would we know that 3 is not correct?

$$\begin{array}{r} 3 \\ 23\,\overline{)674} \\ 69 \end{array}$$

69 ⟵ Since 69 is larger than 67, 3 is too large.

On the other hand, suppose we had chosen 1 for 67 ÷ 23. How would we know that 1 is not correct?

$$\begin{array}{r} 1 \\ 23\,\overline{)674} \\ \underline{23} \\ 44 \end{array}$$

44 ⟵ Since 44 is larger than 23, 1 is too small.

EXAMPLE 3: 8,391 ÷ 64 =

$$
\begin{array}{r@{\,}l}
 & 131 \\
64 & \overline{)8391} \\
 & \underline{64} \\
 & 199 \\
 & \underline{192} \\
 & 71 \\
 & \underline{64} \\
 & 7
\end{array}
$$

So, 8,391 ÷ 64 = 131 R 7

PRACTICE PROBLEM 3: 1,789 ÷ 37 =

Sometimes there is a zero in the quotient. This is illustrated in Example 4.

EXAMPLE 4: 6,939 ÷ 17 =

Solution:

$$
\begin{array}{r@{\,}l}
 & 408 \\
17 & \overline{)6939} \\
 & \underline{68} \\
 & 139 \\
 & \underline{136} \\
 & 3
\end{array}
$$

← After subtracting 68 from 69 and bringing down the 3, we notice that 13 is smaller than 17. We can say that 13 ÷ 17 = 0. Place a 0 next to the 4 in the quotient and bring down the 9 from the dividend. Then divide 139 by 17.

PRACTICE PROBLEM 4: 14,725 ÷ 32 =

Mental Estimation of Quotients

The mental estimation of quotients is handled about the same way as products. However, since the division of the rounded amounts may not be exact, we may have to round the estimated quotients to the nearest unit. Consider the mental estimation of 3,387 ÷ 89. First round each number to the lefthand place value position.

3,387 = 3,000 rounded to the nearest thousand.

88 = 90 rounded to the nearest ten.

To carry out the division mentally, line up the numbers in the place value positions and delete the matching zeros. Deleting the matching zeros means that we are dividing each number by 10.

$$
\begin{array}{r}
3{,}00\not{0} \\
9\not{0} \\
\uparrow
\end{array}
$$

Delete the matching zeros

Then mentally divide 300 by 9 and round to the nearest unit.

300 ÷ 9 = 33 rounded to the nearest unit.

You may believe that it is unrealistic to expect one to master the mental calculation of $300 \div 9$. Like many skills in mathematics, mental division requires practice. With hard work and patience, there is a good chance that what you may consider difficult will become simple and routine. Also, as you practice you may discover other approaches for mental estimation that work better for you.

EXAMPLE 5: Mentally estimate the quotient for $794 \div 31$ by rounding each number to the lefthand place value position. Round the estimate to the nearest unit.

Solution:

$794 = 800$ rounded to the nearest hundred.

$31 = 30$ rounded to the nearest ten.

$800 \div 30 = 80 \div 3 = 27$ rounded to the nearest unit.

PRACTICE PROBLEM 5: Mentally estimate the quotient for $2{,}123 \div 91$ by rounding each number to the lefthand place value position. Round the estimate to the nearest unit.

Be Careful About This

Mental estimation is meant to give us a rough approximation (sometimes referred to as a "ballpark figure") for the exact answer. Care must be taken when using the method illustrated in this section, because the estimate can in some cases vary appreciably from the exact answer. For example, the estimated quotient for $4{,}472 \div 46$ is 80 (the result for $4{,}000 \div 50$). The exact quotient is 97 with a remainder of 10. The significant deviation is caused by the fact that 4,472 is rounded to 4,000. We would get a better estimate if we round 4,472 to the hundreds position. Then the estimated quotient is 90 (the result for $4{,}500 \div 50$).

As you become more proficient with mental estimation of quotients you may want to experiment with rounding to positions that are perhaps one digit to the right of the lefthand position. In any case, always remember that your mental estimate is a rough approximation.

ANSWERS FOR PRACTICE PROBLEMS

1. 9 R 2 **2.** 84 R 7 **3.** 48 R 13 **4.** 460 R 5 **5.** 22

EXERCISE SET 1.6A

Skill Development

In 1 through 30, find each quotient. If the division is not exact, identify the remainder.

1. $4\overline{)96}$ **2.** $8\overline{)764}$ **3.** $7\overline{)483}$ **4.** $5\overline{)6{,}785}$

5. $3\overline{)904}$ **6.** $9\overline{)215}$ **7.** $23\overline{)80}$ **8.** $15\overline{)671}$

9. $34\overline{)451}$ **10.** $20\overline{)560}$ **11.** $41\overline{)12{,}645}$ **12.** $365\overline{)15{,}974}$

13. $76 \div 8 =$

14. $57 \div 2 =$

15. $630 \div 7 =$

16. $121 \div 6 =$

17. $324 \div 3 =$

18. $1{,}245 \div 9 =$

19. $78 \div 13 =$

20. $137 \div 19 =$

21. $597 \div 56 =$

22. $1{,}583 \div 147 =$

23. $2{,}687 \div 200 =$

24. $38{,}350 \div 125 =$

25. $156{,}214 \div 17 =$

26. $587{,}913 \div 62 =$

27. $174{,}000 \div 37 =$

28. $1{,}600{,}000 \div 47 =$

29. $875{,}600 \div 384 =$

30. $976{,}254 \div 89 =$

Applications

31. A lottery prize of $2,400,000 is shared by 8 people. Find each person's share.

32. A car is driven 216 miles and used 8 gallons of gas. How many miles are driven per gallon?

33. Mary Lanier earns $50,868 per year. How much does she earn per month?

34. A case of beer contains 24 cans. At a party 168 cans were consumed. How many cases were consumed?

35. Leonard Carlin distributes his collection of 4,155 baseball cards to his 7 grandchildren. If each child receives the same number of cards, how many cards remain undistributed?

For Some Extra Thought

36. If we divide a three-digit number by a one-digit number can the quotient be more than 900? Explain.

37. If the divisor is 17 what is the largest remainder we can get?

38. If the quotient, remainder, and dividend are, respectively, 2, 3, and 19, find the divisor.

Mental Estimation

In 39 through 45, mentally estimate each quotient by rounding each number to the lefthand place value position. Round each estimate to the nearest unit.

39. $97 \div 17 =$

40. $81 \div 12 =$

41. $324 \div 42 =$

42. $186 \div 17 =$

43. $875 \div 225 =$

44. $3{,}800 \div 620 =$

45. If a math tutor earns $416 for 32 hours of tutoring, mentally estimate the charge per hour.

Review Problems for Maintaining Your Skills

46. Write 87,652,189 in words.

47. Round 12,935 to the nearest thousand.

48. Round 4,609,774 to the nearest hundred thousand.

49. Add the numbers: 321, 897, 48, and 278.

50. Add the numbers: 17, 38, 45, 31, 73, and 89.

51. Subtract 78 from 200.

52. Subtract 645 from 892.

53. $13 \times 782 =$

54. $24 \times 10{,}000 =$

EXERCISE SET 1.6B

Skill Development

In 1 through 30, find each quotient. If the division is not exact, identify the remainder.

1. $6\overline{)78}$ **2.** $5\overline{)185}$ **3.** $9\overline{)507}$ **4.** $2\overline{)5{,}841}$

5. $4\overline{)217}$ **6.** $7\overline{)171}$ **7.** $14\overline{)706}$ **8.** $35\overline{)600}$

9. $21\overline{)676}$ **10.** $50\overline{)9{,}500}$ **11.** $123\overline{)5{,}687}$ **12.** $460\overline{)500}$

13. $89 \div 3 =$

14. $65 \div 7 =$

15. $842 \div 5 =$

16. $453 \div 9 =$

17. $468 \div 2 =$

18. $2{,}874 \div 6 =$

19. $98 \div 56 =$

20. $245 \div 17 =$

21. $900 \div 61 =$

22. $1{,}511 \div 25 =$

23. $22{,}870 \div 54 =$

24. $10{,}000 \div 13 =$

25. $448{,}638 \div 23 =$

26. $2{,}687{,}100 \div 54 =$

27. $216{,}817 \div 19 =$

28. $256{,}781 \div 61 =$

29. $874{,}000 \div 1{,}600 =$

30. $541{,}673 \div 14 =$

Applications

31. Edward Mason registers for 7 credits in a university. If his total tuition is $896, what is the charge per credit?

32. A $108 dinner check is shared by 9 people. What should each person pay?

33. Sam Miller is a math tutor and has earned $752 this month for 47 hours of tutoring. What does Sam charge per hour?

34. A picture album can hold 192 photographs. Pat Winter has 800 photographs. After filling 4 albums, how many photographs does she have for a fifth album?

35. On a 2,968 mile cross-country trip Alan Beck added 8 quarts of oil to his engine. If the car requires the same amount of oil on the return trip, how often (in terms of miles driven) will he have to add oil?

For Some Extra Thought

36. If the dividend is 80 and the divisor is a whole number, what is the largest quotient we can get?

37. Can the remainder be larger than the quotient? Explain.

38. If the quotient, remainder, and divisor are, respectively, 3, 4, and 5, find the dividend.

Mental Estimation

In 39 through 45, mentally estimate each quotient by rounding each number to the lefthand place value position. Round each estimate to the nearest unit.

39. $86 \div 3 =$

40. $76 \div 8 =$

41. $221 \div 67 =$

42. $460 \div 77 =$

43. $690 \div 42 =$

44. $417 \div 105 =$

45. If a $192 restaurant bill is shared by 8 people, mentally estimate each person's share.

Review Problems for Maintaining Your Skills

46. Write 1,000,500,000 in words.

47. Round 489 to the nearest ten.

48. Round 872,967,400 to the nearest ten million.

49. Add the numbers: 2,412, 3,839, 756, 1,955.

50. Add the numbers: 275, 389, 93, 145, 67, and 617.

51. Subtract 45 from 121.

52. Subtract 478 from 1,000.

53. $27 \times 145 =$

54. $326 \times 100{,}000 =$

1.7 MIXED OPERATIONS

Preview Test

1. $8 + 4 \times 5 - 2 =$
2. $6 \times 7 + 9 \times 8 =$
3. $10 + 2 \times 5 - 7 \div 7 =$
4. $40 - (2 + 4 \times 9) =$
5. $54 + 40 \div 8 - 24 \div 8 =$
6. $(17 - 15 \div 5) - (20 - 4 \times 3) =$

When an arithmetic problem involves more than one operation we have to establish a procedure regarding the order in which the operations are performed. Consider the following example.

$$7 + 5 \times 8 - 6 \div 3$$

How to proceed? Should we perform the operations from left to right as they appear in the problem, or should certain operations be performed before others regardless of where they appear? The latter course is the proper course to follow, and it is summarized here.

If no parentheses ( ) or brackets [ ] appear, then the order in which the operations are performed is:

1. multiplication and/or division in the order in which the operations occur from left to right;
2. addition and/or subtraction in the order in which the operations occur from left to right.

Using the above rule for order of operations, the expression $7 + 5 \times 8 - 6 \div 3$ is evaluated as follows.

$$\begin{aligned} & 7 + \underbrace{5 \times 8}_{\text{multiply}} - 6 \div 3 \\ &= 7 + 40 - \underbrace{6 \div 3}_{\text{divide}} \\ &= \underbrace{7 + 40}_{\text{add}} - 2 \\ &= \underbrace{47 - 2}_{\text{subtract}} \\ &= 45 \end{aligned}$$

ANSWERS FOR PREVIEW TEST

1. 26 **2.** 114 **3.** 19 **4.** 2 **5.** 56 **6.** 6

EXAMPLE 1: $9 \times 5 - 4 \times 7 =$

Solution:

$$
\begin{aligned}
&\underbrace{9 \times 5}_{\text{multiply}} - \underbrace{4 \times 7}_{\text{multiply}} \\
&= \underbrace{45 - 28}_{\text{subtract}} \\
&= 17
\end{aligned}
$$

PRACTICE PROBLEM 1: $26 - 16 \div 8 =$

If parentheses or brackets are included, the operations within the parentheses or brackets should be performed first in the correct order: that is, multiplication and/or division from left to right, then addition and/or subtraction from left to right.

For example, to evaluate $(4 + 7 \times 5) \times (20 - 8 \div 4)$ we begin by performing the operations within each set of parentheses in the correct order. The numerical results found within each set of parentheses are multiplied.

$$
\begin{aligned}
&(4 + \underbrace{7 \times 5}_{\text{multiply}}) \times (20 - \underbrace{8 \div 4}_{\text{divide}}) \\
&= \underbrace{(4 + 35)}_{\text{add}} \times \underbrace{(20 - 2)}_{\text{subtract}} \\
&= \underbrace{39 \times 18}_{\text{multiply}} \\
&= 702
\end{aligned}
$$

EXAMPLE 2: $8 + (20 - 4 \times 3) \div (10 - 2 \times 3) =$

Solution:

$$
\begin{aligned}
&8 + (20 - \underbrace{4 \times 3}_{\text{multiply}}) \div (10 - \underbrace{2 \times 3}_{\text{multiply}}) \\
&= 8 + \underbrace{(20 - 12)}_{\text{subtract}} \div \underbrace{(10 - 6)}_{\text{subtract}} \\
&= 8 + \underbrace{8 \div 4}_{\text{divide}} \\
&= \underbrace{8 + 2}_{\text{add}} \\
&\quad 10
\end{aligned}
$$

PRACTICE PROBLEM 2: $7 \times (17 - 63 \div 9 + 2)$

ANSWERS FOR PRACTICE PROBLEMS

1. 24 **2.** 84

EXERCISE SET 1.7A

Skill Development

1. $6 + 7 \times 2 - 8 =$
2. $5 \times 4 + 15 \div 5 - 2 =$
3. $7 - 2 \times 3 + 220 - 18 \div 9 =$
4. $10 + 2 \times 3 + 8 \times 1 - 4 \div 4 =$
5. $100 + 20 \times 4 - 16 \div 2 =$
6. $(30 \times 2 - 7) \times (5 - 16 \div 8) =$
7. $30 - (6 - 2 \times 2) =$
8. $(40 \div 4 \times 2 - 6) \times 3 - (5 + 2 \times 3) =$
9. $(200 \div 40 - 2) \times (5 - 20 \div 10) =$
10. $400 - (7 \times 5) - (8 \times 4) \div (16 \times 2) =$

Applications

11. Ruth Tessler purchases 25 shares of IBM stock for $156 per share and 100 shares of AT&T stock for $27 per share. Find the total cost of the stock.

12. A ticket broker sells 17 twenty-dollar seats and 26 thirty-dollar seats for a concert. Find the total dollar amount of the ticket sales.

13. To find Tim Smith's final average in a course, his midterm grade, 87, is multiplied by 2, his final exam grade, 92, is multiplied by 3, and the sum of the two products is divided by 5. Find his average.

For Some Extra Thought

14. $0 \div 4 + 5 \times 2 =$
15. $1 + 1 \times 1 + 1 \times 1 \times 1 + 1 \times 1 \times 1 \times 1 =$

Review Problems for Maintaining Your Skills

16. Round 4,628,450 to the nearest hundred thousand.
17. Round 2,347 to the nearest ten.
18. Add the numbers: 254, 781, and 2,790.
19. Subtract 374 from 800.
20. $144 \times 10{,}000 =$
21. $48 \times 927 =$

In 22 and 23, find each quotient and remainder.

22. $237 \div 19 =$
23. $4{,}980 \div 71 =$

EXERCISE SET 1.7B

Skill Development

1. $8 + 19 \times 4 - 10 =$
2. $3 \times 6 + 7 \times 8 + 9 \times 1 =$
3. $25 - 2 \times 7 + 12 \div 3 =$
4. $41 - 5 \times 7 + 3 \times 3 - 4 \div 2 =$
5. $25 + 30 \div 6 - 7 \times 2 =$
6. $(25 - 4 \times 5) \times (6 - 8 \div 8) =$
7. $56 - (20 - 12 \div 2) \times 3 =$
8. $(220 - 40 \times 5) \div (100 - 40 \times 2) + 5 =$
9. $500 - (60 \times 3 - 16 \times 10) - 20 \times 20 =$
10. $60 + (17 \times 3 - 6) + 3 - 24 \div 3 =$

Applications

11. From a weight of 350 pounds, Norman Kell lost 15 pounds each week for 6 consecutive weeks, and then 8 pounds each week for 10 consecutive weeks. Find his weight after 16 weeks.

12. Michael Richard's total monthly earnings consist of his \$500 base salary plus \$25 times 82 as commission for the 82 computers he sold. Find his total earnings for the month.

13. Mary Perkins holds two jobs. In her first job she works 23 hours and is paid \$7 per hour. In her second job she works 12 hours and is paid \$9 per hour. Find her total earnings for the two jobs.

For Some Extra Thought

14. $13 \div 13 \times 12 \div 12 \times 11 \div 11 \times 10 \div 10 =$
15. $(3 \times 8 - 6 \times 4) \div 5 =$

Review Problems for Maintaining Your Skills

16. Round 17,892,340 to the nearest ten thousand.
17. Round 45,871 to the nearest hundred.
18. Add the numbers: 483, 145, 79, and 64.
19. Subtract 4,697 from 5,200.

20. 3,247 × 100,000 =

21. 678 × 809 =

In 22 and 23, find each quotient and remainder.

22. 489 ÷ 100 =

23. 3,783 ÷ 42 =

PRACTICE TEST 1A

1. In the number 23,678,507 the 6 is in the ________________ place.
2. Write 204,873 in words.
3. Write the number for seven million, sixty-three thousand, nine hundred four.
4. Round 54,342 to the nearest hundred.
5. Round 6,785,300 to the nearest ten thousand.
6. Add the numbers: 987, 45, 3,470, and 644.
7. The attendance figures at a concert for Friday, Saturday, and Sunday were: 8,356, 9,280, and 7,461. Find the total attendance for the three days.
8. Subtract 23 from 4,000.
9. The price of a car increased from \$15,678 to \$16,764. Find the amount of the increase.
10. $987 \times 43 =$
11. $10{,}000 \times 303 =$

Find the quotient and remainder.

12. $246 \div 13 =$
13. $1{,}307 \div 500 =$
14. Michael Cusak earns \$28,444 per year. How much does he earn per week?
15. $7 \times 5 + 8 \times 6 - 2 \times 7 =$
16. $(80 - 5 \times 7) \times (100 + 2 \times 8) =$

PRACTICE TEST 1B

1. In the number 2,304 the 0 is in the ____________ place.
2. Write 2,000,530 in words.
3. Write the number for eight hundred twenty-three thousand, seventeen.
4. Round 45,687 to the nearest thousand.
5. Round 4,572 to the nearest hundred.
6. Add the numbers: 24, 366, 874, 772, and 95.
7. For the first six games of the season a football team had a total offense in yards of 458, 368, 587, 273, 390, and 588. Find the total yards for the six games.
8. Subtract 457 from 9,125.
9. The tuition at a university increased from \$5,600 to \$7,300 over a ten-year period. Find the amount of the increase.
10. $34 \times 408 =$
11. $100{,}000 \times 41 =$

12. The price per ticket to a baseball game is \$7. Find the cost for 83 tickets.

Find the quotient and remainder.

13. $126 \div 33 =$

14. $1{,}235 \div 347 =$

15. $100 - 8 \times 5 - 4 \times 7 =$

16. $(4 + 2 \times 8) - (30 - 7 \times 4) =$

CHAPTER 2

FRACTIONS AND MIXED NUMBERS

2.1 EQUIVALENT FRACTIONS

2.2 MIXED NUMBERS AND IMPROPER FRACTIONS

2.3 MULTIPLYING FRACTIONS AND MIXED NUMBERS

2.4 DIVIDING FRACTIONS AND MIXED NUMBERS

2.5 ADDING FRACTIONS AND MIXED NUMBERS

2.6 SUBTRACTING FRACTIONS AND MIXED NUMBERS

2.7 MIXED OPERATIONS

There are many situations in which a fraction or mixed number, rather than a whole number, is used.

- The proportion of questions answered correctly on a test: $\frac{22}{25}$
- The increase in the price of a share of stock: $\$1\frac{5}{8}$
- The number of acres cleared for a library building: $\frac{11}{16}$

You may have to add, subtract, multiply, or divide fractions or mixed numbers to answer questions such as:

- If John Brooks owns $\frac{1}{4}$ of the stock in a company and Mary Ortiz owns $\frac{2}{5}$ of the stock, what part of the company do they own together?
- A truck is loaded with $6\frac{3}{4}$ cubic yards of stone. It unloads $3\frac{1}{2}$ cubic yards. How many cubic yards remain in the truck?
- A caterer for a banquet estimates $\frac{2}{5}$ bottles of soda for each guest. How many bottles should the caterer order for 80 guests?
- Virginia Harris paid $5,175 for stock that cost $\$8\frac{5}{8}$ per share. How many shares did she purchase?

2.1 EQUIVALENT FRACTIONS

Preview Test

In questions 1 through 3, reduce each fraction to its lowest terms.

1. $\frac{12}{30}$ **2.** $\frac{20}{65}$ **3.** $\frac{144}{180}$

In questions 4 through 6, convert each fraction to an equivalent fraction with the indicated denominator.

4. $\frac{11}{14}$, 28 **5.** $\frac{17}{20}$, 100 **6.** $\frac{5}{7}$, 56

A **fraction** can be used to indicate:

1. a part of a whole;
2. a ratio of two quantities;
3. one number divided by another number.

ANSWERS FOR PREVIEW TEST

1. $\frac{2}{5}$ **2.** $\frac{4}{13}$ **3.** $\frac{4}{5}$ **4.** $\frac{22}{28}$ **5.** $\frac{85}{100}$ **6.** $\frac{40}{56}$

The "part of a whole" meaning of a fraction will be considered in this section. Ratios are considered in Chapter 5 and fractions indicating division will be handled in section 3 of Chapter 4.

The fraction $\frac{3}{4}$ can identify three of four equal parts. This is illustrated by considering any closed figure divided into four equal parts in which three of the parts are shaded (see Figure 2.1.1).

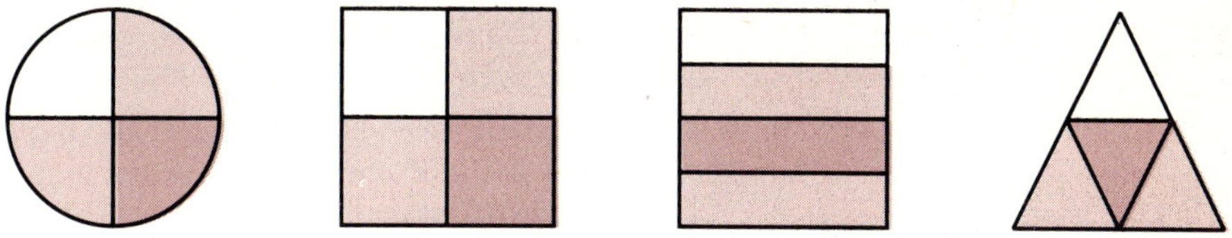

Figure 2.1.1

For the fraction $\frac{3}{4}$ the number on top of the bar is called the **numerator** and the number below the bar is called the **denominator**.

3	←----	Numerator
–	←----	Fraction Bar
4	←----	Denominator

The circle in Figure 2.1.2(a) is divided into four equal parts with three parts shaded. The circle in Figure 2.1.2(b) is divided into eight equal parts with six parts shaded. In each case the same part of the whole is shaded. So it appears that 3 of 4 equal parts equals 6 of 8 equal parts, or

$$\frac{3}{4} = \frac{6}{8}$$

Figure 2.1.2

The fractions $\frac{3}{4}$ and $\frac{6}{8}$ are called **equivalent fractions** because they represent the same value. When $\frac{6}{8}$ is converted to $\frac{3}{4}$ you have reduced $\frac{6}{8}$ to its **lowest terms**. When $\frac{3}{4}$ is converted to $\frac{6}{8}$ you have raised $\frac{3}{4}$ to **higher terms**.

Reducing Fractions

To reduce a fraction to its lowest terms, you must divide the numerator and denominator by all common divisors larger than 1. For example, to reduce $\frac{16}{24}$ to its lowest terms, you divide 16 and 24 by the common divisor 8.

$$\frac{16}{24} = \frac{16 \div 8}{24 \div 8} = \frac{2}{3}$$

If you do not see that 8 is the largest common divisor, but you do realize that 4 is a common divisor, you can first divide by 4 and then by 2.

$$\frac{16}{24} = \frac{16 \div 4}{24 \div 4} = \frac{4}{6} = \frac{4 \div 2}{6 \div 2} = \frac{2}{3}$$

A fraction is reduced to its lowest terms if the numerator and denominator have no common divisors larger than 1. Therefore $\frac{2}{3}$ is reduced to its lowest terms.

To help you find common divisors, the following rules can be used.

1. All even numbers (2, 4, 6, . . .) are divisible by 2.
2. If the sum of the digits is divisible by 3, the number is divisible by 3. For example, the sum of the digits in 594 is $5 + 9 + 4 = 18$. Since 18 is divisible by 3, 594 is divisible by 3. In fact, $594 \div 3 = 198$.
3. If the number formed by the 2 righthand digits is divisible by 4, the number is divisible by 4. For example, in 4,136, the number 36 is divisible by 4. Therefore, 4,136 is divisible by 4.
4. If the units digit is 0 or 5, the number is divisible by 5.
5. If an even number is divisible by 3, it is also divisible by 6. For example, 594 is even and divisible by 3; therefore, 594 is divisible by 6.
6. If the units digit is 0, the number is divisible by 10. This rule can be generalized for divisibility by 100, 1,000, 10,000, and so on for numbers that end with 2, 3, 4, . . . zeros.
7. If the number formed by the 2 righthand digits is 00, 25, 50, or 75, the number is divisible by 25.
8. If the number formed by the 2 righthand digits is 00 or 50, the number is divisible by 50.

EXAMPLE 1: Reduce each fraction to its lowest terms.

a) $\frac{42}{72}$ **b)** $\frac{125}{175}$

Solution:

a) 42 and 72 are both even. They are also divisible by 3 since the sum of the digits for each $(4 + 2 = 6)$ and $(7 + 2 = 9)$ is divisible by 3. Therefore 42 and 72 are divisible by 6.

$$\frac{42}{72} = \frac{42 \div 6}{72 \div 6} = \frac{7}{12}$$

Since 7 and 12 have no common divisors larger than 1, $\frac{7}{12}$ is reduced to its lowest terms.

b) 125 and 175 are divisible by 25.

$$\frac{125}{175} = \frac{125 \div 25}{175 \div 25} = \frac{5}{7}$$

Since 5 and 7 have no common divisors larger than 1, $\frac{5}{7}$ is reduced to its lowest terms.

PRACTICE PROBLEM 1: Reduce each fraction to its lowest terms.

a) $\frac{40}{64}$ **b)** $\frac{150}{450}$

Raising to Higher Terms

To raise a fraction to higher terms, multiply the numerator and denominator by the same nonzero number. For example, to convert $\frac{7}{8}$ to an equivalent fraction whose denominator is 24, multiply the numerator and denominator by 3, because $3 \times 8 = 24$.

$$\frac{7}{8} = \frac{7 \times 3}{8 \times 3} = \frac{21}{24}$$

EXAMPLE 2: Convert each fraction to an equivalent fraction with the indicated denominator.

a) $\frac{7}{10}$, 50 **b)** $\frac{5}{9}$, 54

Solution:

a) Since $10 \times 5 = 50$, multiply 7 and 10 by 5.

$$\frac{7}{10} = \frac{7 \times 5}{10 \times 5} = \frac{35}{50}$$

b) Since $9 \times 6 = 54$, multiply 5 and 9 by 6.

$$\frac{5}{9} = \frac{5 \times 6}{9 \times 6} = \frac{30}{54}$$

PRACTICE PROBLEM 2: Convert each fraction to an equivalent fraction with the indicated denominator.

a) $\frac{1}{3}$, 36 **b)** $\frac{23}{40}$, 200

ANSWERS FOR PRACTICE PROBLEMS

1. a) $\frac{5}{8}$ **b)** $\frac{1}{3}$ **2. a)** $\frac{12}{36}$ **b)** $\frac{115}{200}$

EXERCISE SET 2.1A

Skill Development

Reduce each fraction to its lowest terms.

1. $\frac{42}{44}$ **2.** $\frac{100}{125}$ **3.** $\frac{64}{80}$ **4.** $\frac{124}{240}$ **5.** $\frac{82}{90}$ **6.** $\frac{783}{900}$

Convert each fraction to an equivalent fraction with the indicated denominator.

7. $\frac{3}{4}$, 36 **8.** $\frac{5}{9}$, 63 **9.** $\frac{1}{7}$, 35

10. $\frac{15}{16}$, 32 **11.** $\frac{6}{13}$, 39 **12.** $\frac{3}{11}$, 55

Applications

13. In a class of 32 students 20 plan to attend the local community college. Express the part of the class that plans to attend the local community college as a fraction reduced to its lowest terms.

14. During the 30 days in April this year it rained on 12 days. Express the part of April it rained as a fraction reduced to its lowest terms.

15. Study a map of the United States and determine how many of the 48 states (excluding Alaska and Hawaii) are located completely east of the Mississippi River. Represent this as a fraction reduced to its lowest terms.

16. Ruth Butler averages 4 hits for every 10 times at bat for her softball team. If she comes to bat 80 times, express the number of hits as the numerator of a fraction whose denominator is 80.

For Some Extra Thought

17. Is every fraction whose denominator is one number higher than its numerator reduced to lowest terms?

18. To convert a fraction with a denominator of 19 to an equivalent fraction with a denominator of 19,000, by what number should you multiply the numerator and denominator?

Review Problems for Maintaining Your Skills

19. Round 867,472 to the nearest thousand.

20. Add the numbers: 876, 36, 788, and 51.

21. Subtract 453 from 1,000.

22. $135 \times 100{,}000 =$

23. $78 \times 93 =$

24. For $897 \div 27$ find the quotient and remainder.

25. $7 + 8 \times 4 - 13 =$

EXERCISE SET 2.1B

Skill Development

Reduce each fraction to its lowest terms.

1. $\frac{51}{63}$ **2.** $\frac{26}{52}$ **3.** $\frac{500}{750}$ **4.** $\frac{77}{99}$ **5.** $\frac{54}{81}$ **6.** $\frac{13}{65}$

Convert each fraction to an equivalent fraction with the indicated denominator.

7. $\frac{1}{8}$, 72 **8.** $\frac{23}{100}$, 500 **9.** $\frac{6}{7}$, 56

10. $\frac{21}{25}$, 150 **11.** $\frac{11}{30}$, 120 **12.** $\frac{12}{17}$, 51

Applications

13. To pass an entrance examination for an insurance company, an applicant must correctly answer at least three-fourths of the questions. If the examination contains 60 questions, express the minimum passing grade as the numerator of a fraction whose denominator is 60.

14. Last Tuesday five-eighths of the stock traded on the New York Stock Exchange dropped in price during the day's trading. If 400 stocks were traded that day, express the number of stocks that dropped in price as the numerator of a fraction whose denominator is 400.

15. The women's basketball team won five-sixths of the games they played. If they played 24 games, express the number of games they won as the numerator of a fraction whose denominator is 24.

16. In a history class 12 of 34 students earned an A. Express the part of the class that earned an A as a fraction reduced to its lowest terms.

For Some Extra Thought

17. Is a fraction whose denominator and numerator are both even numbers reduced to its lowest terms?

18. If two fractions have the same denominator and different numerators can they be equivalent?

Review Problems for Maintaining Your Skills

19. Round 796 to the nearest ten.

20. Add the numbers: 1,256, 450, and 872.

21. Subtract 87 from 380.

22. $47 \times 1{,}000 =$

23. $104 \times 36 =$

24. For $59 \div 7$ find the quotient and remainder.

25. $6 \times 9 + 3 \times 7 =$

2.2 MIXED NUMBERS AND IMPROPER FRACTIONS

Preview Test

In questions 1 through 3, convert each improper fraction to a mixed number. Reduce each fraction part to its lowest terms.

1. $\frac{15}{7}$ **2.** $\frac{63}{20}$ **3.** $\frac{100}{30}$

In questions 4 through 6, convert each mixed number to an improper fraction.

4. $5\frac{5}{7}$ **5.** $7\frac{1}{8}$ **6.** $12\frac{2}{5}$

Converting an Improper Fraction to a Mixed Number

A **proper fraction** contains a numerator that is smaller than the denominator. For example, $\frac{7}{11}$, $\frac{13}{20}$, and $\frac{1}{8}$ are proper fractions. An **improper fraction** contains a numerator that is equal to or larger than the denominator. For example, $\frac{13}{6}$, $\frac{22}{17}$, $\frac{9}{3}$, and $\frac{4}{4}$ are improper fractions.

The improper fractions $\frac{9}{3}$ and $\frac{4}{4}$ reduce to the whole numbers 3 and 1, respectively. The improper fractions $\frac{13}{6}$ and $\frac{22}{17}$ do not reduce to whole numbers, but can be expressed in another form by dividing the numerator by the denominator.

$$\frac{13}{6} = 13 \div 6 = 2 \text{ R } 1 = 2\frac{1}{6}$$

This new form, $2\frac{1}{6}$, is an example of a **mixed number**, a number containing a whole number plus a fraction. The equality of $\frac{13}{6}$ and $2\frac{1}{6}$ is illustrated in Figure 2.2.1. Each unit length is divided into six equal parts.

$$\frac{13}{6} = 2 + \frac{1}{6} = 2\frac{1}{6}$$

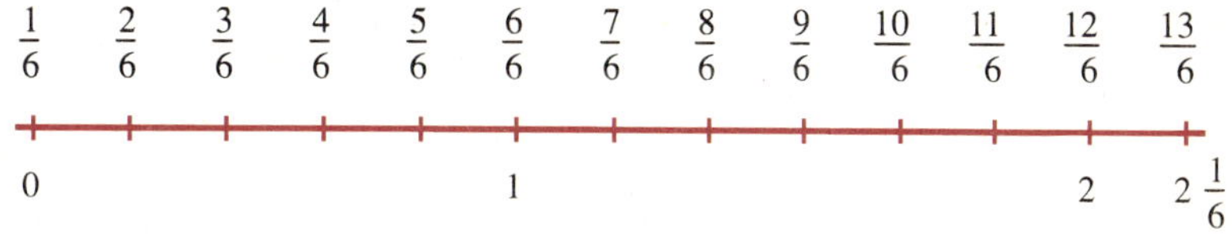

Figure 2.2.1

ANSWERS FOR PREVIEW TEST

1. $2\frac{1}{7}$ **2.** $3\frac{3}{20}$ **3.** $3\frac{1}{3}$ **4.** $\frac{40}{7}$ **5.** $\frac{57}{8}$ **6.** $\frac{62}{5}$

EXAMPLE 1: Convert $\frac{16}{11}$ to a mixed number.

Solution:
If you divide 16 by 11 the quotient is 1 and the remainder is 5. Therefore,

$$\frac{16}{11} = 1\frac{5}{11}$$

PRACTICE PROBLEM 1: Convert $\frac{100}{27}$ to a mixed number.

Converting a Mixed Number to an Improper Fraction

A mixed number can be converted to an improper fraction by the following procedure.

Mixed Number		Improper Fraction
Whole Number $\frac{\text{Numerator}}{\text{Denominator}}$	=	$\frac{\text{Denominator} \times \text{Whole Number} + \text{Numerator}}{\text{Denominator}}$

For example,

$$3\frac{5}{7} = \frac{7 \times 3 + 5}{7} = \frac{26}{7}$$

To help you understand the rule, notice that $3\frac{5}{7}$ can be written as $\frac{3}{1} + \frac{5}{7}$. Then $\frac{3}{1}$ can be expressed with a denominator of 7 by multiplying 3 and 1 by 7. So,

$$\frac{3}{1} + \frac{5}{7} = \frac{7 \times 3}{7 \times 1} + \frac{5}{7} = \frac{21}{7} + \frac{5}{7} = \frac{26}{7}$$

Notice that in this demonstration we have added the two fractions $\frac{21}{7}$ and $\frac{5}{7}$ and obtained as a result one fraction with a denominator of 7. The process of adding fractions with the same denominator is discussed further in Section 2.5.

EXAMPLE 2: Convert $2\frac{1}{5}$ to an improper fraction.

Solution:

$$2\frac{1}{5} = \frac{5 \times 2 + 1}{5} = \frac{11}{5}$$

PRACTICE PROBLEM 2: Convert $15\frac{2}{7}$ to an improper fraction.

ANSWERS FOR PRACTICE PROBLEMS

1. $3\frac{19}{27}$ **2.** $\frac{107}{7}$

EXERCISE SET 2.2A

Skill Development

Convert each improper fraction to a mixed number. Reduce each fraction part to its lowest terms.

1. $\frac{17}{10}$ **2.** $\frac{38}{30}$ **3.** $\frac{120}{7}$

4. $\frac{71}{43}$ **5.** $\frac{200}{155}$ **6.** $\frac{89}{10}$

Convert each mixed number to an improper fraction.

7. $4\frac{1}{5}$ **8.** $1\frac{7}{8}$ **9.** $10\frac{3}{7}$

10. $6\frac{2}{11}$ **11.** $3\frac{5}{9}$ **12.** $2\frac{10}{13}$

Applications

13. An office employee worked 37 hours one week. If each work day is 8 hours, what mixed number represents the number of days worked?

14. If there are 3 feet in 1 yard, express the number of yards in 17 feet as a mixed number.

15. A room is $15\frac{1}{2}$ feet wide. Express its width as an improper fraction.

16. A container of dishwashing fluid contains $1\frac{1}{4}$ pints of a product. Write the number of pints as an improper fraction.

For Some Extra Thought

17. Which has more cheese, a 21-ounce block or a $1\frac{1}{4}$-pound block (1 pound = 16 ounces).

18. A cash register has $\$23\frac{3}{4}$ in quarters. How many quarters does the register contain?

Review Problems for Maintaining Your Skills

19. Round 41,687,300 to the nearest ten thousand.

20. Subtract 156 from 680.

21. $56 \times 907 =$

22. For $451 \div 100$ find the quotient and remainder.

23. $100 - (4 + 5 \times 6) =$

24. Reduce $\frac{52}{65}$ to its lowest terms.

25. Convert $\frac{6}{7}$ to an equivalent fraction with 28 as a denominator.

EXERCISE SET 2.2B

Skill Development

Convert each improper fraction to a mixed number. Reduce each fraction part to its lowest terms.

1. $\frac{25}{9}$ **2.** $\frac{166}{163}$ **3.** $\frac{50}{11}$

4. $\frac{55}{3}$ **5.** $\frac{300}{200}$ **6.** $\frac{42}{30}$

Convert each mixed number to an improper fraction.

7. $3\frac{2}{3}$ **8.** $8\frac{20}{21}$ **9.** $1\frac{2}{5}$

10. $5\frac{5}{8}$ **11.** $4\frac{3}{7}$ **12.** $15\frac{4}{7}$

Applications

13. The value of a home has increased $\frac{3}{2}$ from its purchase price to its selling price. What mixed number represents its increase in value as a multiplier?

14. A boat is 22 feet long. What is its length in yards as a mixed number? (1 yard = 3 feet)

15. Computer paper is $11\frac{3}{4}$ inches wide. What is its width expressed as an improper fraction?

16. A case of oil contains 26 1-quart cans. How many gallons of oil is this expressed as a mixed number? (1 gallon = 4 quarts)

For Some Extra Thought

17. Which bottle has more liquid, one with 43 ounces or one with $1\frac{1}{2}$ quarts? (1 quart = 32 fluid ounces)

18. A cash register has \$$15\frac{1}{2}$ in half dollars. How many half dollars does the register contain?

Review Problems for Maintaining Your Skills

19. Round 28,549 to the nearest hundred.

20. Add the numbers: 46, 853, 92, and 500.

21. $905 \times 10 =$

22. For $546 \div 79$ find the quotient and remainder.

23. $96 - (5 \times 3 + 4 \times 9) =$

24. Reduce $\frac{45}{63}$ to its lowest terms.

25. Convert $\frac{3}{5}$ to an equivalent fraction with a denominator of 200.

2.3 MULTIPLYING FRACTIONS AND MIXED NUMBERS

Preview Test

Find each product. Reduce each product to its lowest terms.

1. $\frac{12}{25} \times \frac{2}{15} =$ **2.** $\frac{1}{3} \times \frac{3}{4} \times \frac{2}{9} =$

3. $5 \times 2\frac{5}{7} =$ **4.** $3\frac{1}{2} \times 4\frac{4}{5} =$

Multiplying Fractions

To multiply two fractions, multiply the respective numerators and denominators. For $\frac{4}{5} \times \frac{2}{3}$ we have

$$\frac{4}{5} \times \frac{2}{3} = \frac{4 \times 2}{5 \times 3} = \frac{8}{15}$$

We can illustrate this multiplication by considering a bar that is divided into 15 equal parts (see Figure 2.3.1).

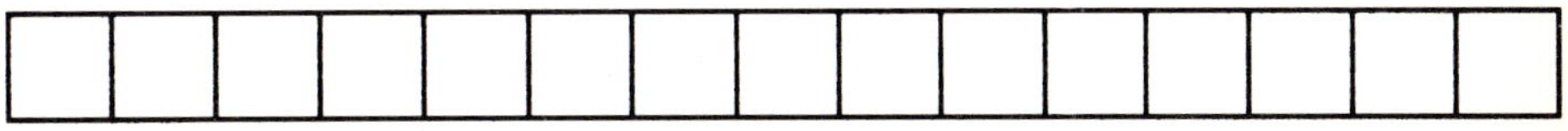

Figure 2.3.1

Of the 15 equal parts, 10 represent $\frac{2}{3}$ of the original length, since $\frac{10}{15} = \frac{2}{3}$ (see Figure 2.3.2).

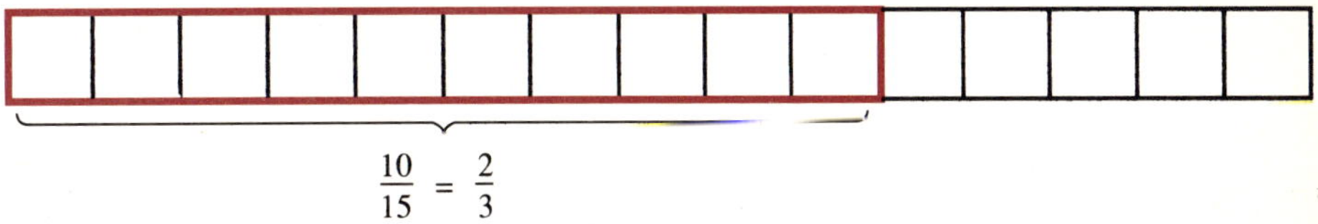

Figure 2.3.2

Now to find $\frac{4}{5}$ of $\frac{2}{3}$, we shade 8 of the 10 parts identified in Figure 2.3.2, since $\frac{8}{10} = \frac{4}{5}$ (see Figure 2.3.3).

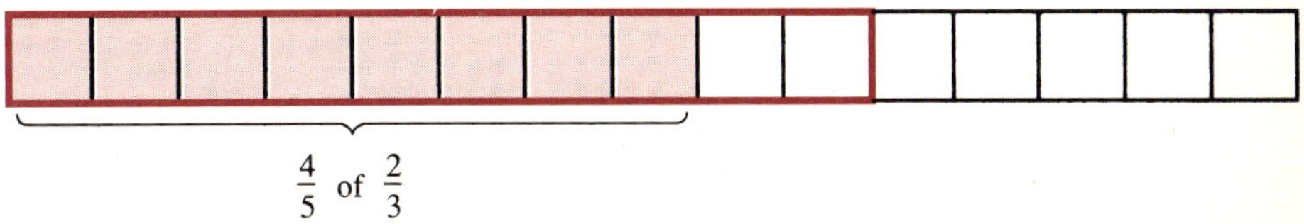

Figure 2.3.3

ANSWERS FOR PREVIEW TEST

1. $\frac{8}{125}$ **2.** $\frac{1}{18}$ **3.** $13\frac{4}{7}$ **4.** $16\frac{4}{5}$

Since the 8 shaded boxes represent 8 of 15 parts, it appears that

$$\frac{4}{5} \text{ of } \frac{2}{3} = \frac{4}{5} \times \frac{2}{3} = \frac{4 \times 2}{5 \times 3} = \frac{8}{15}$$

Consider $\frac{14}{15} \times \frac{3}{28} = \frac{14 \times 3}{15 \times 28} = \frac{42}{420} = \frac{1}{10}$

An alternate approach involves the reduction of fractions before multiplying. This can be done by dividing a numerator and denominator by a common divisor. Notice that 14 and 28 are divided by 14 and 3 and 15 are divided by 3.

$$\frac{14}{15} \times \frac{3}{28} = \frac{\overset{1}{\cancel{14}}}{\underset{5}{\cancel{15}}} \times \frac{\overset{1}{\cancel{3}}}{\underset{2}{\cancel{28}}} = \frac{1}{10}$$

You can see from this example that the numerator and denominator you are reducing do not have to be in the same fraction.

Be Careful About This

Make sure that when you reduce, you always reduce a number in a numerator with a number in a denominator. The numbers can be in the same fraction or in different fractions. However, you cannot reduce a numerator with a numerator (such as 3 and 6 in the multiplication problem $\frac{3}{5} \times \frac{6}{7}$) or two denominators with each other (such as 5 and 10 in the multiplication problem $\frac{2}{5} \times \frac{7}{10}$).

Products of more than two fractions offer no additional difficulty. First reduce the fractions and then multiply the respective numerators and denominators. This is illustrated in part (b) of Example 1.

EXAMPLE 1: Multiply and reduce each product to its lowest terms.

a) $\frac{9}{14} \times \frac{21}{22}$ **b)** $8 \times \frac{1}{14} \times \frac{21}{27}$

Solution:

a) Divide 14 and 21 by 7.

$$\frac{9}{\underset{2}{\cancel{14}}} \times \frac{\overset{3}{\cancel{21}}}{22} = \frac{9 \times 3}{2 \times 22} = \frac{27}{44}$$

b) Express 8 as $\frac{8}{1}$. Divide 21 and 27 by 3. Divide 7 and 14 by 7. Divide 8 and 2 by 2.

$$\frac{\overset{4}{\cancel{8}}}{1} \times \frac{1}{\underset{\underset{1}{\cancel{2}}}{\cancel{14}}} \times \frac{\overset{\overset{1}{\cancel{7}}}{\cancel{21}}}{\underset{9}{\cancel{27}}} = \frac{4 \times 1 \times 1}{1 \times 1 \times 9} = \frac{4}{9}$$

PRACTICE PROBLEM 1: Multiply and reduce each product to its lowest terms.

a) $\frac{12}{17} \times \frac{7}{18}$ **b)** $\frac{8}{9} \times 4 \times \frac{15}{16}$

Be Careful About This

The reason we have suggested that you write the whole number multiplier with a denominator of 1 is to avoid the common error of multiplying both the numerator and denominator by the whole number. That is, *do not do the following.*

$$2 \times \frac{3}{7} = \frac{2 \times 3}{2 \times 7} = \frac{6}{14}$$

By writing 2 as $\frac{2}{1}$, the above error would not be made.

$$2 \times \frac{3}{7} = \frac{2}{1} \times \frac{3}{7} = \frac{2 \times 3}{1 \times 7} = \frac{6}{7}$$

Multiplication Involving Mixed Numbers

To multiply mixed numbers, change the mixed numbers to improper fractions and multiply the respective numerators and denominators.

EXAMPLE 2: Multiply and reduce each product to its lowest terms.

a) $2\frac{4}{5} \times 1\frac{7}{10}$ **b)** $5 \times 3\frac{2}{5}$

Solution:

a) $2\frac{4}{5} \times 1\frac{7}{10} = \frac{\overset{7}{\cancel{14}}}{5} \times \frac{17}{\underset{5}{\cancel{10}}} = \frac{7 \times 17}{5 \times 5} = \frac{119}{25} = 4\frac{19}{25}$

b) $5 \times 3\frac{2}{5} = \frac{\overset{1}{\cancel{5}}}{1} \times \frac{17}{\underset{1}{\cancel{5}}} = 17$

PRACTICE PROBLEM 2: Multiply and reduce each product to its lowest terms.

a) $3\frac{5}{7} \times 1\frac{1}{2}$ **b)** $\frac{2}{3} \times 6\frac{1}{7}$

Applications involving multiplication of fractions are usually stated with the word "of" playing the role of the word "times." Recall in our original illustration with the bar that we found $\frac{4}{5}$ of $\frac{2}{3}$ by calculating $\frac{4}{5} \times \frac{2}{3}$.

ANSWERS FOR PRACTICE PROBLEMS

1 a) $\frac{14}{51}$ **b)** $3\frac{1}{3}$ **2 a)** $5\frac{4}{7}$ **b)** $4\frac{2}{21}$

EXERCISE SET 2.3A

Skill Development

Find each product. Reduce all results to lowest terms.

1. $\frac{3}{4} \times \frac{2}{9} =$
2. $\frac{7}{8} \times \frac{3}{4} =$
3. $\frac{5}{6} \times \frac{1}{10} =$
4. $\frac{12}{13} \times \frac{5}{12} =$
5. $\frac{20}{21} \times \frac{15}{16} =$
6. $\frac{10}{11} \times \frac{3}{14} =$
7. $\frac{1}{3} \times \frac{3}{4} \times \frac{2}{5} =$
8. $\frac{7}{8} \times \frac{10}{21} \times \frac{3}{7} =$
9. $\frac{25}{26} \times \frac{12}{13} \times \frac{4}{5} =$
10. $2\frac{2}{5} \times 1\frac{3}{4} =$
11. $5\frac{1}{2} \times 8\frac{4}{5} =$
12. $10 \times 3\frac{1}{4} =$
13. $1\frac{2}{5} \times \frac{7}{8} =$
14. $4\frac{4}{5} \times 3\frac{2}{5} =$
15. $3\frac{1}{3} \times \frac{5}{6} =$

Applications

16. From the high school graduating class $\frac{9}{10}$ of the lower $\frac{3}{5}$ plan to attend a two-year college. What fractional part of the graduating class plans to attend a two-year college?

17. In a small town $\frac{5}{6}$ of the eighteen year olds are registered to vote and $\frac{3}{7}$ of the registered eighteen year olds actually voted in the national election. What fractional part of the eighteen year olds actually voted?

18. John Smith owns $\frac{3}{4}$ acres of land. He plants corn on $\frac{7}{8}$ of his land. How many acres are planted with corn?

19. An insurance company agrees to pay $\frac{4}{5}$ of $\frac{9}{10}$ of the cost of repairing a house damaged by fire. What fractional part of the cost is the insurance company paying?

For Some Extra Thought

20. Can the product of two mixed numbers be less than 1? If your answer is yes, give an example? If your answer is no, explain why.

21. Two-thirds times what number equals 2?

Review Problems for Maintaining Your Skills

22. Round 1,582,743 to the nearest million.

23. $4{,}000 - 3{,}723 =$

24. $47 \times 89 =$

25. For $89 \div 17$ find the quotient and remainder.

26. $6 \times 7 - 5 \times 8 =$

27. Reduce $\frac{75}{85}$ to lowest terms.

28. Convert $\frac{11}{15}$ to an equivalent fraction with a denominator of 60.

29. Convert $\frac{19}{13}$ to a mixed number.

30. Convert $4\frac{5}{7}$ to an improper fraction.

EXERCISE SET 2.3B

Skill Development

Find each product. Reduce all results to lowest terms.

1. $\frac{2}{5} \times \frac{7}{9} =$

2. $\frac{12}{13} \times \frac{8}{9} =$

3. $\frac{1}{6} \times \frac{5}{8} =$

4. $\frac{3}{4} \times \frac{11}{15} =$

5. $\frac{7}{8} \times \frac{2}{9} =$

6. $\frac{14}{15} \times \frac{10}{21} =$

7. $4 \times \frac{4}{9} \times \frac{3}{10} =$

8. $\frac{6}{7} \times \frac{5}{8} \times \frac{1}{6} =$

9. $\frac{3}{8} \times \frac{8}{9} \times \frac{6}{7} =$

10. $4\frac{1}{2} \times 6\frac{3}{5} =$

11. $7\frac{2}{3} \times 1\frac{4}{7} =$

12. $8 \times 5\frac{1}{4} =$

13. $2\frac{1}{2} \times \frac{3}{4} =$

14. $4\frac{1}{2} \times 7\frac{1}{2} =$

15. $1\frac{3}{4} \times \frac{8}{9} =$

Applications

16. A health insurance plan provides for reimbursement for $\frac{4}{5}$ of $\frac{95}{100}$ of a person's medical expenses. What fractional part of the expense do they pay?

17. Neal Hamilton buys 20 shares of stock at a price of \$$4\frac{1}{8}$ per share. Find the total cost of the 20 shares.

18. Mary Lasker buys $\frac{1}{2}$ cord of wood for her fireplace. Over a four-week period she burns $\frac{3}{5}$ of her supply. How much wood does she burn in four weeks?

19. Two-thirds of History 101 deals with the twentieth century and one-third of the twentieth century work deals with World War II. What fractional part of the course deals with World War II?

For Some Extra Thought

20. Will the product $\frac{89}{90} \times 2$ be greater than or less than 2?

21. Seven-eighths times what fraction equals 1?

Review Problems for Maintaining Your Skills

22. Round 786 to the nearest ten.

23. $560 - 378 =$

24. $897 \times 10{,}000 =$

25. For $139 \div 12$ find the quotient and remainder.

26. $89 - 4 \times 3 + 7 \times 2 =$

27. Reduce $\frac{44}{60}$ to its lowest terms.

28. Convert $\frac{6}{13}$ to an equivalent fraction with a denominator of 52.

29. Convert $\frac{22}{9}$ to a mixed number.

30. Convert $8\frac{3}{5}$ to an improper fraction.

2.4 DIVIDING FRACTIONS AND MIXED NUMBERS

Preview Test

Find each quotient. Reduce each quotient to its lowest terms.

1. $\frac{3}{4} \div \frac{7}{8}$ **2.** $\frac{9}{11} \div 5$ **3.** $2 \div \frac{1}{3}$ **4.** $3\frac{3}{4} \div 1\frac{1}{2}$

Division Involving Fractions

To divide by a fraction, invert the divisor and use the rule for multiplying fractions. For $\frac{5}{9} \div \frac{3}{4}$ we have

Invert the divisor

$$\frac{5}{9} \div \frac{3}{4} = \frac{5}{9} \times \frac{4}{3} = \frac{20}{27}$$

Division to multiplication

Notice that we've done two activities in one step (inverted the divisor and changed to multiplication). To help you understand why inverting and multiplying works we will write the division as

$$\frac{\frac{5}{9}}{\frac{3}{4}}$$

← five-ninths
← divided by
← three-quarters

If we multiply $\frac{5}{9}$ and $\frac{3}{4}$ by the same fraction we are not changing the value of the quotient, since we are multiplying by an equivalent of 1. Also, by choosing $\frac{4}{3}$ (the inverted form of $\frac{3}{4}$) as the multiplier, something nice happens; the denominator turns out to be 1. This is illustrated below.

$$\frac{\frac{5}{9}}{\frac{3}{4}} \times \frac{\frac{4}{3}}{\frac{4}{3}} = \frac{\frac{5}{9} \times \frac{4}{3}}{\frac{3}{4} \times \frac{4}{3}} = \frac{\frac{20}{27}}{\frac{12}{12}} = \frac{\frac{20}{27}}{1} = \frac{20}{27}$$

Since the process of multiplying both fractions by the inverted form of the divisor will always give us 1 in the denominator, we can abbreviate the above procedure and use the rule stated at the beginning of this section.

ANSWERS FOR PREVIEW TEST

1. $\frac{6}{7}$ **2.** $\frac{9}{55}$ **3.** 6 **4.** $2\frac{1}{2}$

Be Careful About This

It is always the divisor that is inverted. Try to avoid these common errors.

1. Inverting the dividend. $\frac{3}{4} \div \frac{5}{7} = \frac{4}{3} \times \frac{5}{7}$
2. Inverting both fractions. $\frac{3}{4} \div \frac{5}{7} = \frac{4}{3} \times \frac{7}{5}$
   Of course, if the division is done correctly we have

$$\frac{3}{4} \div \frac{5}{7} = \frac{3}{4} \times \frac{7}{5} = \frac{21}{20} = 1\frac{1}{20}$$

EXAMPLE 1: Divide and reduce each quotient to its lowest terms.

a) $\frac{7}{10} \div \frac{3}{4}$ **b)** $7 \div \frac{2}{3}$ **c)** $\frac{4}{9} \div 5$

Solution:

Invert the divisor

a) $\frac{7}{10} \div \frac{3}{4} = \frac{7}{10} \times \frac{4}{3}$

Division to multiplication

Now reduce by dividing 4 and 10 by 2.

$$\frac{7}{\overset{}{\underset{5}{\cancel{10}}}} \times \frac{\overset{2}{\cancel{4}}}{3} = \frac{7 \times 2}{5 \times 3} = \frac{14}{15}$$

b) The whole number 7 can be written as $\frac{7}{1}$. Therefore,

$$7 \div \frac{2}{3} = \frac{7}{1} \div \frac{2}{3} = \frac{7}{1} \times \frac{3}{2} = \frac{7 \times 3}{1 \times 2} = \frac{21}{2} = 10\frac{1}{2}$$

c) Notice that 5 is the divisor. After writing it as $\frac{5}{1}$, invert it to $\frac{1}{5}$.

$$\frac{4}{9} \div 5 = \frac{4}{9} \div \frac{5}{1} = \frac{4}{9} \times \frac{1}{5} = \frac{4 \times 1}{9 \times 5} = \frac{4}{45}$$

Be Careful About This

Do not reduce until you have inverted and converted the division to a multiplication operation. For example, *do not do the following.*

$$\frac{7}{9} \div \frac{6}{7} = \frac{\overset{1}{\cancel{7}}}{\underset{3}{\cancel{9}}} \div \frac{\overset{2}{\cancel{6}}}{\underset{1}{\cancel{7}}} = \frac{1}{3} \times \frac{1}{2} = \frac{1}{6}$$

The correct approach is: $\frac{7}{9} \div \frac{6}{7} = \frac{7}{9} \times \frac{7}{6} = \frac{7 \times 7}{9 \times 6} = \frac{49}{54}$

PRACTICE PROBLEM 1: Divide and reduce each quotient to its lowest terms.

a) $\frac{13}{18} \div \frac{5}{6}$ **b)** $4 \div \frac{5}{7}$ **c)** $\frac{7}{8} \div 2$

Division Involving Mixed Numbers

To divide mixed numbers, change the mixed numbers to improper fractions. Then invert the divisor and multiply.

EXAMPLE 2: Divide and express each quotient in its lowest terms.

a) $5\frac{2}{3} \div 7\frac{1}{2}$ **b)** $8\frac{4}{5} \div 3$

Solution:

a) $5\frac{2}{3} \div 7\frac{1}{2} = \frac{17}{3} \div \frac{15}{2} = \frac{17}{3} \times \frac{2}{15} = \frac{17 \times 2}{3 \times 15} = \frac{34}{45}$

b) $8\frac{4}{5} \div 3 = \frac{44}{5} \div \frac{3}{1} = \frac{44}{5} \times \frac{1}{3} = \frac{44 \times 1}{5 \times 3} = \frac{44}{15} = 2\frac{14}{15}$

PRACTICE PROBLEM 2: Divide and express each quotient in its lowest terms.

a) $6\frac{3}{5} \div 2\frac{1}{2}$ **b)** $5\frac{2}{3} \div \frac{7}{9}$

ANSWERS FOR PRACTICE PROBLEMS

1. a) $\frac{13}{15}$ **b)** $5\frac{3}{5}$ **c)** $\frac{7}{16}$ **2. a)** $2\frac{16}{25}$ **b)** $7\frac{2}{7}$

EXERCISE SET 2.4A

Skill Development

Divide and reduce each quotient to its lowest terms. Express any quotient that is an improper fraction as a mixed number.

1. $\frac{1}{8} \div \frac{3}{5} =$
2. $\frac{12}{13} \div \frac{5}{3} =$
3. $\frac{1}{4} \div \frac{4}{15} =$
4. $\frac{17}{18} \div \frac{2}{5} =$
5. $\frac{6}{11} \div \frac{9}{5} =$
6. $\frac{12}{7} \div \frac{12}{5} =$
7. $\frac{6}{7} \div 2 =$
8. $\frac{1}{12} \div \frac{20}{19} =$
9. $10 \div \frac{5}{14} =$
10. $6\frac{2}{7} \div 1\frac{3}{4} =$
11. $8\frac{1}{9} \div 2\frac{5}{6} =$
12. $\frac{3}{4} \div 6\frac{5}{7} =$
13. $7\frac{2}{5} \div 3 =$
14. $12 \div 3\frac{1}{3} =$
15. $5\frac{3}{5} \div \frac{6}{7} =$

Applications

16. How many $\frac{3}{4}$ acre plots are there in 15 acres?

17. Each hamburger requires $\frac{1}{6}$ pound of meat. How many hamburgers can we make from $\frac{2}{3}$ pounds of meat?

18. Ten people share $\frac{1}{2}$ the cost of uniforms for a softball team. What is each person's fractional share of the cost?

19. Mike spends $17\frac{1}{2}$ hours per week delivering groceries. Each delivery takes him about $\frac{5}{8}$ hours. How many deliveries can he make in one week?

For Some Extra Thought

20. Two-thirds divided by what number equals two?

21. If a whole number is divided by one-seventh, is the quotient a whole number? Explain.

Review Problems for Maintaining Your Skills

22. Round 1,682 to the nearest hundred.

23. Add the numbers: 478, 763, and 1,094.

24. For $123 \div 50$ find the quotient and remainder.

25. $100 - (45 + 8 \times 6) =$

26. Reduce $\frac{56}{72}$ to its lowest terms.

27. Convert $\frac{3}{4}$ to an equivalent fraction with a denominator of 48.

28. Convert $\frac{9}{5}$ to a mixed number.

29. Convert $5\frac{6}{7}$ to an improper fraction.

In 30 through 32, multiply and reduce each result to its lowest terms.

30. $\frac{5}{7} \times \frac{13}{15} =$

31. $5\frac{1}{2} \times 6 =$

32. $2\frac{3}{4} \times 3\frac{1}{4} =$

EXERCISE SET 2.4B

Skill Development

Divide and reduce each quotient to its lowest terms. Express any quotient that is an improper fraction as a mixed number.

1. $\frac{2}{5} \div \frac{7}{8} =$
2. $\frac{3}{8} \div \frac{10}{11} =$
3. $\frac{5}{7} \div \frac{1}{4} =$
4. $\frac{7}{8} \div \frac{4}{3} =$
5. $\frac{1}{3} \div \frac{17}{15} =$
6. $\frac{5}{6} \div \frac{6}{5} =$
7. $2 \div \frac{6}{5} =$
8. $\frac{9}{14} \div \frac{4}{5} =$
9. $\frac{4}{7} \div 8 =$
10. $5\frac{1}{4} \div 1\frac{4}{5} =$
11. $3\frac{1}{3} \div 2\frac{5}{7} =$
12. $7\frac{1}{4} \div 1\frac{2}{3} =$
13. $5\frac{2}{3} \div \frac{8}{9} =$
14. $5 \div 1\frac{1}{4} =$
15. $9\frac{1}{4} \div 4 =$

Applications

16. How many $\frac{2}{3}$ acre plots of land are there in 20 acres?

17. Each bookshelf requires $\frac{1}{10}$ pint of paint. How many bookshelves can we paint with $\frac{3}{5}$ pints of paint?

18. Seven people share $\frac{2}{3}$ of a lottery prize equally. What is each person's fractional share of the prize?

19. A builder divides $6\frac{1}{2}$ acres of land into 10 lots for houses. Find the number of acres for each building.

For Some Extra Thought

20. Three-fifths divided by what number equals six?

21. If an even whole number is divided by two-thirds, is the quotient a whole number? Explain.

Review Problems for Maintaining Your Skills

22. Round 45,872 to the nearest thousand.

23. Add the numbers: 245, 72, 144, and 37.

24. For $67 \div 13$ find the quotient and remainder.

25. $200 - (5 \times 6 + 7 \times 10) =$

26. Reduce $\frac{39}{65}$ to its lowest terms.

27. Convert $\frac{11}{14}$ to an equivalent fraction with a denominator of 56.

28. Convert $\frac{33}{7}$ to a mixed number.

29. Convert $4\frac{3}{8}$ to an improper fraction.

In 30 through 32, multiply and reduce each result to its lowest terms.

30. $\frac{12}{13} \times \frac{3}{16} =$

31. $5\frac{1}{2} \times \frac{15}{22} =$

32. $1\frac{1}{2} \times 1\frac{1}{2} =$

2.5 ADDING FRACTIONS AND MIXED NUMBERS

Preview Test

In questions 1 and 2, find the least common multiple.

1. 7 and 9 **2.** 5, 8, and 12

In questions 3 through 8, find each sum. Express each sum that is an improper fraction as a mixed number.

3. $\frac{3}{8} + \frac{2}{8}$ **4.** $\frac{4}{5} + \frac{1}{8} + \frac{1}{2}$ **5.** $3 + \frac{5}{8} + \frac{1}{2}$

6. $5\frac{1}{2} + 2\frac{2}{5}$ **7.** $1\frac{2}{7} + 3\frac{3}{4} + \frac{7}{8}$ **8.** $4\frac{3}{5} + 3 + 1\frac{2}{5}$

Finding the Least Common Multiple

To add two fractions with like denominators, add the numerators but do not change the denominators. For example,

$$\frac{5}{9} + \frac{2}{9} = \frac{7}{9}$$

If the denominators are not alike, we must identify the **least common multiple** of the denominators (referred to as the **least common denominator** or **L.C.D.**) and raise each fraction whose denominator is not the L.C.D. to higher terms.

To help you understand the meaning of L.C.D., we will find the least common multiple of 5, 8, and 10 using a method that is clear, but rather lengthy for universal application. The multiples of 5, 8, and 10 are found by multiplying each by 1, 2, 3, 4, and so on. The smallest number that appears in all three rows is the least common multiple of 5, 8, and 10. That number is 40, which is circled.

Multiples of 5: 5, 10, 15, 20, 25, 30, 35, (40), 45, 50, . . .
Multiples of 8: 8, 16, 24, 32, (40), 48, 56, . . .
Multiples of 10: 10, 20, 30, (40), 50, 60, 70, . . .

Another method for finding the least common multiple of a set of numbers is far more efficient. It requires the use of prime numbers.

A **prime number** is a whole number larger than 1 which is only divisible by itself and 1. The first ten primes are 2, 3, 5, 7, 11, 13, 17, 19, 23, and 29. The number 4 is not prime, since it is divisible by a number other than itself and 1, namely 2.

To find the least common multiple of a set of numbers, complete the following steps.

Step 1: If the numbers have no common prime number divisors, the least common multiple is the product of the numbers. For example, since 3 and 8 have no common prime number divisors, the least common multiple of 3 and 8 is 24.

ANSWERS FOR PREVIEW TEST

1. 63 **2.** 120 **3.** $\frac{5}{8}$ **4.** $1\frac{17}{40}$ **5.** $4\frac{1}{8}$ **6.** $7\frac{9}{10}$

7. $5\frac{51}{56}$ **8.** 9

Step 2: If at least two of the numbers have a common prime number divisor, arrange the numbers in a row and carry out the division where it is exact. When the division is exact, bring down the quotient to the next row. When the division is not exact do not divide, but instead, bring the number down to the next row. For example, in 3, 8, and 10, the prime number 2 is a common divisor of 8 and 10.

Divisor ⟶ $2 \,|\, 3 \quad 8 \quad 10$

Since 3 is not divisible by 2, bring down the 3. ⟶ $3 \quad 4 \quad 5$

$8 \div 2 = 4$ (↑ under 4); $10 \div 2 = 5$ (↑ under 5)

Step 3: Continue the process with the new row until there are no common prime number divisors. The least common multiple is the product of the divisor(s) and the numbers in the bottom row. In the illustration in step 2 the numbers in the bottom row, 3, 4, and 5 have no common prime number divisors. Therefore, the least common multiple is

$$\underset{\substack{\uparrow \\ \text{Prime divisor}}}{2} \times \underbrace{3 \times 4 \times 5}_{\text{Numbers in the bottom row}} = 120$$

EXAMPLE 1: Find the least common multiple of each set.

a) 5 and 13 **b)** 6, 15, and 20

Solution:

a) Since 5 and 13 have no common prime number divisors, the least common multiple is $5 \times 13 = 65$.

b)

Divide 6 and 20 by 2:	2	6	15	20
Divide 3 and 15 by 3:	3	3	15	10
Divide 5 and 10 by 5:	5	1	5	10
		1	1	2

The least common multiple is $2 \times 3 \times 5 \times 1 \times 1 \times 2 = 60$.

PRACTICE PROBLEM 1: Find the least common multiple of each set.

a) 7 and 12 **b)** 8, 18, and 24

Adding Fractions or Mixed Numbers

To add fractions or mixed numbers with unlike denominators complete the following steps.

Step 1: Find the L.C.D.

Step 2: Change each fraction whose denominator is not the L.C.D. to an equivalent fraction with the L.C.D. as the denominator.

Step 3: Add the fractions by placing the sum of the numerators over the L.C.D.

EXAMPLE 2: Add **a)** $4 + \frac{7}{8} + \frac{1}{3}$ **b)** $5\frac{6}{7} + 2\frac{4}{9}$

Solution:

a) Express 4 as the fraction $\frac{4}{1}$. For $\frac{4}{1} + \frac{7}{8} + \frac{1}{3}$ the L.C.D. is 24.

$$\frac{4}{1} = \frac{4 \times 24}{1 \times 24} = \frac{96}{24} \qquad \frac{7}{8} = \frac{7 \times 3}{8 \times 3} = \frac{21}{24} \qquad \frac{1}{3} = \frac{1 \times 8}{3 \times 8} = \frac{8}{24}$$

The equivalent fractions are now added.

$$\frac{96}{24} + \frac{21}{24} + \frac{8}{24} = \frac{125}{24} = 5\frac{5}{24}$$

b) The L.C.D. is 63.

Convert $\frac{6}{7}$ and $\frac{4}{9}$ to equivalent fractions with denominators of 63. Then add the whole numbers and fractions separately.

$$5\frac{6 \times 9}{7 \times 9} + 2\frac{4 \times 7}{9 \times 7} = 5\frac{54}{63} + 2\frac{28}{63} = 7\frac{82}{63}$$

For $7\frac{82}{63}$, since $\frac{82}{63}$ is an improper fraction we must convert $\frac{82}{63}$ to the mixed number $1\frac{19}{63}$ and add this mixed number to 7.

$$7\frac{82}{63} = 7 + 1\frac{19}{63} = 8\frac{19}{63}$$

PRACTICE PROBLEM 2: Add **a)** $\frac{4}{9} + 5 + \frac{2}{5}$ **b)** $4\frac{3}{4} + 5\frac{7}{8}$

Be Careful About This

After mastering the skill of multiplying fractions, which calls for the multiplication of the numerators and denominators, it is quite natural to make the error of adding the numerators and denominators when you are adding fractions. Therefore, try to avoid this common error.

$$\frac{3}{8} + \frac{2}{8} = \frac{3+2}{8+8} = \frac{5}{16}$$

ANSWERS FOR PRACTICE PROBLEMS

1. a) 84 **b)** 72 **2. a)** $5\frac{38}{45}$ **b)** $10\frac{5}{8}$

EXERCISE SET 2.5A

Skill Development

Find the least common multiple of each set.

1. 24 and 15 **2.** 2, 7, 21, and 28 **3.** 15, 40, and 60

Find each sum. Reduce all results to their lowest terms and express improper fractions as mixed numbers.

4. $\frac{2}{7} + \frac{3}{7} =$

5. $\frac{2}{5} + \frac{1}{5} + \frac{3}{5} =$

6. $\frac{11}{25} + \frac{13}{25} =$

7. $\frac{1}{2} + \frac{3}{4} =$

8. $\frac{1}{5} + \frac{7}{9} =$

9. $\frac{7}{25} + \frac{1}{40} =$

10. $\frac{16}{45} + \frac{7}{20} =$

11. $\frac{1}{5} + \frac{2}{15} =$

12. $\frac{13}{18} + \frac{14}{27} =$

13. $\frac{3}{7} + \frac{1}{2} + \frac{5}{6} =$

14. $\frac{2}{15} + \frac{3}{10} + \frac{1}{20} =$

15. $\frac{2}{3} + \frac{2}{5} + \frac{2}{7} =$

16. $2 + \frac{1}{2} + \frac{1}{3} =$

17. $\frac{3}{8} + 5 + \frac{2}{5} =$

18. $\frac{1}{7} + \frac{3}{8} + 3 =$

19. $1\frac{3}{5} + 4\frac{7}{15} =$

20. $5\frac{1}{2} + 7\frac{9}{16} =$

21. $10\frac{1}{2} + 8\frac{1}{8} =$

22. $4 + 1\frac{3}{5} + 3\frac{1}{2} =$

23. $\frac{2}{5} + 1\frac{3}{5} + 5\frac{1}{9} =$

24. $2 + 5\frac{1}{6} + \frac{2}{9} =$

Applications

25. A contract for a construction project calls for payment of $\frac{1}{3}$ of the total price when the contract is signed and $\frac{1}{15}$ of the total price per month for ten consecutive months. What fractional part of the cost is paid after the second monthly payment?

26. Edward Martin writes a will leaving $\frac{1}{4}$ of his estate to his wife and $\frac{1}{6}$ to each of his three children. How much of the estate is he leaving to his wife and children?

27. Ken Lyman's lawn mower uses $\frac{2}{9}$ gallons of gas to mow the Ingram's lawn and $\frac{1}{6}$ gallon to mow the Miller's lawn. How many gallons are used to mow the two lawns?

28. In a math class $\frac{2}{5}$ of the class earned an A and $\frac{2}{7}$ earned a B. What fractional part of the class earned an A or a B?

For Some Extra Thought

29. Six-seventeenths plus what fraction equals one?

30. Find five fractions whose sum equals one-fourth.

Review Problems for Maintaining Your Skills

31. Round 56,379 to the nearest thousand.

32. $5{,}600 - 3{,}709 =$

33. $123 \times 64 =$

34. For $674 \div 37$ find the quotient and remainder.

35. Reduce $\frac{48}{72}$ to its lowest terms.

36. Convert $\frac{7}{10}$ to an equivalent fraction with a denominator of 200.

37. Convert $\frac{19}{5}$ to a mixed number.

38. Convert $5\frac{3}{7}$ to an improper fraction.

In 39 through 42, reduce each result to its lowest terms and express any result that is an improper fraction as a mixed number.

39. $\frac{7}{9} \times \frac{3}{7} =$

40. $1\frac{2}{5} \times 3\frac{3}{4} =$

41. $\frac{1}{2} \div \frac{1}{7} =$

42. $5 \div 5\frac{2}{3} =$

EXERCISE SET 2.5B

Skill Development

Find the least common multiple of each set.

1. 7 and 12 **2.** 3, 6, and 8 **3.** 12, 15, 20, and 24

Find each sum. Reduce all results to their lowest terms and express improper fractions as mixed numbers.

4. $\frac{4}{9} + \frac{1}{9} =$

5. $\frac{2}{13} + \frac{1}{13} + \frac{3}{13} =$

6. $\frac{17}{35} + \frac{13}{35} =$

7. $\frac{1}{4} + \frac{3}{8} =$

8. $\frac{2}{3} + \frac{5}{12} =$

9. $\frac{5}{21} + \frac{1}{14} + \frac{3}{7} =$

10. $\frac{13}{40} + \frac{4}{25} =$

11. $\frac{1}{2} + \frac{1}{3} + \frac{1}{4} =$

12. $\frac{17}{20} + \frac{10}{13} =$

13. $\frac{5}{6} + \frac{2}{3} + \frac{1}{4} =$

14. $\frac{24}{25} + \frac{11}{75} =$

15. $\frac{2}{5} + \frac{5}{6} + \frac{4}{9} =$

16. $3 + \frac{1}{2} + \frac{3}{5} =$

17. $\frac{12}{13} + \frac{4}{5} + 10 =$

18. $\frac{5}{24} + 7 + \frac{1}{2} =$

19. $3\frac{1}{4} + 1\frac{2}{3} =$

20. $4\frac{5}{7} + 6\frac{4}{9} =$

21. $1\frac{5}{6} + 1\frac{3}{4} =$

22. $5\frac{2}{3} + 3 + 1\frac{1}{2} =$

23. $7\frac{3}{5} + \frac{4}{7} + 1\frac{1}{2} =$

24. $3\frac{2}{5} + 5 + 3\frac{1}{4} =$

Applications

25. A fruit salad calls for $\frac{1}{4}$ pound of sliced peaches, $\frac{1}{3}$ pound of melon cubes, and $\frac{1}{5}$ pound of strawberries. How many pounds of fruit are in the salad?

26. In a presidential election the Democratic candidate received $\frac{23}{50}$ of the votes and the Republican candidate received $\frac{9}{20}$ of the votes. What fractional part of the vote did the two candidates get between them?

27. A brake pad wears down $\frac{1}{64}$ inches during one long trip. On another trip the brake wear is $\frac{3}{128}$ inches. How many inches of brake pad are worn on the two trips?

28. A bicycle is shipped in a wooden crate. The crate weighs $8\frac{1}{4}$ pounds without the bicycle and the bicycle weighs $28\frac{5}{16}$ pounds. How much do the bicycle and crate weigh together?

For Some Extra Thought

29. Nine thirty-sevenths plus what mixed number equals two?

30. In an election is it possible for the two candidates to receive $\frac{3}{7}$ and $\frac{5}{8}$ of the votes, respectively? Explain.

Review Problems for Maintaining Your Skills

31. Round 4,189,733 to the nearest hundred.

32. $720 - 633 =$

33. $10{,}000 \times 45 =$

34. For $719 \div 100$ find the quotient and remainder.

35. Reduce $\frac{125}{175}$ to its lowest terms.

36. Convert $\frac{7}{15}$ to an equivalent fraction with a denominator of 75.

37. Convert $\frac{37}{17}$ to a mixed number.

38. Convert $3\frac{1}{8}$ to an improper fraction.

In 39 through 42, reduce each result to its lowest terms and express any result that is an improper fraction as a mixed number.

39. $\frac{11}{15} \times \frac{3}{4} =$

40. $2\frac{4}{5} \times \frac{3}{7} =$

41. $\frac{2}{5} \div \frac{1}{5} =$

42. $1\frac{2}{7} \div 3\frac{1}{2} =$

2.6 SUBTRACTING FRACTIONS AND MIXED NUMBERS

Preview Test

In questions 1 through 6, find each difference. Express each difference that is an improper fraction as a mixed number.

1. $\frac{7}{9}-\frac{2}{9}$ **2.** $\frac{15}{16}-\frac{3}{8}$ **3.** $\frac{5}{12}-\frac{4}{15}$

4. $5\frac{1}{2}-2\frac{3}{8}$ **5.** $12\frac{5}{9}-3\frac{2}{3}$ **6.** $7-1\frac{6}{7}$

To subtract one fraction from another with like denominators, subtract the numerators and place the difference over the common denominator. For example,

$$\frac{13}{20}-\frac{6}{20}=\frac{7}{20}$$

For subtraction involving fractions or mixed numbers with unlike denominators, use the same three-step procedure you used for addition.

Step 1: Find the L.C.D.

Step 2: Change each fraction whose denominator is not the L.C.D. to an equivalent fraction with the L.C.D. as the denominator.

Step 3: Subtract by placing the difference of the numerators over the L.C.D.

EXAMPLE 1: Subtract **a)** $\frac{5}{24}-\frac{1}{20}$ **b)** $6\frac{3}{4}-2\frac{1}{8}$

c) $2\frac{3}{4}-1\frac{13}{16}$ **d)** $13-5\frac{1}{7}$

Solution:

a) The L.C.D. is 120.

$$\frac{5}{24}=\frac{5\times 5}{24\times 5}=\frac{25}{120} \qquad \frac{1}{20}=\frac{1\times 6}{20\times 6}=\frac{6}{120}$$

$$\frac{25}{120}-\frac{6}{120}=\frac{19}{120}$$

b) The L.C.D. is 8.

First convert $6\frac{3}{4}$ to $6\frac{6}{8}$.

Then subtract the fraction parts and whole number parts. Therefore,

$$6\frac{6}{8}-2\frac{1}{8}=4\frac{5}{8}$$

ANSWERS FOR PREVIEW TEST

1. $\frac{5}{9}$ **2.** $\frac{9}{16}$ **3.** $\frac{3}{20}$ **4.** $3\frac{1}{8}$ **5.** $8\frac{8}{9}$ **6.** $5\frac{1}{7}$

c) The L.C.D. is 16.

First convert $2\frac{3}{4}$ to $2\frac{12}{16}$. Since we cannot subtract $\frac{13}{16}$ from $\frac{12}{16}$ you must borrow 1 from 2, rename 1 as $\frac{16}{16}$ and add $\frac{16}{16}$ to $\frac{12}{16}$. So,

$$2\frac{12}{16} = 1 + \frac{16}{16} + \frac{12}{16} = 1\frac{28}{16}$$

Then, $1\frac{28}{16} - 1\frac{13}{16} = \frac{15}{16}$.

d) First convert 13 to $12\frac{7}{7}$ by borrowing 1 from 13. Then

$$12\frac{7}{7} - 5\frac{1}{7} = 7\frac{6}{7}$$

PRACTICE PROBLEM 1: Subtract

a) $\frac{17}{25} - \frac{19}{40}$ **b)** $12\frac{2}{5} - 4\frac{1}{3}$ **c)** $7\frac{1}{8} - 4\frac{5}{6}$ **d)** $20 - 12\frac{3}{8}$

Be Careful About This

Borrowing is only necessary if the fraction in the subtrahend is larger than the fraction in the minuend. So borrowing is necessary for $4\frac{2}{5} - 2\frac{3}{5}$, but is not necessary for $6\frac{7}{9} - 2\frac{5}{9}$.

ANSWERS FOR PRACTICE PROBLEMS

1. **a)** $\frac{41}{200}$ **b)** $8\frac{1}{15}$ **c)** $2\frac{7}{24}$ **d)** $7\frac{5}{8}$

EXERCISE SET 2.6A

Skill Development

Find each difference. Reduce all results to their lowest terms.

1. $\frac{24}{40} - \frac{11}{40} =$

2. $\frac{15}{26} - \frac{3}{26} =$

3. $\frac{85}{100} - \frac{74}{100} =$

4. $\frac{1}{2} - \frac{1}{8} =$

5. $\frac{7}{10} - \frac{1}{12} =$

6. $\frac{20}{21} - \frac{15}{28} =$

7. $\frac{19}{20} - \frac{8}{15} =$

8. $\frac{3}{8} - \frac{1}{7} =$

9. $\frac{15}{400} - \frac{7}{200} =$

10. $6\frac{3}{4} - 1\frac{1}{4} =$

11. $3\frac{4}{5} - 2\frac{5}{8} =$

12. $17\frac{5}{6} - 13\frac{3}{4} =$

13. $10\frac{5}{6} - 3\frac{8}{15} =$

14. $8\frac{1}{4} - 3\frac{1}{9} =$

15. $18\frac{6}{7} - 9\frac{3}{5} =$

16. $12 - 9\frac{3}{13} =$

17. $5\frac{1}{7} - 4 =$

18. $16 - 15\frac{15}{16} =$

Applications

19. The thickness of a new brake pad is $\frac{3}{4}$ inches. If 10,000 miles of driving wear out $\frac{5}{16}$ inches, how many inches of pad remain after the first 10,000 miles of driving?

20. The price of a stock drops from $\$\frac{7}{8}$ per share to $\$\frac{13}{16}$ per share. Find the amount of the drop.

21. A car with 4 passengers weighs 2 tons. If the passengers weigh $\frac{3}{10}$ tons, find the weight of the car in tons.

22. A mailbox post is $7\frac{3}{4}$ feet long. If it is set $1\frac{11}{16}$ feet into the ground, how high is the post above the ground?

For Some Extra Thought

23. Find two fractions whose difference is four-fifths.

24. Five-thirteenths plus what fraction equals one?

Review Problems for Maintaining Your Skills

25. Round 6,807, 466 to the nearest ten thousand.

26. For 127 ÷ 11, find the quotient and remainder.

27. Reduce $\frac{24}{38}$ to its lowest terms.

28. Convert $\frac{5}{12}$ to an equivalent fraction with a denominator of 84.

29. Convert $\frac{39}{5}$ to a mixed number.

30. Convert $6\frac{3}{8}$ to an improper fraction.

In 31 through 36, reduce each result to its lowest terms and express any result that is an improper fraction as a mixed number.

31. $\frac{12}{13} \times \frac{5}{24} =$

32. $\frac{1}{2} \div \frac{3}{4} =$

33. $8\frac{2}{5} \div 2\frac{1}{5} =$

34. $\frac{2}{7} + \frac{3}{4} =$

35. $\frac{3}{8} + \frac{1}{2} + \frac{3}{4} =$

36. $3\frac{5}{7} + 2 =$

EXERCISE SET 2.6B

Skill Development

Find each difference. Reduce all results to their lowest terms.

1. $\frac{7}{8} - \frac{2}{8} =$
2. $\frac{17}{20} - \frac{12}{20} =$
3. $\frac{9}{31} - \frac{5}{31} =$
4. $\frac{3}{8} - \frac{1}{4} =$
5. $\frac{5}{6} - \frac{2}{3} =$
6. $\frac{17}{20} - \frac{1}{5} =$
7. $\frac{6}{7} - \frac{2}{5} =$
8. $\frac{9}{10} - \frac{1}{4} =$
9. $\frac{14}{15} - \frac{3}{5} =$
10. $5\frac{5}{6} - 2\frac{1}{6} =$
11. $12\frac{4}{5} - 8\frac{2}{3} =$
12. $7\frac{3}{4} - 6\frac{5}{9} =$
13. $10\frac{5}{7} - 1\frac{2}{3} =$
14. $6\frac{3}{4} - 2\frac{2}{5} =$
15. $20\frac{1}{2} - 15\frac{3}{4} =$
16. $12 - 8\frac{4}{5} =$
17. $8\frac{7}{9} - 5 =$
18. $3\frac{4}{5} - \frac{7}{8} =$

Applications

19. A golf course orders $\frac{3}{4}$ tons of fertilizer for their golf greens. If they use $\frac{2}{3}$ tons, how many tons remain?

20. In a 1-mile race the lead runner has completed $\frac{11}{32}$ miles. How far is the lead runner from the finish line?

21. The combined weight of a bike and its shipping carton is 31 pounds. If the bike weighs $22\frac{3}{4}$ pounds find the weight of the carton.

22. The price of a stock drops from $\$167\frac{1}{4}$ per share to $\$164\frac{7}{8}$ per share. Find the amount of the drop in price.

For Some Extra Thought

23. Find two mixed numbers whose difference is 1.

24. Three and two-fifths plus what number equals four?

Review Problems for Maintaining Your Skills

25. Round 739 to the nearest ten.

26. For $89 \div 9$ find the quotient and remainder.

27. Reduce $\frac{150}{750}$ to lowest terms.

28. Convert $\frac{17}{20}$ to an equivalent fraction with a denominator of 300.

29. Convert $\frac{78}{23}$ to a mixed number.

30. Convert $1\frac{7}{9}$ to an improper fraction.

In 31 through 36, reduce each result to its lowest terms and express any result that is an improper fraction as a mixed number.

31. $\frac{5}{6} \times \frac{3}{10} =$

32. $\frac{7}{10} \times \frac{2}{5} =$

33. $3\frac{1}{4} \div \frac{5}{8} =$

34. $\frac{6}{7} + \frac{2}{5} =$

35. $\frac{1}{5} + \frac{1}{3} + \frac{1}{4} =$

36. $3\frac{2}{5} + 4\frac{5}{6} =$

2.7 MIXED OPERATIONS

Preview Test

1. $\frac{2}{3} \times \frac{1}{4} + \frac{1}{6} =$

2. $\frac{3}{5} - \frac{1}{3} \div \frac{2}{3} =$

3. $4\frac{1}{2} - \left(\frac{1}{3} + \frac{2}{3} \times \frac{1}{2}\right) =$

4. $7 \div \left(\frac{1}{4} + \frac{1}{8} \times 3\right) =$

The order in which operations should be performed is the same for fractions and mixed numbers as it was for whole numbers.

EXAMPLE 1: $\frac{1}{2} \times \frac{3}{4} + \frac{1}{4} \times \frac{2}{5} =$

$$\underbrace{\frac{1}{2} \times \frac{3}{4}}_{\text{Multiply}} + \underbrace{\frac{1}{4} \times \frac{2}{5}}_{\text{Multiply}}$$

$$\frac{3}{8} + \frac{1}{10}$$

The LCD = 40.

$$= \frac{3 \times 5}{8 \times 5} + \frac{1 \times 4}{10 \times 4}$$

$$= \frac{15}{40} + \frac{4}{40} = \frac{19}{40}$$

PRACTICE PROBLEM 1: $\frac{7}{8} - \frac{3}{5} \div \frac{4}{5} =$

EXAMPLE 2: $7\frac{1}{2} - \left(4\frac{1}{2} + \frac{1}{3} \div \frac{1}{3}\right) =$

Solution:

$$7\frac{1}{2} - \left(4\frac{1}{2} + \underbrace{\frac{1}{3} \div \frac{1}{3}}_{\text{Divide}}\right)$$

$$= 7\frac{1}{2} - \underbrace{\left(4\frac{1}{2} + 1\right)}_{\text{Add}}$$

$$= 7\frac{1}{2} - 5\frac{1}{2} = 2$$

ANSWERS FOR PREVIEW TEST

1. $\frac{1}{3}$ **2.** $\frac{1}{10}$ **3.** $3\frac{5}{6}$ **4.** $11\frac{1}{5}$

PRACTICE PROBLEM 2: $\left(\frac{9}{10} - \frac{1}{5}\right) \times \left(\frac{2}{3} - \frac{1}{6}\right) =$

ANSWERS FOR PRACTICE PROBLEMS

1. $\frac{1}{8}$ **2.** $\frac{7}{20}$

EXERCISE SET 2.7A

Skill Development

1. $\frac{3}{4} \times \frac{7}{8} + \frac{1}{2} =$

2. $\frac{5}{6} - \frac{1}{2} \times \frac{1}{2} =$

3. $\frac{2}{3} \times \frac{1}{2} + \frac{2}{3} \times \frac{1}{2} =$

4. $\frac{4}{5} + \frac{2}{3} \div \frac{1}{3} + \frac{2}{5} =$

5. $\frac{1}{3} \times \frac{2}{5} + \frac{4}{5} \times \frac{2}{3} + \frac{1}{5} \times \frac{1}{2} =$

6. $4\frac{1}{2} + 2\frac{1}{2} \times 1\frac{1}{2} =$

7. $\frac{4}{5} - \left(\frac{1}{5} + \frac{2}{3} \times \frac{3}{4}\right) =$

8. $2 - \left(\frac{1}{2} + \frac{3}{4} \div \frac{4}{5}\right) =$

9. $\left(\frac{1}{2} \times \frac{2}{3} + \frac{1}{3}\right) \div \left(\frac{2}{3} - \frac{4}{5} \times \frac{1}{2}\right) =$

10. $\left(\frac{3}{4} \times \frac{2}{3} - \frac{1}{2} \times \frac{1}{6}\right) \times 3\frac{1}{4} - 1 =$

For Some Extra Thought

11. $4\frac{1}{2} \div 4\frac{1}{2} \div 4\frac{1}{2} =$

12. $0 \times 5\frac{4}{5} \times 6\frac{2}{3} \div 1\frac{1}{2} =$

Review Problems for Maintaining Your Skills

13. Round 2,145,837 to the nearest hundred thousand.

14. $230 - 189 =$

15. $52 \times 405 =$

16. Reduce $\frac{20}{65}$ to its lowest terms.

17. Convert $\frac{11}{13}$ to an equivalent fraction with a denominator of 78.

18. Convert $\frac{74}{73}$ to a mixed number.

19. Convert $1\frac{2}{9}$ to an improper fraction.

In 20 through 28, reduce each fraction to its lowest terms and express any result that is an improper fraction as a mixed number.

20. $\frac{20}{21} \times \frac{3}{5} =$

21. $3\frac{1}{4} \times 5\frac{2}{3} =$

22. $\frac{3}{8} \div 1\frac{1}{2} =$

23. $\frac{3}{4} + \frac{4}{5} =$

24. $2\frac{2}{3} + \frac{5}{6} =$

25. $4 + 1\frac{2}{9} =$

26. $\frac{10}{11} - \frac{3}{22} =$

27. $12 - 3\frac{5}{8} =$

28. $6\frac{1}{2} - 1\frac{3}{5} =$

EXERCISE SET 2.7B

Skill Development

1. $\frac{1}{2} \div \frac{3}{4} + \frac{5}{6} =$

2. $\frac{11}{16} - \frac{5}{16} \times \frac{3}{8} =$

3. $\frac{4}{5} \times \frac{1}{2} + \frac{3}{4} \times \frac{5}{6} =$

4. $\frac{3}{4} - \frac{1}{2} \div \frac{5}{6} + \frac{2}{3} =$

5. $\frac{2}{3} \times \frac{3}{4} + \frac{1}{2} \times \frac{2}{3} + \frac{1}{3} \times \frac{3}{4} =$

6. $5\frac{1}{3} - 2\frac{1}{2} \times 1\frac{2}{3} =$

7. $\frac{5}{6} + \left(\frac{2}{5} - \frac{1}{2} \times \frac{1}{2}\right) =$

8. $4 \times \left(\frac{2}{3} + \frac{1}{2} \times \frac{3}{4}\right) =$

9. $\left(\frac{2}{3} \times \frac{1}{2} - \frac{1}{4}\right) \div \left(\frac{1}{4} + \frac{3}{4} \times \frac{2}{3}\right) =$

10. $\left(\frac{2}{5} \times \frac{1}{2} + \frac{3}{5} \times \frac{1}{2}\right) \div 4\frac{1}{2} + 2 =$

For Some Extra Thought

11. $\frac{2}{5} \div \frac{2}{5} \div \frac{2}{5} =$

12. $0 \div \left(\frac{1}{2} \times \frac{5}{6} + \frac{2}{3}\right) =$

Review Problems for Maintaining Your Skills

13. Round 32,755 to the nearest hundred.

14. $1{,}200 - 943 =$

15. $3{,}422 \times 100{,}000 =$

16. Reduce $\frac{64}{96}$ to its lowest terms.

17. Convert $\frac{7}{10}$ to an equivalent fraction with a denominator of 40.

18. Convert $\frac{47}{33}$ to a mixed number.

19. Convert $5\frac{3}{7}$ to an improper fraction.

In 20 through 28, reduce each fraction to its lowest terms and express any result that is an improper fraction as a mixed number.

20. $\frac{12}{13} \times \frac{2}{3} =$

21. $2\frac{3}{4} \times 6\frac{1}{2} =$

22. $7 \div 3\frac{1}{2} =$

23. $\frac{5}{6} + \frac{3}{4} =$

24. $3\frac{3}{5} + 1\frac{2}{3} =$

25. $3\frac{2}{5} + \frac{10}{13} =$

26. $\frac{5}{8} - \frac{5}{12} =$

27. $10 - 6\frac{5}{9} =$

28. $7\frac{3}{4} - 6\frac{4}{5} =$

PRACTICE TEST 2A

1. Reduce to their lowest terms.

a) $\frac{125}{350}$ **b)** $\frac{74}{78}$ **c)** $\frac{64}{192}$

2. Convert each fraction to an equivalent fraction with the indicated denominator.

a) $\frac{7}{9}$, 54 **b)** $\frac{7}{16}$, 64

3. Convert each improper fraction to a mixed number. Reduce each fraction part to its lowest terms.

a) $\frac{19}{7}$ **b)** $\frac{100}{16}$ **c)** $\frac{51}{20}$

4. Convert each mixed number to an improper fraction.

a) $5\frac{4}{9}$ **b)** $12\frac{2}{15}$ **c)** $1\frac{5}{7}$

5. Find each product. Reduce all results to their lowest terms.

a) $\frac{4}{9} \times \frac{3}{10} =$ **b)** $5\frac{1}{2} \times 3\frac{4}{5} =$

6. Divide and reduce each quotient to its lowest terms. Express any quotient that is an improper fraction as a mixed number.

a) $\frac{2}{5} \div \frac{3}{5} =$ **b)** $5 \div \frac{1}{4} =$ **c)** $6\frac{3}{4} \div 2\frac{2}{5} =$

7. At $\$\frac{3}{4}$ per share how many shares of stock can you purchase for \$600?

8. Find the least common multiple of 8, 12, and 20.

9. Find each sum. Reduce all results to their lowest terms.

a) $\frac{1}{6} + \frac{2}{5} =$ **b)** $5\frac{2}{5} + \frac{3}{10} =$ **c)** $10\frac{4}{7} + 3\frac{1}{4} =$

10. Find each difference. Reduce all results to their lowest terms.

a) $\frac{13}{15} - \frac{5}{12} =$ **b)** $7\frac{2}{3} - 5 =$ **c)** $12\frac{1}{2} - 3\frac{2}{5} =$

11. The fractional part of a senior class that attends the local community college has increased from $\frac{1}{5}$ to $\frac{1}{4}$. Find the increase as a fraction.

12. $\frac{7}{8} - \frac{3}{4} \times \frac{1}{2} =$

13. $\left(\frac{1}{2} + \frac{1}{3}\right) \times \left(\frac{1}{2} - \frac{1}{3}\right) =$

PRACTICE TEST 2B

1. Reduce to their lowest terms.

a) $\frac{44}{48}$ **b)** $\frac{300}{850}$ **c)** $\frac{12}{200}$

2. Convert each fraction to an equivalent fraction with the indicated denominator.

a) $\frac{3}{8}$, 72 **b)** $\frac{13}{100}$, 1,000

3. Convert each improper fraction to a mixed number. Reduce each fraction part to its lowest terms.

a) $\frac{100}{40}$ **b)** $\frac{43}{31}$ **c)** $\frac{75}{55}$

4. Convert each mixed number to an improper fraction.

a) $4\frac{13}{15}$ **b)** $13\frac{1}{10}$ **c)** $2\frac{7}{8}$

5. Find each product. Reduce all results to their lowest terms.

a) $\frac{11}{12} \times \frac{3}{5} =$ **b)** $4\frac{1}{2} \times 6\frac{1}{3}$

6. Divide and reduce each quotient to its lowest terms. Express any quotient that is an improper fraction as a mixed number.

a) $\frac{4}{7} \div \frac{1}{7} =$ **b)** $5\frac{3}{8} \div 2 =$ **c)** $10\frac{1}{2} \div 3\frac{4}{5}$

7. Three-quarters of a ton of coal is to be divided into 5 equal parts. Find the number of tons in each part.

8. Find the least common multiple of 5, 20, and 30.

9. Find each sum. Reduce all results to their lowest terms.

a) $\frac{11}{15} + \frac{3}{10} =$ **b)** $6\frac{1}{2} + \frac{2}{15} =$ **c)** $9\frac{1}{3} + 2\frac{1}{8} =$

10. Find each difference. Reduce all results to their lowest terms.

a) $\frac{3}{4} - \frac{1}{5} =$ **b)** $8 - 7\frac{1}{4} =$ **c)** $13\frac{1}{4} - 5\frac{3}{5} =$

11. Two-fifths of the athletic budget is for football and one-fourth is for basketball. What fraction of the athletic budget is assigned to the two sports?

12. $\frac{3}{5} + \frac{1}{2} \times \frac{3}{4} =$ **13.** $\frac{19}{20} \div \left(\frac{1}{2} + \frac{3}{4} - \frac{1}{5}\right) =$

CUMULATIVE REVIEW EXERCISES FOR CHAPTERS 1 AND 2

1. In the number 2,543,789 the 4 is in the ______________ place.
2. Write 21,341,000 in words.
3. Write the number for one million, eighty thousand, two hundred.
4. Round 23,478 to the nearest hundred.
5. Add the numbers: 1,235, 148, and 2,400
6. Subtract 567 from 3,840.
7. Find the total price for 34 concert tickets if each ticket costs $28.
8. Find the quotient and remainder for $684 \div 27$.
9. $6 \times 34 - 10 \times 2$
10. Reduce $\frac{120}{145}$ to its lowest terms.
11. Convert $\frac{3}{8}$ to an equivalent fraction with denominator 48.
12. Convert $\frac{78}{25}$ to a mixed number.
13. Convert $4\frac{5}{7}$ to an improper fraction.

In 14 through 18, reduce each fraction to its lowest terms and express any result that is an improper fraction as a mixed number.

14. $\frac{10}{11} \times \frac{22}{25} =$
15. $2\frac{1}{2} \div 1\frac{1}{2} =$
16. $3\frac{2}{5} + 4\frac{1}{2} =$
17. $\frac{6}{7} - \frac{1}{3} =$
18. $\frac{1}{3} \times \frac{1}{4} + \frac{3}{4} \times \frac{1}{3} =$

CHAPTER 3

DECIMALS

3.1 PLACE VALUE

3.2 ROUNDING

3.3 ADDITION AND SUBTRACTION

3.4 MULTIPLICATION

3.5 DIVISION

3.6 MIXED OPERATIONS

There are many situations where a decimal is used.

- The price of a computer disk: \$1.57
- A baseball player's batting average: 0.317
- Your grade point average: 3.24
- A normal body temperature: 98.6 degrees

You may have to add, subtract, multiply, or divide decimal numbers to answer questions such as:

- At three consecutive stops for gas for your car you buy 9.67 gallons, 10.32 gallons, and 7.19 gallons. Find the total number of gallons purchased.
- How many miles did you ride if your bike odometer changes from 437.2 to 456.8?
- At \$28.75 for one share of AT&T stock, what is the cost for 100 shares?
- If 14 tickets to a concert cost \$332.50, what is the price for 1 ticket?

3.1 PLACE VALUE

Preview Test

Write the name of the place and the value of the underlined digit.

1. $3.8\underline{5}24$ **2.** $0.045\underline{3}79$ **3.** $74.\underline{6}83$

Write each number in words.

4. 4.62

5. 0.0795

6. 34.009

Identify the number that is written in words.

7. two and five hundred thirty-seven ten-thousandths.

8. sixty-five thousandths.

9. twenty and five hundred-thousandths.

Place Value Names

With decimal numbers, we have place values to the right of the units position. A period, which we call a decimal point, separates the units position from the tenths

ANSWERS FOR PREVIEW TEST

1. hundredths, $\frac{5}{100}$ **2.** ten-thousandths, $\frac{3}{10,000}$ **3.** tenths, $\frac{6}{10}$

4. four and sixty-two hundredths

5. seven hundred ninety-five ten-thousandths

6. thirty-four and nine thousandths

7. 2.0537 **8.** 0.065 **9.** 20.00005

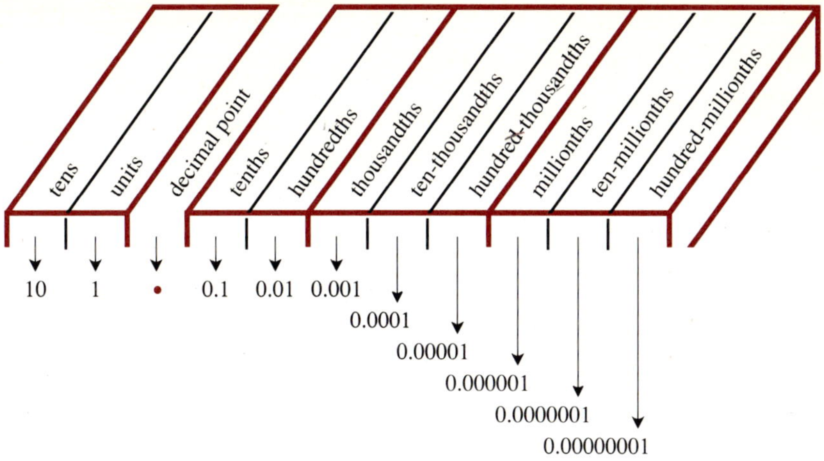

Figure 3.1.1

position. The names of the places from tens through hundred-millionths, along with their numerical values, is shown in Figure 3.1.1

If you know the names for the whole number place values, then it is easy to learn the names for the decimal place values. The whole number place value names end with "s" (units, tens, hundreds, thousands, . . .). The decimal place value names end with "ths" (tenths, hundredths, thousandths, . . .).

EXAMPLE 1: Write the name of the place and the value of each underlined digit.

a) $2.3\underline{4}86$ **b)** $45.087\underline{1}6$ **c)** $0.00006\underline{7}$

Solution:

	Place	Value
a)	hundredths	$\frac{4}{100}$
b)	ten-thousandths	$\frac{1}{10,000}$
c)	millionths	$\frac{7}{1,000,000}$

PRACTICE PROBLEM 1: Write the name of the place and the value of each underlined digit.

a) $1.65\underline{8}2$ **b)** $0.0000003\underline{1}$ **c)** $47.\underline{6}42$

Names for Decimal Numbers

To write a decimal number in words write the set of numbers to the right of the decimal point as a whole number followed by the name of the rightmost position.

0.0042 in words is forty-two ten-thousandths

If the decimal number has no whole number part, then it is customary to write a zero to the left of the decimal point as we did for 0.0042.

If the decimal number has a whole number part, then use the word "and" for the decimal point.

5.07 in words is five and seven hundredths

EXAMPLE 2: Write each number in words.

a) 2.003 **b)** 0.00782

Solution:

a) two and three thousandths

b) seven hundred eighty-two hundred-thousandths

PRACTICE PROBLEM 2: Write each number in words.

a) 0.0004003 **b)** 62.078

EXAMPLE 3: Identify the number that is written in words.

a) six and seventeen hundred-thousandths

b) four thousand five ten-millionths

Solution:

a) 6.00017 **b)** 0.0004005

PRACTICE PROBLEM 3: Identify the number that is written in words.

a) sixty-two and nine tenths

b) forty-three millionths

Be Careful About This

Be very careful with zeros in decimal numbers. For example, 0.7 is not equal to 0.07, because

$$0.7 = \frac{7}{10} \quad \text{and} \quad 0.07 = \frac{7}{100}$$

However, 0.07 is equal to 0.070, because

$$0.07 = \frac{7}{100} \quad \text{and} \quad 0.070 = \frac{70}{1{,}000}, \text{ which reduces to } \frac{7}{100}.$$

Although 0.07 equals 0.070, these numbers have different names. The name for 0.07 is seven hundredths and the name for 0.070 is seventy thousandths.

ANSWERS FOR PRACTICE PROBLEMS

1. **a)** thousandths, $\frac{8}{1{,}000}$ **b)** hundred-millionths, $\frac{1}{100{,}000{,}000}$

   c) tenths, $\frac{6}{10}$

2. **a)** four thousand three ten-millionths

   b) sixty-two and seventy-eight thousandths

3. **a)** 62.9 **b)** 0.000043

EXERCISE SET 3.1A

Skill Development

Write the name of the place and the value of each underlined digit.

1. $2.57\underline{8}$
2. $0.0046\underline{2}6$
3. $78.\underline{9}041$
4. $0.000\underline{8}$
5. $67.504\underline{7}$
6. $9.40\underline{6}1$
7. $0.0470\underline{3}$
8. $480.6\underline{7}42$

Write each number in words.

9. 7.6023
10. 0.007023
11. 75.0981
12. 0.000531
13. 5.9
14. 0.000006
15. 345.2273
16. 3.4063

Identify the number that is written in words.

17. four and seven tenths
18. thirty-five ten-thousandths
19. four millionths
20. six ten-millionths
21. two and twenty-three hundred-thousandths
22. sixteen millionths
23. forty-two and one tenth
24. six and thirty-two ten-thousandths

Applications

25. Last year the rainfall in Denmore County was 29.472 inches. What is the place name and value of the 7 in this measurement?

26. Arlene McCall's batting average for her softball team was three hundred forty-five thousandths. Write her batting average as a decimal number.

27. The average family size in Camden County is 2.63. Write this statistic in words.

For Some Extra Thought

28. In some countries the roles of the comma and decimal point are reversed. That is, one thousand two hundred and thirty-five hundredths is written as 1.200,35. Using this system write 23.462,007 in words.

29. Refer to 28 and write the number for six and forty ten-thousandths.

Review Problems for Maintaining Your Skills

30. Round 21,687 to the nearest thousand.

31. For 185 ÷ 17 find the quotient and remainder.

32. $5 \times 8 - 7 \times 2 =$

33. Reduce $\frac{48}{72}$ to its lowest terms.

34. Convert $\frac{17}{7}$ to a mixed number.

In 35 through 40, reduce each fraction to its lowest terms and express any result that is an improper fraction as a mixed number.

35. $\frac{3}{5} \times \frac{1}{3} =$

36. $5\frac{2}{3} \div \frac{3}{5} =$

37. $\frac{17}{20} + \frac{3}{10} =$

38. $12\frac{1}{4} - 1\frac{2}{3} =$

39. $\frac{1}{4} + \frac{2}{3} \times \frac{8}{9} =$

40. $5 \times \frac{4}{5} - 3\frac{1}{2} =$

EXERCISE SET 3.1B

Skill Development

Write the name of the place and the value of each underlined digit.

1. $5.6\underline{8}43$
2. $0.\underline{3}0477$
3. $3.1087\underline{3}$
4. $38.40\underline{5}3$
5. $45.804\underline{2}$
6. $0.000000\underline{6}$
7. $13.7\underline{1}8$
8. $2.413\underline{7}$

Write each number in words.

9. 6.8044
10. 0.074633
11. 56.8047
12. 4.8026
13. 27.64
14. 32.0004
15. 4.168421
16. 0.017

Identify the number that is written in words.

17. six and eighteen hundredths
18. fifty-six and seven ten-thousandths
19. seventy-two thousandths
20. thirty and four tenths
21. twenty-one and five hundred sixty hundred-thousandths
22. six ten-thousandths
23. eighty-two and thirty-three hundredths
24. ninety-four hundred-thousandths

Applications

25. The normal body temperature for a human is 98.6 degrees Fahrenheit. What is the name and value of the 8 in this measurement?

26. Ed Walsh had a lifetime earned run average of one and eighty-two hundredths. Write his earned run average as a decimal number.

27. Harold Taylor's grade point average is 3.29. Write his grade point average in words.

For Some Extra Thought

28. After the hundred-millionths place, the place value names for the next three positions are billionths, ten-billionths, and hundred-billionths. Write the number for thirty ten-billionths.

29. Refer to 28 and write 0.0000000006 in words.

Review Problems for Maintaining Your Skills

30. Round 8,764 to the nearest ten.

31. For 478 ÷ 123 find the quotient and remainder.

32. $4 \times 9 + 3 \times 7 =$

33. Reduce $\frac{125}{175}$ to its lowest terms.

34. Convert $\frac{33}{17}$ to a mixed number.

In 35 through 40, reduce each fraction to its lowest terms and express any result that is an improper fraction as a mixed number.

35. $\frac{13}{15} \times \frac{5}{13} =$

36. $4\frac{1}{3} \div 2\frac{2}{5} =$

37. $\frac{3}{4} + \frac{1}{12} =$

38. $15\frac{1}{4} - 4 =$

39. $\frac{3}{4} - \frac{1}{3} \times \frac{4}{5} =$

40. $7 \times 3\frac{1}{2} - 3\frac{1}{2} =$

3.2 ROUNDING

> **Preview Test**
>
> **1.** Round 0.5743 to the nearest hundredth.
>
> **2.** Round 8.07531 to the nearest ten-thousandth.
>
> **3.** Round 1.00007494 to the nearest millionth.
>
> **4.** Round 0.85 to the nearest tenth.
>
> **5.** Round $7.4173 to the nearest cent.

To round 7.38 to the nearest tenth you have to answer the question, "Is 7.38 closer to 7.3 or 7.4?" Since 7.35 is halfway between 7.3 and 7.4, it is easy to see in Figure 3.2.1 that 7.38 is closer to 7.4. So, 7.38 rounded to the nearest tenth is 7.4.

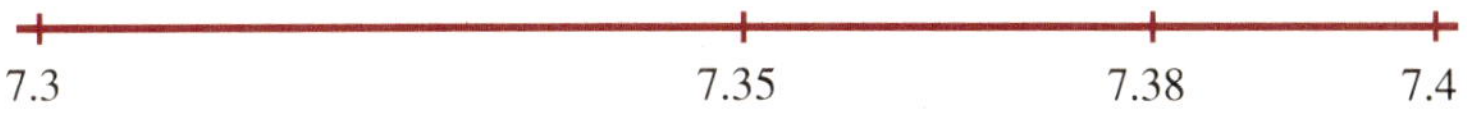

Figure 3.2.1

The alternate "Digit to the right method" which was used in Section 1.2 works as well for decimal numbers. The only difference is we delete the numbers after completing Step 2.

To round a decimal number to a place value position

Step 1: Look at the digit to the right of the place value rounding position.
Step 2: If the digit in Step 1 is 0, 1, 2, 3, or 4, then the digit in the rounding position is not changed. If the digit in Step 1 is 5, 6, 7, 8, or 9, then the digit in the rounding position is increased by 1.
Step 3: Delete all digits to the right of the place value rounding position.

Lct's illustrate this method for rounding 7.38 to the nearest tenth.

Step 1: The digit that lies to the right of the tenths position is 8.
Step 2: Add 1 to the 3 in the tenths position.
Step 3: Delete the 8.
Therefore, 7.38 rounded to the nearest tenth equals 7.4.

> **Be Careful About This**
>
> The last sentence written above ends with "nearest tenth equals 7.4." The period after the 4 is the period for the end of the sentence. It is not a decimal point. The decimal point is the period symbol between the 7 and 4.
>
> 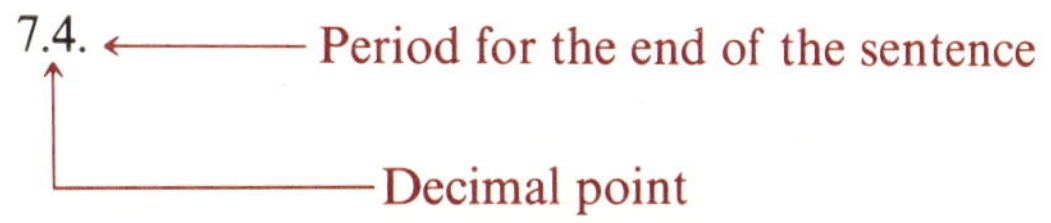
>

ANSWERS FOR PREVIEW TEST

1. 0.57 **2.** 8.0753 **3.** 1.000075 **4.** 0.9 **5.** $7.42

EXAMPLE 1: Round each decimal number.

a) 3.4728 to the nearest hundredth

b) 0.047968 to the nearest hundred-thousandth

Solution:

a) **Step 1:** The digit to the right of the hundredths position is 2.
Step 2: Do not change the 7 in the hundredths position.
Step 3: Delete the 2 and 8.
Therefore, 3.4728 rounded to the nearest hundredth equals 3.47.

b) **Step 1:** The digit to the right of the hundred-thousandths position is 8.
Step 2: Add 1 to the 6 in the hundred-thousandths position.
Step 3: Delete the 8.
Therefore, 0.047968 rounded to the nearest hundred-thousandth equals 0.04797.

PRACTICE PROBLEM 1: Round each decimal number.

a) 48.63492 to the nearest thousandth.

b) 1.99 to the nearest tenth.

For money amounts, the cents position is in the hundredths position. Let's round $4.563 to the nearest cent.

Step 1: The digit to the right of the cents (hundredths) position is 3.
Step 2: Do not change the 6 in the cents position.
Step 3: Delete the 3.
Therefore, $4.563 rounded to the nearest cent equals $4.56.

EXAMPLE 2: Round $13.8074 to the nearest cent.

Solution:

Step 1: The digit to the right of the cents position is 7.
Step 2: Add 1 to the 0 in the cents position.
Step 3: Delete the 7 and 4.
Therefore, $13.8074 rounded to the nearest cent equals $13.81.

PRACTICE PROBLEM 2: Round $8.3647 to the nearest cent.

ANSWERS FOR PRACTICE PROBLEMS

1. **a)** 48.635 **b)** 2.0 **2.** $8.36

EXERCISE SET 3.2A

Skill Development

1. Round 45.236 to the nearest tenth.
2. Round 6.904 to the nearest hundredth.
3. Round 12.2504678 to the nearest ten-thousandth.
4. Round 1.2576331 to the nearest hundred-thousandth.
5. Round 14.555555 to the nearest hundredth.
6. Round 0.070905 to the nearest thousandth.
7. Round 43.64284573 to the nearest millionth.
8. Round 0.143789547 to the nearest ten-millionth.
9. Round 5.4455 to the nearest tenth.
10. Round $4.5637 to the nearest cent.
11. Round $33.257 to the nearest cent.
12. Round $3.87493 to the nearest cent.

Applications

13. In May the rainfall measured 2.637 inches. Round this measurement to the nearest hundredth.

14. Jack Watson's grade point average is 3.096. Round his grade point average to the nearest tenth.

15. With a precision pressure gauge, the pressure in a bike tire was measured at 84.3714 pounds. Round the pressure to the nearest thousandth.

16. The exchange rate is 2,143.714 pesos per dollar. Round this exchange rate to the nearest tenth of a peso.

For Some Extra Thought

17. Round $4.05273 to the nearest tenth of a cent.
18. Is 1.0172 closer to 1.01 or 1.02?

Review Problems for Maintaining Your Skills

19. Round 43,586,733 to the nearest hundred thousand.
20. $530 - 429 =$
21. $56 \times 1{,}000 =$
22. Convert $\frac{3}{8}$ to an equivalent fraction with a denominator of 56.
23. Convert $6\frac{3}{5}$ to an improper fraction.

In 24 through 29, reduce each fraction to its lowest terms and express any result that is an improper fraction as a mixed number.

24. $\frac{14}{15} \times \frac{3}{28} =$ **25.** $\frac{7}{8} \div 2\frac{1}{2} =$ **26.** $\frac{2}{5} + \frac{3}{8} =$

27. $13 - 6\frac{4}{9} =$ **28.** $\frac{2}{3} + \frac{3}{5} \times \frac{5}{4} =$ **29.** $5\frac{1}{4} - 4 \times \frac{1}{2} =$

30. Write the place and value of 6 in the decimal number 2.0761.

31. Write 1.478 in words.

32. Identify the number for seventeen and sixty-two millionths.

EXERCISE SET 3.2B

Skill Development

1. Round 5.807 to the nearest tenth.
2. Round 14.876912 to the nearest hundred-thousandth.
3. Round 0.47833 to the nearest thousandth.
4. Round 23.781322 to the nearest ten-thousandth.
5. Round 8.7777777 to the nearest hundredth.
6. Round 0.004545454545 to the nearest millionth.
7. Round 3.170899 to the nearest hundred-thousandth.
8. Round 5.2783 to the nearest thousandth.
9. Round 0.746238645 to the nearest ten-millionth.
10. Round $6.9745 to the nearest cent.
11. Round $14.37806 to the nearest cent.
12. Round $8.34165 to the nearest cent.

Applications

13. The thickness of a book is 3.5784 centimeters. Round the thickness to the nearest tenth.

14. The snowfall measured 12.5784 inches. Round the snowfall to the nearest hundredth.

15. The winner of the 100-meter dash had a time of 10.0487 seconds. Round the time to the nearest hundredth.

16. The average number of children per family in Calley County is 2.637. Round this number to the nearest tenth.

For Some Extra Thought

17. Round $0.006728 to the nearest hundredth of a cent.
18. Is 2.0346 closer to 2.03 or 2.04?

Review Problems for Maintaining Your Skills

19. Round 487 to the nearest hundred.
20. Subtract 47 from 120.
21. $62 \times 143 =$
22. Convert $\frac{3}{5}$ to an equivalent fraction with a denominator of 110.
23. Convert $10\frac{5}{7}$ to an improper fraction.

In 24 through 29, reduce each fraction to its lowest terms and express any result that is an improper fraction as a mixed number.

24. $1\frac{1}{4} \times 2\frac{3}{5} =$ **25.** $\frac{2}{5} \div \frac{3}{5} =$ **26.** $5\frac{3}{8} + \frac{2}{5} =$

27. $\frac{12}{13} - \frac{1}{12} =$ **28.** $4 \times \frac{1}{2} - \frac{2}{3} =$ **29.** $\frac{3}{8} - \frac{1}{8} \times \frac{4}{5} =$

30. Write the place and value of the 4 in the decimal number 6.408.

31. Write 5.0004 in words.

32. Identify the number for thirty-eight hundred-thousandths.

3.3 ADDITION AND SUBTRACTION

Preview Test

1. $15.8 + 24.1 =$
2. $7.96 + 13.71 =$
3. $14.6 - 8.45 =$
4. $20 - 0.67 =$
5. Add the following numbers: 213, 35.245, 4.9, and 10.06
6. Subtract 0.003 from 10.

Adding Decimal Numbers

Adding decimal numbers is much like adding whole numbers. We use the same basic facts, but have to adjust for place values. You may have already practiced adding decimal numbers without realizing it. You add decimal numbers every time you make several purchases and total your bill.

The key to adding is to match place values, that is, match units to units, tenths to tenths, and so on. This matching will occur automatically if you line up decimal points. Consider, for example, adding 1.4 to 3.5. Write the numbers one over another, lining up the decimal points.

decimal points lined up
↓

1	.	4
3	.	5
↑		↑
units place		tenths place

Notice that when the decimal points are lined up, the place values for the two numbers will match up. We can now add the two numbers as if they were whole numbers.

$$\begin{array}{rcr} 1 & . & 4 \\ +3 & . & 5 \\ \hline 4 & & 9 \end{array}$$

We still need to locate the decimal point in the answer. This is done by writing the decimal point under the decimal points for 1.4 and 3.5.

ANSWERS FOR PREVIEW TEST

1. 39.9 **2.** 21.67 **3.** 6.15 **4.** 19.33 **5.** 263.205 **6.** 9.997

```
  1 . 4
 +3 . 5
 ------
  4 . 9
    ↑
```

The decimal point in the answer is lined up under the other decimal points.

Thus, 1.4 added to 3.5 equals 4.9.

EXAMPLE 1: Add each set of numbers.

a) 6.47, 1.32, and 7.89

b) 8, 5.9, and 1.76

Solution:

a) Line up the decimal points and add.

```
  6.47
  1.32
  7.89
 -----
 15.68
```

b) The decimal point for the whole number 8 is positioned on its right.

8 = 8. ←—— Decimal point positioned here.

The decimal points are now lined up.

```
8.
5.9
1.76
```

As an optional step to make the addition clearer you can fill out the empty positions with zeros. Then add.

```
  8.00
  5.90
  1.76
 -----
 15.66
```

PRACTICE PROBLEM 1: Add each set of numbers.

a) 12.6, 4.8, 1.3, and 0.8

b) 14.32, 10, 1.678, and 7.9

Subtracting Decimal Numbers

Lining up decimal points works when subtracting decimal numbers, as the following example shows.

EXAMPLE 2: 9.32 − 6.75 =

Solution:

Write the numbers one over the other with the decimal points lined up. Make sure that you place 9.32 on top.

```
9.32
6.75
```

Subtract as if the numbers were whole numbers and place the decimal point in the answer under the two decimal points.

$$\begin{array}{r} 9.32 \\ -6.75 \\ \hline 2.57 \end{array}$$

Notice that this subtraction problem requires borrowing. If you've forgotten how to borrow, go back to Section 4 of Chapter 1 and review. You'll not be able to do decimal number subtraction if you don't remember your whole number subtraction! We'll also need to use some vocabulary we defined in Chapter 1. The names of the numbers in our subtraction problem are identified below. Hopefully, the names will look familiar!

$$\begin{array}{rl} 9.32 & \text{Minuend} \\ -6.75 & \text{Subtrahend} \\ \hline 2.57 & \text{Difference} \end{array}$$

PRACTICE PROBLEM 2: 14.6 − 8.9 =

EXAMPLE 3: Subtract 4.514 from 12.7.

Solution:
Read the directions in this problem carefully. The 12.7 must be written first because of the "subtract from" wording.

$$\begin{array}{lcrl} 12.7 & \longrightarrow & 12.700 & \text{Add 2 zeros to make 3 decimal places} \\ & & -\ 4.514 & \\ \hline & & 8.186 & \longleftarrow \text{Subtract then write in the decimal point} \end{array}$$

PRACTICE PROBLEM 3: Subtract 3.762 from 6.6.

Mental Estimation of Sums and Differences

Have you ever been the lucky one at a restaurant who is asked to check the accuracy of a dinner bill? Consider the check shown in Figure 3.3.1.

You can mentally estimate the accuracy of this dinner check by rounding each number to the nearest dollar and adding the rounded numbers.

\$ 4.50 = \$ 5 to the nearest dollar
\$ 1.30 = \$ 1 to the nearest dollar
\$ 4.40 = \$ 4 to the nearest dollar
\$13.50 = \$14 to the nearest dollar
\$17.00 = \$17 to the nearest dollar
\$ 9.75 = \$10 to the nearest dollar
\$ 3.00 = \$ 3 to the nearest dollar

Then, adding mentally

$$\$5 + \$1 + \$4 + \$14 + \$17 + \$10 + \$3 = \$54$$

This tells you that the \$53.45 total on the dinner check is reasonably accurate.

GUEST CHECK				
DATE	TABLE NUMBER	NO. PERSONS	SERVER	NO. 97752
9/18	3	4	1	

QUANTITY	ITEM	AMOUNT	
2	Shrimp Cocktails	4	50
2	Tomato Juice	1	30
4	Veg. Soup	4	40
1	12 oz. Steak	13	50
2	Roast Chicken	17	00
1	Salmon Steak	9	75
4	Coffee	3	00
	Total	53	45

Figure 3.3.1

EXAMPLE 4: Mentally estimate the sum for 34.95, 19.13, 26.581, and 10.6 by rounding each number to the nearest unit.

Solution:
The rounded numbers are, respectively:

$$35, 19, 27, \text{ and } 11$$

The estimated sum is

$$35 + 19 + 27 + 11 = 92$$

PRACTICE PROBLEM 4: Mentally estimate the sum for 53.8, 26.21, 40.05, and 18.7 by rounding each number to the nearest unit.

EXAMPLE 5: Mentally estimate the difference between \$1.21 and \$0.48 by rounding each amount to the nearest ten cents.

Solution:
The rounded amounts are, respectively:

$$\$1.20 \text{ and } \$0.50$$

The estimated difference is

$$\$1.20 - \$0.50 = \$0.70$$

PRACTICE PROBLEM 5: Mentally estimate the difference between \$12.68 and \$7.13 by rounding each amount to the nearest ten cents.

ANSWERS FOR PRACTICE PROBLEMS

1. a) 19.5 **b)** 33.898 **2.** 5.7 **3.** 2.838 **4.** 139 **5.** \$5.60

EXERCISE SET 3.3A

Skill Development

1. 13.6 + 15.9 =
2. 265.3 + 12.9 =
3. 7.54 − 1.91 =
4. 51.9 − 13.1 =
5. 2.73 + 1.5 =
6. 7.9 + 3.641 =
7. 9.34 − 6.172 =
8. 28 − 19.35 =
9. 15.16 − 8 =
10. 17.01 − 9.00654 =
11. 13.2 + 18.65 + 13.4 + 23 =
12. 24 + 19.81 + 0.3 + 1.56 =
13. Add the numbers: 15.2, 2.56, 0.478, and 1.25
14. Add the numbers: 175.06, 125, 450.3367, and 200
15. Subtract 12.3 from 19.1.
16. Subtract 14.26 from 25.9.
17. Subtract 1.667 from 5.
18. Subtract 0.03 from 0.0514.

Applications

19. From 1983 through 1987, the yearly costs for repairing Arlene Ferro's car were \$17.32, \$55.00, \$124.50, \$500.60, and \$978.12, respectively. Find the total cost for repairs over the 5-year period.

20. The bimonthly gas and electrical bills for the Tessler family were \$125.76, \$235.60, \$198.76, \$297.42, \$329.33, and \$367.91. Find the total cost for the year.

21. Between 6 a.m. and 2 p.m. the temperature increased from 17.2 degrees Celsius to 25.1 degrees Celsius. Find the rise in temperature during the 8-hour period.

22. The diameter of a living organism observed under a microscope increased from 0.00932 cm. to 0.0159 cm. over a two-month period. Find the amount of increase.

For Some Extra Thought

23. Find two numbers whose sum is 0.001.

24. A brake pad must be replaced when its thickness measures 0.09 inches or less. If the pad now measures 0.1 inches, and loses 0.005 inches per month, in how many months will the pad need replacement?

25. Find eight numbers whose sum is 0.33333333.

Mental Estimation

26. Estimate the sum for 65.8, 17.52, and 13.6 mentally by rounding each number to the nearest unit.

27. Estimate the difference between 89.22 and 35.701 mentally by rounding each number to the nearest unit.

28. Estimate the sum for \$7.25, \$9.65, \$12.25, and \$7.90 mentally by rounding each amount to the nearest dollar.

29. Estimate the sum for \$7.25, \$9.65, \$12.25, and \$7.90 mentally by rounding each the nearest ten cents.

Review Problems for Maintaining Your Skills

30. Write 2,008,345 in words.

31. Find the quotient and remainder for $179 \div 34$.

32. Reduce $\frac{28}{63}$ to its lowest terms.

33. Convert $\frac{57}{11}$ to a mixed number.

In 34 through 36, reduce each fraction to its lowest terms and express any result that is an improper fraction as a mixed number.

34. $\frac{5}{8} + \frac{1}{4} + \frac{2}{5} =$

35. $10 - 6\frac{3}{7} =$

36. $\frac{11}{14} \div \frac{5}{7} =$

37. Write the place and value of 5 in the decimal number 0.0005.

38. Write 5.0496 in words.

39. Round 6.04863 to the nearest ten-thousandth.

40. Round \$2.4583 to the nearest cent.

EXERCISE SET 3.3B

Skill Development

1. $23.7 + 5.1 =$
2. $78.12 + 45.66 =$
3. $14.65 - 7.23 =$
4. $56.1 - 9.7 =$
5. $1.89 + 0.365 =$
6. $5.6 + 2.48 =$
7. $10.6 - 5.89 =$
8. $32 - 14.61 =$
9. $21.3 - 4 =$
10. $45.05 - 23.00652 =$
11. $17.3 + 1.23 + 5.701 + 20 =$
12. $12.03 + 9 + 0.054 + 5.461 + 6.95 =$
13. Add the numbers: 2.03, 5.6, 1.405, 0.008, and 7.

14. Add the numbers: 12, 15.6, 9.05, 17.66, and 9.0535.

15. Subtract 15.6 from 25.68.

16. Subtract 2.47 from 5.3.

17. Subtract 0.0356 from 1.

18. Subtract 0.106 from 0.2789.

Applications

19. From Monday through Friday the sales credited to Tom Brett at the Dalton Shoe Store were \$423.17, \$389.22, \$526.80, \$779.50 and \$750.62, respectively. Find his total sales for the 5-day period.

20. On April 18, Alice Slater purchased 100 shares of AT&T stock for \$2,784.19. On July 25, she sold the 100 shares for \$2,813.70. Find her gain on the sale of the stock.

21. According to a map, the estimated travel time is 4.58 hours from Springfield to St. Louis, 5.83 hours from St. Louis to Louisville, and 2.17 hours from Louisville to Cincinnati. What is the total estimated travel time from Springfield to Cincinnati, if you must pass through St. Louis?

22. The weekly rainfall for a 4-week period in April was 1.32 inches, 0.019 inches, 2.3 inches, and 1.981 inches. Find the total rainfall for the four weeks.

For Some Extra Thought

23. Find two numbers whose difference is 0.005.

24. If we subtract 0.1 from 1 and round the difference to the nearest unit, the answer is 1. Explain the reason for what some people may believe is a surprising result.

25. Find nine numbers whose sum is 0.123456789.

Mental Estimation

26. Estimate the sum for 32.9, 10.02 and 4.33 mentally by rounding each number to the nearest unit.

27. Estimate the difference between 56.77 and 12.4 mentally by rounding each number to the nearest unit.

28. Estimate the sum for \$1.22, \$0.89, \$4.60, and \$0.32 mentally by rounding each amount to the nearest dollar.

29. Estimate the difference between \$17.62 and \$10.85 by rounding each amount to the nearest ten cents.

Review Problems for Maintaining Your Skills

30. Write 12,347 in words.

31. Find the quotient and remainder for $67 \div 13$.

32. Reduce $\frac{25}{85}$ to its lowest terms.

33. Convert $\frac{41}{20}$ to a mixed number.

In 34 through 36, reduce each fraction to its lowest terms and express any result that is an improper fraction as a mixed number.

34. $\frac{2}{5} + \frac{4}{7} + \frac{1}{5} =$

35. $12 - 3\frac{2}{5} =$

36. $\frac{3}{4} \times \frac{8}{9} =$

37. Write the place and value of 3 in the decimal number 0.4235.

38. Write 1.24087 in words.

39. Round 0.007744 to the nearest ten-thousandth.

40. Round \$12.7805 to the nearest cent.

3.4 MULTIPLICATION

Preview Test

Find each product.

1.	$\begin{array}{r} 356 \\ \times\ 1.9 \\ \hline \end{array}$	**2.**	$\begin{array}{r} 1,230 \\ \times\ 3.17 \\ \hline \end{array}$	**3.**	$\begin{array}{r} 56.7 \\ \times\ 8.3 \\ \hline \end{array}$
4.	$\begin{array}{r} 2.38 \\ \times 0.09 \\ \hline \end{array}$	**5.**	$\begin{array}{r} 0.009 \\ \times\ \ 0.06 \\ \hline \end{array}$	**6.**	$\begin{array}{r} 79.365 \\ \times\ \ 1.46 \\ \hline \end{array}$

7. $3 \times 23.78 =$ **8.** $1.2 \times 6.09 =$

9. $0.26 \times 10{,}000 =$ **10.** $0.01 \times 5{,}589 =$

We can use repeated addition to multiply a decimal number by a whole number. Let's consider 3×4.52.

$$3 \times 4.52 = 4.52 + 4.52 + 4.52 = 13.56$$

To find the product for 3×4.52 without using repeated addition, we can multiply using the same procedure as we used in Section 1.5 to multiply two whole numbers. However, once the multiplication is complete, we have to find the proper place for the decimal point. The rule to place the decimal point is:

> There must be as many digits to the right of the decimal point in the product as the total number of digits to the right of the decimal points in the factors.

Let's use this rule for 3×4.52. Notice that we get the same result (13.56) as we got when repeated addition was used.

$$\begin{array}{r} 4.52 \\ \times\ \ 3 \\ \hline 13.56 \end{array}$$

4.52 ← A total of 2 digits to the right of the decimal point.

↑ Position the decimal point here so there are 2 digits to the right of the decimal point.

ANSWERS FOR PREVIEW TEST

1. 676.4 **2.** 3,899.1 **3.** 470.61 **4.** 0.2142 **5.** 0.00054
6. 115.8729 **7.** 71.34 **8.** 7.308 **9.** 2,600 **10.** 55.89

EXAMPLE 1: Find the product for 14 × 27.9

Solution:

```
 27.9  ←—  A total of 1 digit to the
 × 14      right of the decimal point.
 ----
 1116
 279
 -----
 390.6
   ↑
```

Position the decimal point here so there is 1 digit to the right of the decimal point.

PRACTICE PROBLEM 1: Find the product for 3.08 × 37

EXAMPLE 2: Find the product for 2.7 × 9.05

Solution:

```
   9.05  ←—  A total of 3 digits to the
  × 2.7  ←—  right of the decimal point.
  -----
   6335
  1810
  ------
  24.435
    ↑
```

Position the decimal point here so there are 3 digits to the right of the decimal point.

PRACTICE PROBLEM 2: Find the product for 3.2 × 12.16

In Example 2 we found that two and seven tenths times nine and five hundredths equals twenty-four and four hundred thirty-five thousandths. It looks like a decimal number with tenths multiplying a decimal number with hundredths gives us a decimal number with thousandths as a product. The fact that tenths times hundredths gives us thousandths is consistent with the multiplication of place value numbers. With place value whole numbers

$10 \times 10 = 100$	ten × ten = one hundred
$10 \times 100 = 1{,}000$	ten × one hundred = one thousand
$10 \times 1{,}000 = 10{,}000$	ten × one thousand = ten thousand

and with place value decimal numbers

$0.1 \times 0.1 = 0.01$	tenths × tenths = hundredths
$0.1 \times 0.01 = 0.001$	tenths × hundredths = thousandths
$0.1 \times 0.001 = 0.0001$	tenths × thousandths = ten-thousandths

The above list is only presented to give you an intuitive understanding of the placement of the decimal point in the product. You can still use the rule to get the result.

tenths × hundreds = thousandths

0.1 × 0.0 1 = 0.0 0 1

↑ ↑ ↑

A total of 3 digits to the right of the decimal point. — Position the decimal point here so there are 3 digits to the right of the decimal point.

Sometimes we have to add zeros to our answer so that we can have enough digits to the right of the decimal point. Suppose, for example, the product before placing the decimal point is 43 and we want 5 digits to the right of the decimal point. We would have to add 3 zeros.

This zero is the customary zero to the left of the decimal point to show that the whole number is 0.

↓

0.0 0 0 4 3

Add 3 zeros to give us 5 digits to the right of the decimal point.

Notice that we wrote the 3 zeros *before* the 4. If we had written the zeros after the 3 in 43 (0.43000) we would have gotten the wrong answer.

EXAMPLE 3: Find the product for 0.007 × 1.2

Solution:

```
        1.2   ←—— A total of 4 digits to the
  × 0.0 0 7   ←—— right of the decimal point.
  ---------
  0.0 0 8 4
```

Customary zero showing that the whole number is zero ——→ 0.0084

Add 2 zeros before the 8 so that there are 4 digits to the right of the decimal point.

PRACTICE PROBLEM 3: Find the product for 0.06 × 0.07

When the multiplier is a whole number power of ten (10, 100, 1,000, and so on) we can find the product simply by moving the decimal point. Count the number of zeros to the right of the 1 in the power of 10 and move the decimal point that many places to the right. For example,

10 × 3.25 = 3.2.5 = 32.5 ←—— 1 zero, so move the decimal point 1 place to the right.

100 × 3.25 = 3.25. = 325. ←—— 2 zeros, so move the decimal point 2 places to the right.

1,000 × 3.25 = 3.250. = 3,250 ←—— 3 zeros, so move the decimal point 3 places to the right (which means you have to add 1 zero).

10,000 × 3.25 = 3.2500. = 32,500. ←—— 4 zeros, so move the decimal point 4 places to the right (which means you have to add 2 zeros).

When the multiplier is a decimal power of ten (0.1, 0.01, 0.001 and so on) we can also find the product simply by moving the decimal point. Count the number of digits to the right of the decimal point in the decimal power of ten and move the decimal point that many places to the left. For example,

$0.1 \times 3.25 = 0.3.25 = 0.325$ ⟵ 1 digit, so move the decimal point 1 place to the left.

$0.01 \times 3.25 = 0.03.25 = 0.0325$ ⟵ 2 digits, so move the decimal point 2 places to the left (which means you must add 1 zero to the left of 3).

$0.001 \times 3.25 = 0.003.25 = 0.00325$ ⟵ 3 digits, so move the decimal point 3 places to the left (which means you must add 2 zeros to the left of 3).

EXAMPLE 4: Find the product for 100×6.248

Solution:
Since 100 has 2 zeros we'll move the decimal point 2 places to the right.

$$100 \times 6.248 = 6.24.8 = 624.8$$

PRACTICE PROBLEM 4: Find the product for 10×0.023

EXAMPLE 5: Find the product for 0.001×6.248

Solution:
Since 0.001 has 3 digits to the right of the decimal point, we'll move the decimal point in 6.248 3 places to the left.

$$0.001 \times 6.248 = 0.006.248 = 0.006248$$

PRACTICE PROBLEM 5: Find the product for 0.0001×689.5

Be Careful About This

When you multiply a power of ten by a whole number, remember that the whole number has an understood decimal point to the right of the units position. For example, to multiply 0.001×45, we can write the 45 with the decimal point after the 5. Then

$$0.001 \times 45. = 0.045$$

Mental Estimation of Products

The product for 27.86×3.2 can be estimated mentally by rounding 27.86 to 30 and 3.2 to 3. That is, each number is rounded to the lefthand place value position. The estimated product is then

$$30 \times 3 = 90$$

EXAMPLE 6: Mentally estimate the product for 7.8×42.25 by rounding each number to the lefthand place value position.

Solution:

$$7.8 = 8 \text{ to the nearest unit}$$
$$42.25 = 40 \text{ to the nearest ten}$$

Estimated product $= 8 \times 40 = 320$

PRACTICE PROBLEM 6: Mentally estimate the product for 91.6×58.7 by rounding each number to the lefthand place value position.

ANSWERS FOR PRACTICE PROBLEMS

1. 113.96 **2.** 38.912 **3.** 0.0042 **4.** 0.23
5. 0.06895 **6.** 5,400

EXERCISE SET 3.4A

Skill Development

In questions 1 through 20, find each product.

1. $\begin{array}{r} 64 \\ \times 1.2 \\ \hline \end{array}$ **2.** $\begin{array}{r} 850 \\ \times 6.7 \\ \hline \end{array}$ **3.** $\begin{array}{r} 1{,}295 \\ \times 1.08 \\ \hline \end{array}$

4. $\begin{array}{r} 75.6 \\ \times 2.3 \\ \hline \end{array}$ **5.** $\begin{array}{r} 8.41 \\ \times 0.6 \\ \hline \end{array}$ **6.** $\begin{array}{r} 42.750 \\ \times 0.08 \\ \hline \end{array}$

7. $\begin{array}{r} 1.268 \\ \times 1.09 \\ \hline \end{array}$ **8.** $\begin{array}{r} 265.73 \\ \times 0.018 \\ \hline \end{array}$ **9.** $\begin{array}{r} 0.0057 \\ \times 2.026 \\ \hline \end{array}$

10. $25.7 \times 6.08 =$

11. $683.45 \times 12.2 =$

12. $3.908 \times 45 =$

13. $0.36 \times 1.5 =$

14. $8.05 \times 0.006 =$

15. $100 \times 2.3874 =$

16. $1{,}000 \times 1.2 =$

17. $10 \times 0.0035 =$

18. $0.001 \times 56.7 =$

19. $0.1 \times 0.0023 =$

Applications

20. Find the cost of 5 boxes of computer disks if each box costs $17.52.

21. If each credit of B+ work is worth 3.3 quality points, find the number of quality points for 12 credits of B+ work.

22. The area of a rectangle is the product of its length and width. Find the area in square feet of a rectangle whose length and width are, respectively, 5.23 feet and 4.17 feet.

23. The sales tax is 0.0625 times the purchase price. Find the sales tax for a men's suit with a price of $189.50. Round the result to the nearest cent.

24. At $0.895 per gallon find, to the nearest cent, the cost for 12.62 gallons of gas.

For Some Extra Thought

25. Find the product of five thousandths and seventeen ten-thousandths.

26. If the product of two numbers is 4 and one of the numbers is 8, find the other number.

27. One thousand times what decimal number equals one?

Mental Estimation

In 28 through 31, mentally estimate each product by rounding as follows: (a) numbers larger than 1 round to the lefthand place value position and (b) numbers less than 1 round to the place value of the leftmost nonzero digit.

28. $8.2 \times 3.9 =$

29. $12.25 \times 0.87 =$

30. $0.021 \times 0.709 =$

31. $3{,}100 \times 62.5 =$

Review Problems for Maintaining Your Skills

32. Write the number for fifty-six billion, twenty million.

33. $4{,}500 - 3{,}578 =$

34. $51 \times 100 =$

35. Reduce $\frac{35}{49}$ to its lowest terms.

36. Convert $4\frac{3}{8}$ to an improper fraction.

In 37 through 39, reduce each fraction to its lowest terms and express any result that is an improper fraction as a mixed number.

37. $3\frac{2}{5} + 4\frac{1}{2} =$

38. $\frac{13}{15} - \frac{3}{10} =$

39. $1\frac{1}{4} \div 2\frac{3}{4} =$

40. Write 22.568 in words.

41. Round 4.50897 to the nearest hundredth.

42. Round $5.12736 to the nearest cent.

43. $5.9 + 7.45 + 8 =$

44. $2.18 - 0.304 =$

EXERCISE SET 3.4B

Skill Development

In questions 1 through 20, find each product.

1. $\begin{array}{r} 78 \\ \times 4.6 \\ \hline \end{array}$

2. $\begin{array}{r} 564 \\ \times 8.09 \\ \hline \end{array}$

3. $\begin{array}{r} 2{,}657 \\ \times 0.02 \\ \hline \end{array}$

4. $\begin{array}{r} 456.3 \\ \times 1.54 \\ \hline \end{array}$

5. $\begin{array}{r} 876.45 \\ \times 0.75 \\ \hline \end{array}$

6. $\begin{array}{r} 65.72 \\ \times 1.20 \\ \hline \end{array}$

7. $\begin{array}{r} 38.014 \\ \times 2.091 \\ \hline \end{array}$

8. $\begin{array}{r} 0.000008 \\ \times 0.007 \\ \hline \end{array}$

9. $\begin{array}{r} 1.2371 \\ \times 0.85 \\ \hline \end{array}$

10. $345 \times 1.26 =$
11. $36.8 \times 1.7 =$
12. $12.45 \times 0.003 =$
13. $4{,}500 \times 6.25 =$
14. $8.05 \times 0.0027 =$
15. $1{,}000 \times 4.5 =$
16. $10 \times 0.0032 =$
17. $100{,}000 \times 0.000678 =$
18. $0.01 \times 0.005 =$
19. $0.1 \times 45{,}780 =$

Applications

20. Find the cost of 7 cases of soda if the price per case is $5.23.

21. If each credit of C+ work is worth 2.3 quality points, find the number of quality points for 15 credits of C+ work.

22. The monthly interest charge for a charge account is 0.015 times the outstanding balance. Find the interest charge if the outstanding balance is \$165.80. Round the result to the nearest cent.

23. Mike Nelson's new weekly salary will be 1.08 times his present weekly salary, which is \$489.53. Find his new weekly salary to the nearest cent.

24. Find, to the nearest cent, the cost for 9.31 gallons of gas if it is selling for \$0.749 per gallon.

For Some Extra Thought

25. Find the product of twenty-six hundred-thousandths and ten thousand.

26. If the product of two numbers is 0.7 and one of the numbers is 0.1, find the other number.

27. One million times what decimal number equals one thousand?

Mental Estimation

In 28 through 31, mentally estimate each product by rounding as follows: (a) numbers larger than 1 round to the lefthand place value position and (b) numbers less than 1 round to the place value of the leftmost nonzero digit.

28. $54.6 \times 67.8 =$

29. $8.6 \times 0.205 =$

30. $0.0018 \times 0.627 =$

31. $2{,}300 \times 104.67 =$

Review Problems for Maintaining Your Skills

32. Write the number for twenty-three million, eight hundred five.

33. $1{,}850 - 1{,}243 =$

34. $34 \times 87 =$

35. Reduce $\frac{34}{51}$ to its lowest terms.

36. Convert $13\frac{2}{7}$ to an improper fraction.

In 37 through 39, reduce each fraction to its lowest terms and express any result that is an improper fraction as a mixed number.

37. $2\frac{3}{4} + 1\frac{1}{3} =$

38. $\frac{5}{8} - \frac{2}{5} =$

39. $3\frac{1}{2} \times 4\frac{1}{2} =$

40. Write 0.00527 in words.

41. Round 2.706 to the nearest tenth.

42. Round \$4.3419 to the nearest cent.

43. $1.5 + 3 + 0.47 =$

44. $10 - 6.73 =$

3.5 DIVISION

Preview Test

In questions 1 through 6, find each quotient.

1. $4\overline{)15.8}$ **2.** $25\overline{)1.6}$ **3.** $2.9\overline{)4.292}$

4. $45 \div 0.01 =$ **5.** $6.82 \div 0.001 =$ **6.** $76.982 \div 100 =$

In questions 7 through 9, find and round each quotient to the nearest tenth.

7. $26.09 \div 7 =$ **8.** $4 \div 0.13 =$ **9.** $2.03 \div 1.2 =$

In questions 10 through 12, find and round each quotient to the nearest hundredth.

10. $45.3 \div 17 =$ **11.** $1.9 \div 0.07 =$ **12.** $0.85 \div 13.4 =$

We can divide a decimal number by a whole number using essentially the same method as we used in Section 1.6 to divide two whole numbers. For example, consider a dinner check for $76.72 that is shared by 8 people. To find each person's share we have to divide $76.72 by 8. We place a decimal point in the quotient directly above the decimal point in the dividend.

Place a decimal point here above the decimal point in 76.72.

↓

$$\;\;\;\;.$$
$$8\overline{)76.72}$$

Now carry out the division.

$$\begin{array}{r} 9.59 \\ 8\overline{)76.72} \\ \underline{72} \\ 47 \\ \underline{40} \\ 72 \\ \underline{72} \end{array}$$

So, each person's share is $9.59.

ANSWERS FOR PREVIEW TEST

1. 3.95 **2.** 0.064 **3.** 1.48 **4.** 4,500 **5.** 6,820 **6.** 0.76982 **7.** 3.7 **8.** 30.8 **9.** 1.7 **10.** 2.66 **11.** 27.14 **12.** 0.06

EXAMPLE 1: Divide 538.5 by 15.

Solution:
Place a decimal point for the quotient directly above the decimal point in the dividend, 538.5. Then divide.

```
     3 5.9
1 5 |5 3 8.5
     4 5
       8 8
       7 5
       1 3 5
       1 3 5
```

PRACTICE PROBLEM 1: Divide 74.04 by 12.

The problems we've done so far have been exact (that is, the remainder is zero). If the division is not exact, we may be directed to round the quotient to a designated position. This is illustrated in the next example.

EXAMPLE 2: Divide 86.2 by 7 and round the quotient to the nearest tenth.

Solution:
To round to the tenths place, we'll need to know the digit to the right, in the hundredths place. Therefore, we'll divide to the hundredths position.

```
   1 2.3 1
7 |8 6.2 0  <----  Add this zero so we can
   7               carry the quotient to
   1 6             the hundredths place.
   1 4
     2 2
     2 1
       1 0
         7
         3
```

The quotient, 12.31, rounded to the nearest tenth is 12.3.

PRACTICE PROBLEM 2: Divide 8.96 by 3 and round the quotient to the nearest hundredth.

If the divisor is a decimal we have to convert it to a whole number before dividing. To do this

Step 1: Move the decimal point to the right of the righthand digit in the divisor.
Step 2: Move the decimal point in the dividend the same number of places to the right.
Step 3: Finally, place the decimal point for the quotient above the moved decimal point in the dividend and divide.

For example, to divide 3.553 by 1.7 we need to move the decimal point in 1.7 one place to the right to make the whole number 17. We'll also have to move the decimal point in the dividend 3.553 one place to the right.

```
1.7⟌3.5.5 3
```

Move the decimal point one place to the right.

Now, place the decimal point in the quotient above the decimal point in the dividend and divide.

```
        2.0 9
1.7⟌3.5.5 3
    3 4
    ---
      1 5 3
      1 5 3
      -----
```

You may wonder how we can be sure that moving the decimal point to the right will not change the answer. Consider the following division examples.

$$8 \div 2 = 4$$
$$80 \div 20 = 4$$
$$800 \div 200 = 4$$

Starting with $8 \div 2$, we can multiply the dividend and divisor by 10 and the quotient does not change.

$$8 \div 2 = (8 \times 10) \div (2 \times 10) = 80 \div 20 = 4$$

Starting again with $8 \div 2$, we can multiply the divisor and dividend by 100 and the quotient still does not change.

$$8 \div 2 = (8 \times 100) \div (2 \times 100) = 800 \div 200 = 4$$

When we move the decimal point to the right in the divisor and dividend we are multiplying both numbers by the same power of 10. For 3.553 divided by 1.7 we have

$$3.553 \div 1.7 = (3.553 \times 10) \div (1.7 \times 10) = 35.53 \div 17$$

EXAMPLE 3: Divide 2.7 by 9.54 and round the quotient to the nearest hundredth.

Solution:

Since 954 is larger than 270 place a 0 here to show that the whole number part is 0.

```
           0.2 8 3
9.5 4⟌2.7 0.0 0 0
         1 9 0 8
         -------
           7 9 2 0
           7 6 3 2
           -------
             2 8 8 0
             2 8 6 2
             -------
                 1 8
```

Move the decimal point 2 places to the right. →

← Since we are rounding to the hundredth we need 3 zeros to right of the decimal point.

The quotient, 0.283, rounded to the nearest hundredth is 0.28.

PRACTICE PROBLEM 3: Divide 5.74 by 6.9 and round the quotient to the nearest thousandth.

EXAMPLE 4: Divide 23 by 0.09 and round the quotient to the nearest unit.

Solution:

```
         2 5 5.5
0.0 9 | 2 3.0 0.0
        1 8
          5 0
          4 5
            5 0
            4 5
              5 0
              4 5
                5
```

The quotient, 255.5, rounded to the nearest unit is 256.

PRACTICE PROBLEM 4: Divide 176 by 3.7 and round the quotient to the nearest unit.

EXAMPLE 5: Divide 0.6 by 79.34 and round the quotient to the nearest thousandth.

Solution:

```
           0.0 0 7 5
7 9.3 4 | 0.6 0.0 0 0 0
          5 5 5 3 8
            4 4 6 2 0
            3 9 6 7 0
              4 9 5 0
```

The quotient, 0.0075, rounded to the nearest thousandth is 0.008.

PRACTICE PROBLEM 5: Divide 0.19 by 8.6 and round the quotient to the nearest thousandth.

Just as we have a shortcut for multiplying by powers of 10, we have one for dividing by powers of 10. When the divisor is a whole number power of ten (10, 100, 1,000, and so on) we can count the number of zeros to the right of the 1 in the power of 10 and find the quotient by moving the decimal point that many places to the left. For example,

$897 \div 10 = 89.7$ ⟵ 1 zero, so move the decimal point 1 place to the left.

$897 \div 100 = 8.97$ ⟵ 2 zeros, so move the decimal point 2 places to the left.

$897 \div 1{,}000 = 0.897$ ⟵ 3 zeros, so move the decimal point 3 places to the left.

When the divisor is a decimal power of ten (0.1, 0.01, 0.001, and so on) we can count the number of digits to the right of the decimal point in the decimal power of 10 and find the quotient by moving the decimal point that many places to the right. For example,

$3.42 \div 0.1 = 34.2$ ⟵ 1 digit, so move the decimal point 1 place to the right.

$3.42 \div 0.01 = 342.$ ⟵ 2 digits, so move the decimal point 2 places to the right.

$3.42 \div 0.001 = 3{,}420.$ ⟵ 3 digits, so move the decimal point 3 places to the right.

EXAMPLE 6: Divide 45.6 by 100.

Solution:
There are 2 zeros in 100, so move the decimal point in 45.6 two places to the left.

$$45.6 \div 100 = 0.45.6 = 0.456$$

PRACTICE PROBLEM 6: Divide 32.14 by 10.

EXAMPLE 7: Divide 8.73 by 0.001.
0.001 has 3 digits to the right of the decimal point. The decimal point in 8.73 is moved 3 places to the right. Notice that a zero has to be added as a placeholder.

$$8.73 \div 0.001 = 8.730. = 8{,}730$$

PRACTICE PROBLEM 7: Divide 0.002 by 0.001.

Mental Estimation of Quotients

For $81.3 \div 6.8$, the quotient can be estimated mentally by rounding 81.3 to 80 and 6.8 to 7. That is, just as with products of numbers larger than 1, each number is rounded to the lefthand place value position. Then the estimated quotient is

$$80 \div 7 = 11 \text{ to the nearest unit}$$

EXAMPLE 8: Mentally estimate the quotient for $385.25 \div 6.75$ and round the estimate to the nearest unit.

Solution:

$$385.25 = 400 \text{ to the nearest hundred.}$$
$$6.75 = 7 \text{ to the nearest unit.}$$
$$\text{Estimated quotient} = 400 \div 7 = 57 \text{ to the nearest unit}$$

PRACTICE PROBLEM 8: Mentally estimate the quotient for $17.2 \div 1.9$ and round the estimate to the nearest unit.

Mental estimation of quotients involving numbers less than 1 can be carried out by rounding each number less than 1 to the place value of the leftmost nonzero digit. Then convert the divisor to a whole number. This reads as a very complicated procedure, but it is really not that difficult. Consider $7.93 \div 0.021$.

$$7.93 = 8 \text{ to the nearest unit}$$
$$0.021 = 0.02 \text{ to the nearest hundredth}$$

Now move the decimal point in 0.02 and 8 two places to the right and divide. So, the estimated quotient is

$$800 \div 2 = 400$$

EXAMPLE 9: Mentally estimate the quotient for 0.074 ÷ 0.0059 and round the estimate to the nearest unit.

Solution:

0.074 = 0.07 to the nearest hundredth

0.0059 = 0.006 to the nearest thousandth

Estimated quotient = 0.07 ÷ 0.006 = 70 ÷ 6 = 12 to the nearest unit

PRACTICE PROBLEM 9: Mentally estimate the quotient for 19.5 ÷ 0.0391 and round the estimate to the nearest unit.

ANSWERS FOR PRACTICE PROBLEMS

1. 6.17 **2.** 2.99 **3.** 0.832 **4.** 48

5. 0.022 **6.** 3.214 **7.** 2 **8.** 10 **9.** 500

EXERCISE SET 3.5A

Skill Development

In questions 1 through 9, find each quotient.

1. $3\overline{)26.7}$ **2.** $13\overline{)107.9}$ **3.** $1.7\overline{)8.891}$

4. $19.38 \div 1.02 =$ **5.** $28.53 \div 3.17 =$

6. $4.62 \div 100 =$ **7.** $94 \div 1{,}000 =$

8. $0.035 \div 0.01 =$ **9.** $6 \div 0.1 =$

In questions 10 and 11, find each quotient and round to the nearest unit.

10. $4.7 \div 2.06 =$ **11.** $36 \div 7.9 =$

In questions 12 and 13, find each quotient and round to the nearest tenth.

12. $36.07 \div 1.9 =$ **13.** $0.145 \div 0.7 =$

In questions 14 and 15, find each quotient and round to the nearest hundredth.

14. $1.26 \div 0.47 =$ **15.** $62.3 \div 2.9 =$

In questions 16 and 17, find each quotient and round to the nearest thousandth.

16. $3.45 \div 1.9 =$ **17.** $76.13 \div 1.72 =$

Applications

18. Nine box seats for a baseball game cost $74.25. What is the price of one box seat?

19. The cost for 17.4 pounds of grass seed is $84.39. Find the cost per pound.

20. John Gregory worked 37.5 hours and earned \$254.25. What does John earn per hour?

21. Larry Jensen receives a dividend of \$118.35 for 45 shares of MPT common stock. What is the dividend for each share?

For Some Extra Thought

22. Find the quotient for eighteen thousandths divided by twenty-four thousandths.

23. How can we know before dividing that the quotient for 0.057 divided by 0.063 is less than 1?

24. If the divisor is 0.02 and the quotient is 10, what is the dividend?

Mental Estimation

In 25 through 28, mentally estimate each quotient by rounding as follows: (a) numbers larger than 1 round to the lefthand place value position and (b) numbers less than 1 round to the place value of the leftmost nonzero digit.

25. $387.5 \div 52.5 =$

26. $8.6 \div 0.205 =$

27. $81.56 \div 3.95 =$

28. $0.98 \div 1.02 =$

Review Problems for Maintaining Your Skills

29. Round 5,143,872 to the nearest hundred thousand.

30. Find the quotient and remainder for $127 \div 13$.

31. Reduce $\frac{20}{55}$ to lowest terms.

32. Convert $\frac{39}{17}$ to a mixed number.

In 33 through 35, reduce each fraction to its lowest terms and express any result that is an improper fraction as a mixed number.

33. $\frac{12}{13} + \frac{3}{10} =$ **34.** $2\frac{1}{2} - 1\frac{3}{8} =$ **35.** $5 \div 3\frac{1}{5} =$

36. Round 2.4058 to the nearest hundredth.

37. Round $3.672 to the nearest cent.

38. $6.52 + 7 + 0.7 =$ **39.** $10.86 - 7.413 =$

40. $2.45 \times 100 =$ **41.** $5.1 \times 6.37 =$

EXERCISE SET 3.5B

Skill Development

In questions 1 through 12, find the quotient.

1. $7\overline{)187.6}$ **2.** $19\overline{)39.52}$ **3.** $1.2\overline{)75.768}$

4. $31.496 \div 6.2 =$ **5.** $31.672 \div 8.56 =$

6. $235 \div 100 =$ **7.** $6{,}784 \div 10 =$

8. $2.5 \div 0.01 =$ **9.** $0.5 \div 0.0001 =$

In questions 10 and 11, find each quotient and round to the nearest unit.

10. $6.21 \div 2.3 =$ **11.** $678 \div 0.87 =$

In questions 12 and 13, find each quotient and round to the nearest tenth.

12. $87.6 \div 13.09 =$ **13.** $54 \div 0.65 =$

In questions 14 and 15, find each quotient and round to the nearest hundredth.

14. $5.02 \div 0.34 =$ **15.** $97.54 \div 56.7 =$

In questions 16 and 17, find each quotient and round to the nearest thousandth.

16. $3.04 \div 8.22 =$ **17.** $45.6 \div 24.59 =$

Applications

18. The cost of 15 cases of soda is \$88.35. Find the cost per case.

19. The cost for 16 gallons of gas is \$14.32. Find the price per gallon.

20. Alice Ingram worked 28.6 hours and earned \$278.85. What does she earn per hour?

21. Jane Lance receives an interest payment of \$637 for a quantity of city bonds which she owns. If each bond pays \$22.75, how many bonds does she own?

For Some Extra Thought

22. Find the quotient for twenty-four thousandths divided by six.

23. How can we know before dividing that the quotient for $0.005 \div 0.00007$ is larger than 1?

24. If the dividend is 0.01 and the quotient is 10, what is the divisor?

Mental Estimation

In 25 through 28, mentally estimate each quotient by rounding as follows: (a) numbers larger than 1 round to the lefthand place value position and (b) numbers less than 1 round to the place value of the leftmost nonzero digit.

25. $78 \div 1.95 =$

26. $21.5 \div 0.51 =$

27. $912.75 \div 49.3 =$

28. $2.02 \div 0.004 =$

Review Problems for Maintaining Your Skills

29. Round 7,854 to the nearest ten.

30. Find the quotient and remainder for $79 \div 41$.

31. Reduce $\frac{46}{68}$ to its lowest terms.

32. Convert $\frac{13}{5}$ to a mixed number.

In 33 through 35, reduce each fraction to its lowest terms and express any result that is an improper fraction as a mixed number.

33. $\frac{3}{5} + \frac{4}{15} =$ **34.** $8\frac{2}{5} - 5\frac{1}{3} =$ **35.** $2\frac{2}{3} \times 3\frac{1}{4} =$

36. Round 0.00074 to the nearest ten-thousandth.

37. Round $0.41633 to the nearest cent.

38. $5.43 + 4.1 + 3.92 =$ **39.** $56 - 40.807 =$

40. $36 \times 0.001 =$ **41.** $4.52 \times 1.9 =$

3.6 MIXED OPERATIONS

Preview Test

1. $8.1 + 4.9 \times 2 =$
2. $0.3 \times 0.5 + 1.8 \div 2 =$
3. $0.24 \div 0.02 - 1.3 \times 0.9 =$
4. $25.2 - (7.3 + 4.9 \times 0.05) =$
5. $(7.1 - 1.3 \times 1.5) \times (2 - 1.2 \div 2) =$
6. $(4 + 0.7 \div 0.7) \times (0.8 \div 0.4 \times 0.2) =$

The order in which operations should be performed is the same for decimal numbers as it was for whole numbers and fractions. For your convenience, we'll repeat the rules here for order of operations.

If no parentheses ( ) or brackets [ ] appear, the order in which the operations are performed is

1. multiplication and/or division in the order in which the operations occur from left to right;
2. addition and/or subtraction in the order in which the operations occur from left to right.

If parentheses or brackets are included, the operations should be performed within the parentheses or brackets first in the correct order. That is, multiplication and/or division from left to right and then addition and/or subtraction from left to right.

EXAMPLE 1: $2.3 \times 9.5 - 6.2 \times 1.3 =$

Solution:
This calculation requires multiplication and subtraction. Following the order of operations, multiply first, then subtract.

$$\underbrace{2.3 \times 9.5}_{\text{multiply}} - \underbrace{6.2 \times 1.3}_{\text{multiply}}$$

$$= \underbrace{21.85 \quad - \quad 8.06}_{\text{subtract}}$$

$$=$$

PRACTICE PROBLEM 1: $32.75 - 9.6 \times 2.75 =$

EXAMPLE 2: $(200 - 25 \div 0.5) \times (0.12 \div 0.25 - 0.4) =$

Solution:
Perform the operations within the parentheses first. Since division has priority over subtraction, do it before the subtraction. Then multiply the final numbers in the two parentheses.

ANSWERS FOR PREVIEW TEST

1. 17.9 **2.** 1.05 **3.** 10.83 **4.** 17.655 **5.** 7.21 **6.** 2

$$\begin{aligned}
&(200 - \underbrace{25 \div 0.5}_{\text{divide}}) \times (\underbrace{0.12 \div 0.25}_{\text{divide}} - 0.4) \\
= \; &\underbrace{(200 - 50)}_{\text{subtract}} \times \underbrace{(0.48 - 0.4)}_{\text{subtract}} \\
= \; &\underbrace{150 \times 0.08}_{\text{multiply}} \\
= \; &12
\end{aligned}$$

PRACTICE PROBLEM 2: $12.5 - (28.6 - 2.3 \times 10) =$

ANSWERS FOR PRACTICE PROBLEMS

1. 6.35 **2.** 6.9

EXERCISE SET 3.6A

Skill Development

1. $5.4 + 6.3 \times 7.25 =$
2. $9.1 \times 5.4 + 3.6 \times 4.1 =$
3. $25.5 - 3.6 \times 5.25 + 1.7 =$
4. $36 \div 1.2 + 2.4 \div 0.4 =$
5. $0.006 \times 1{,}000 - 50 \times 0.1 =$
6. $(2.2 \times 4.5 - 6.2) \times (17.25 - 3.1 \times 5.05) =$
7. $30 - (5.6 + 2.25 \times 4) - (6 - 2.1 \times 2.1) =$
8. $(20 \div 0.1 - 1 \div 0.01) \times 100 =$
9. $(5.8 - 1.5 \times 1.5) \times (1 - 1 \times 0.9) =$
10. $8 + (2 \div 0.25 + 8 \div 0.25) \times (4 - 35 \times 0.1) =$

Applications

11. Mary Lasker's total monthly earnings consist of her $1,500 base salary plus 0.02 times her sales. If her sales amount to $60,000 in September, find her monthly earnings.

12. Michael Kelly's grade point average for the semester is calculated by dividing the sum of 6×3.3 and 9×3.7 by 15. Find his grade point average.

13. A ticket broker sells eighteen $9.75 seats and twenty-six $12.50 seats for a concert. Find the total amount of the ticket sales.

For Some Extra Thought

14. $0.01 \times 100 \times 0.01 \times 100 \times 0.01 \times 100 \times 0.01 \times 100 =$
15. $(1.2 \times 10 - 0.12 \times 100) \div 100 =$

Review Problems for Maintaining Your Skills

16. Write 1,026,800 in words.

17. $608 - 79 =$

18. $536 \times 703 =$

19. Reduce $\frac{27}{63}$ to its lowest terms.

20. Convert $7\frac{3}{4}$ to an improper fraction.

In 21 through 23, reduce each fraction to its lowest terms and express any result that is an improper fraction as a mixed number.

21. $3\frac{3}{5} + 1\frac{1}{2} =$

22. $\frac{17}{20} - \frac{1}{15} =$

23. $\frac{3}{8} \div \frac{1}{3} =$

24. Round 2.047355 to the nearest ten-thousandth.

25. Round \$5.87463 to the nearest cent.

26. $3.5 + 6.9 + 4 + 1.33 =$

27. $87.6 - 20.708 =$

28. $5.42 \times 100 =$

29. $678.3 \times 2.3 =$

30. $32.67 \div 1{,}000 =$

31. $0.158 \div 0.00001 =$

32. Divide 47.8 by 1.33 and round the quotient to the nearest hundredth.

33. Divide 2.709 by 1.14 and round the quotient to the nearest tenth.

EXERCISE SET 3.6B

Skill Development

1. $6.25 \times 4 - 7.5 \times 3 =$
2. $1.5 \times 1.5 + 2.6 \times 3.2 =$
3. $20 - 16 \times 0.1 - 8 \times 0.2 - 30 \times 0.5 =$
4. $42.5 + 6.7 \times 8.5 - 3.9 \times 2.75 =$
5. $0.07 \times 100 + 0.06 \times 100 + 0.13 \times 100 =$
6. $(8 - 0.5 \times 3.5) \div (6 - 5.9) =$
7. $50 - (6 \times 5.5 + 7.2 \times 2.3) =$
8. $(8.8 + 2.6 \times 1.25) \times (12.6 - 3.2 \times 3.2) =$
9. $(30 - 2.6 \times 9.85) - (6.2 - 2.2 \times 1.25) =$
10. $(40.2 + 3.5 \div .01) \div (7.6 - 5 \times 1.5) =$

Applications

11. Richard Unger's total weekly earnings consist of his $525 base salary plus 0.075 times his sales. If his sales amount to $12,000, find his weekly earnings.

12. Marcie Barker's grade point average for the semester is calculated by dividing the sum of 3×3.3, 3×3.7, and 4×4.0 by 10. Find her grade point average.

13. Larry Perkins holds two jobs. In his first job he works 17.5 hours and earns $6.40 per hour. In his second job he works 8.2 hours and earns $8.65 per hour. Find his total earnings for the two jobs.

For Some Extra Thought

14. $0.00001 - 1 \div 100{,}000 =$
15. $(4 \div 10 \times 10 - 4) \times 4 =$

Review Problems for Maintaining Your Skills

16. Write 2,453,078,100 in words.

17. $340 - 298 =$

18. $47 \times 805 =$

19. Reduce $\frac{44}{64}$ to its lowest terms.

20. Convert $5\frac{2}{7}$ to an improper fraction.

In 21 through 23, reduce each fraction to its lowest terms and express any result that is an improper fraction as a mixed number.

21. $4\frac{2}{5} + 5\frac{4}{7} =$ **22.** $13 - 6\frac{7}{8} =$ **23.** $\frac{5}{9} \div \frac{5}{12} =$

24. Round 0.00478322 to the nearest hundred-thousandth.

25. Round $12.048273 to the nearest cent.

26. $12 + 4.06 + 7.9 =$ **27.** $63.57 - 12.8 =$

28. $47.33 \times 0.001 =$ **29.** $4.51 \times 1.06 =$

30. $102 \div 100 =$ **31.** $5.4076 \div 0.0001 =$

32. Divide 5.8 by 1.45 **33.** Divide 45.81 by 2.7 and round the quotient to the thousandth.

PRACTICE TEST 3A

1. Write 4.503 in words.

2. Write the number for thirty-five and seventeen ten-thousandths.

3. Round 4.5087 to the nearest tenth.

4. Round 0.0782864 to the nearest hundred-thousandth.

5. Round $41.23706 to the nearest cent.

6. $5.4 + 4.72 + 0.8 + 6 =$

7. $4.52 - 1.673 =$

8. From increased effort, Robert Lyman's grade point average increased from 2.16 to 2.74. Find the amount of the increase.

9. $2.47 \times 1.6 =$

10. $100 \times 0.0468 =$

11. At $1.289 per gallon find, to the nearest cent, the cost for 8.4 gallons.

12. Divide 270.88 by 3.2.

13. Divide 45.7 by 17.39 and round the quotient to the nearest hundredth.

14. The cost for 10.6 ounces of cheese is $3.83. Find the cost per ounce to the nearest cent.

15. $7.4 \times 3.5 - 2.4 \times 5.6 =$

PRACTICE TEST 3B

1. Write 23.0672 in words.

2. Write the number for seventy-six millionths.

3. Round 2.40673 to the nearest hundredth.
4. Round 364.2525 to the nearest tenth.
5. Round \$1.342777 to the nearest cent.
6. $6.13 + 5 + 4.1 =$ 7. $10 - 5.642 =$
8. In a one-hour period the temperature decreased from 12.67 degrees Celsius to 9.78 degrees Celsius. Find the amount of the decrease.

9. $34.6 \times 0.08 =$ 10. $0.01 \times 625 =$
11. At \$1,540 per acre, find the cost for 7.3 acres.
12. Divide 516 by 6.45.
13. Divide 9.312 by 1.57 and round the quotient to the nearest thousandth.
14. The cost for 500 sheets of bond paper is \$9.56. Find the cost per sheet to the nearest cent.
15. $12.5 - 2.3 \times 1.6 =$

CUMMULATIVE REVIEW EXERCISES FOR CHAPTERS 1 THROUGH 3

1. Write 3,067,121 in words.

2. Round 12,547 to the nearest hundred.
3. Subtract 123 from 1,000.
4. Find the quotient and remainder for $65 \div 14$.
5. Reduce $\frac{72}{84}$ to its lowest terms.
6. Convert $6\frac{7}{8}$ to an improper fraction.
7. Convert $\frac{100}{7}$ to a mixed number.

In 8 through 11, reduce each fraction to its lowest terms and express any result that is an improper fraction as a mixed number.

8. $\frac{6}{7} \times \frac{11}{18} =$
9. $5\frac{3}{4} \div 1\frac{1}{3} =$
10. $\frac{5}{8} + \frac{1}{6} =$
11. $14 - 3\frac{5}{7} =$
12. Write 4.105 in words.

13. Write the number for seven millionths.

14. Round 12.057423 to the nearest ten-thousandth.
15. Round \$1.23804 to the nearest cent.
16. $1.3 + 84.2 + 4.16 =$
17. $2.03 - 1.007 =$
18. $648 \times 0.01 =$
19. $2.47 \times 1.3 =$
20. Divide 6.46 by 1.7.
21. Divide 143.45 by 1.1 and round the quotient to the nearest tenth.
22. $6.4 \times 8.3 + 10.1 =$

C H A P T E R 4

PERCENTS

4.1 PERCENT/DECIMAL AND PERCENT/FRACTION CONVERSIONS

4.2 FINDING THE AMOUNT

4.3 FINDING THE RATE

4.4 FINDING THE BASE

4.5 FINDING THE AMOUNT, RATE, OR BASE

4.6 PERCENT INCREASE AND DECREASE

Suppose a loan company advertises that they charge \$180 per year for a \$2,000 loan. Another loan company offers \$5,000 loans at an annual charge of \$440. If you want to compare the charges of the two companies it is helpful to calculate what each one charges for every \$100 of loan.

First Company: In \$2,000 there are twenty units of \$100. So the charge per \$100 is \$180/20 = \$9.

Second Company: In \$5,000 there are fifty units of \$100. So the charge per \$100 is \$440/50 = \$8.80.

It is customary to express the charge per \$100 in a special notation called **percent notation**. In percent notation a special symbol "%" which means "per hundred" is used. Since the first loan company charges \$9 per \$100 of loan, they can advertise that their charge is 9%. The second company charges 8.80%.

By using percent notation it is easy to compare the charges of two competing loan companies. This also applies to comparing different rates of interest that banks pay you for money you have on deposit with them. The bank that advertises a 7.62% interest rate for a one-year certificate of deposit is paying \$7.62 for every \$100 you deposit. So, if you deposit \$1,000 the bank pays 10 × \$7.62 or \$76.20 interest for the year.

In this chapter you will learn how to use percents to solve a variety of practical problems.

4.1 PERCENT/DECIMAL AND PERCENT/FRACTION CONVERSIONS

Preview Test

1. Convert 17.2% to a decimal.
2. Convert 500% to a decimal.
3. Convert 0.0575 to a percent.
4. Convert 2 to a percent.
5. Convert 13% to a fraction.
6. Convert 5.3% to a fraction.
7. Convert $\frac{37}{50}$ to a percent.
8. Convert $\frac{19}{47}$ to the nearest tenth of a percent.

The symbol % is read **percent** and means "per hundred." For example, 7% is read "seven percent." Since 7% means 7 per hundred

$$7\% = \frac{7}{100} = 0.07$$

ANSWERS FOR PREVIEW TEST

1. 0.172 **2.** 5 **3.** 5.75% **4.** 200%

5. $\frac{13}{100}$ **6.** $\frac{53}{1{,}000}$ **7.** 74% **8.** 40.4%

This suggests a method for converting a

1. percent to a decimal
2. decimal to a percent
3. percent to a fraction
4. fraction to a percent

Converting a Percent to a Decimal

To convert a percent to a decimal, divide the number to the left of the percent symbol by 100 and delete the percent symbol. The effect of dividing by 100 is the movement of the decimal point two places to the left. An abbreviated rule that you will find useful is

> Move the decimal point two places to the left and delete the percent symbol.

EXAMPLE 1: Convert each percent to a decimal.

a) 47% **b)** 0.6% **c)** $7\frac{3}{8}\%$

Solution:

a) Since 47 is a whole number, the decimal point is understood to be to the right of 7. So,

$$47\% = 47.\% = 0.47.\% = 0.47$$

b) $0.6\% = 0.00.6\% = 0.006$

c) First convert $\frac{3}{8}$ to the decimal 0.375.

Then $7\frac{3}{8}\% = 7.375\% = 0.07.375\% = 0.07375$

PRACTICE PROBLEM 1: Convert each percent to a decimal.

a) 525% **b)** 0.08% **c)** $6\frac{1}{4}\%$

Converting a Decimal or Whole Number to a Percent

To convert a decimal or whole number to a percent, multiply by 100 to move the decimal point two places to the right and place a percent symbol after the number.

EXAMPLE 2: Convert each number to a percent.

a) 0.094 **b)** 3

Solution:

a) $0.094 = 0.09.4\% = 9.4\%$

Notice that the two zeros to the left of the 9 are dropped.

b) $3 = 3.00.\% = 300\%$

PRACTICE PROBLEM 2: Convert each number to a percent.

a) 0.72 **b)** 1.4

Converting a Percent to a Fraction

To convert a percent to a fraction, place the number to the left of the percent symbol over 100 and delete the percent symbol. If the numerator is not an integer, multiply the numerator and denominator by the appropriate power of ten (10, 100, 1,000, . . .) to convert the numerator to an integer.

EXAMPLE 3: Convert each percent to a fraction and reduce the fractions to their lowest terms.

a) 24% **b)** 8.7% **c)** $5\frac{3}{4}\%$

Solution:

a) $24\% = \frac{24}{100} = \frac{6}{25}$

b) $8.7\% = \frac{8.7}{100} = \frac{8.7 \times 10}{100 \times 10} = \frac{87}{1{,}000}$

c) $5\frac{3}{4}\% = 5.75\% = \frac{5.75}{100} = \frac{5.75 \times 100}{100 \times 100} = \frac{575}{10{,}000} = \frac{23}{400}$

PRACTICE PROBLEM 3: Convert each percent to a fraction and reduce the fractions to their lowest terms.

a) 7% **b)** 5.4% **c)** $2\frac{1}{2}\%$

Converting a Fraction to a Percent

To convert a fraction to a percent, first change the fraction to a decimal by dividing the numerator by the denominator. Then proceed with the rule for converting a decimal to a percent.

EXAMPLE 4: Convert each fraction to a percent.

a) $\frac{17}{25}$

b) $\frac{8}{19}$, round to the nearest hundredth of a percent.

Solution:

a) $\frac{17}{25} = 0.68 = 0.68.\% = 68\%$

b) $\frac{8}{19} = 0.42105 = 0.42.105\% = 42.11\%$

Notice that you have to carry the quotient to five decimal places in order to express the result to the nearest hundredth of a percent.

PRACTICE PROBLEM 4: Convert each fraction or mixed number to a percent. Round to the nearest hundredth of a percent.

a) $\frac{24}{43}$ **b)** $1\frac{3}{7}$

ANSWERS FOR PRACTICE PROBLEMS

1. a) 5.25 **b)** 0.0008 **c)** 0.0625 **2. a)** 72% **b)** 140%

3. a) $\frac{7}{100}$ **b)** $\frac{27}{500}$ **c)** $\frac{1}{40}$ **4. a)** 55.81% **b)** 142.86%

EXERCISE SET 4.1A

Skill Development

Convert each percent to a decimal.

1. 29%	**2.** 36%	**3.** 5.3%	**4.** 78.14%	**5.** 2.9%
6. 0.7%	**7.** 0.019%	**8.** $4\frac{1}{4}\%$	**9.** $15\frac{3}{5}\%$	**10.** $\frac{99}{100}\%$

Convert each number to a percent.

11. 0.47	**12.** 0.09	**13.** 0.25	**14.** 0.076	**15.** 0.0612
16. 0.007	**17.** 0.159	**18.** 3.2	**19.** 5	**20.** 41

Convert each percent to a fraction and reduce the fraction to its lowest terms.

21. 31%	**22.** 74%	**23.** 9%	**24.** 9.3%	**25.** 4.1%
26. 3.25%	**27.** 0.7%	**28.** $4\frac{3}{4}\%$	**29.** $10\frac{1}{2}\%$	**30.** $\frac{3}{8}\%$

Convert each fraction or mixed number to a percent. If the result is not a whole percent, round to the nearest tenth of a percent.

31. $\frac{41}{50}$	**32.** $\frac{17}{20}$	**33.** $\frac{4}{5}$	**34.** $\frac{93}{100}$	**35.** $\frac{1}{4}$
36. $\frac{19}{37}$	**37.** $\frac{3}{13}$	**38.** $\frac{14}{53}$	**39.** $2\frac{1}{6}$	**40.** $7\frac{5}{9}$

Applications

41. A 1.5% defective rate for gaskets has been determined from quality control procedures. Write this percent as a decimal.

42. A hospital estimates that 0.008 of the average hospital bill is for supplies. Write this decimal as a percent.

43. Eighty-six percent of a salesperson's mileage was for business purposes. Write this percent as a fraction.

44. An architect's contract specifies that she design an office building with one-eighth of the square footage to be employee lounge areas. Change this fraction to a percent.

For Some Extra Thought

45. A hospital has a staff of 2,000. On a particular day, six employees request sick leave. What percent of the employees have requested sick leave?

46. A city water company attaches a $\frac{3}{4}\%$ finance charge and a $\frac{1}{8}\%$ interest charge on delinquent bills. What fraction represents the total penalty?

Review Problems for Maintaining Your Skills

47. Round 4,563 to the nearest ten.

48. Find the quotient and remainder for $52 \div 9$.

In 49 through 51, reduce each fraction to its lowest terms and express any result that is an improper fraction as a mixed number.

49. $2\frac{3}{5} + 3\frac{1}{4} =$ 50. $\frac{13}{20} - \frac{1}{10} =$ 51. $\frac{3}{8} \div \frac{2}{5} =$

52. Round 4.2016 to the nearest thousandth.

53. Round $0.204813 to the nearest cent.

54. $5.3 + 4.12 + 8 =$ 55. $201.3 - 145.67 =$

56. $4.5 \times 6.3 =$ 57. $0.2 \div 0.001 =$

58. Divide 4.56 by 1.7 and round the quotient to the nearest hundredth.

59. $5.3 + 3.2 \times 1.9 =$ 60. $10 - 3.4 \times 1.6 =$

EXERCISE SET 4.1B

Skill Development

Convert each percent to a decimal.

1. 37% **2.** 81% **3.** 7.2% **4.** 43.61% **5.** 4.3%

6. 0.04% **7.** 0.039% **8.** $5\frac{1}{2}\%$ **9.** $23\frac{2}{5}\%$ **10.** $\frac{63}{100}\%$

Convert each number to a percent.

11. 0.56 **12.** 0.14 **13.** 0.50 **14.** 0.083 **15.** 0.04

16. 0.002 **17.** 0.273 **18.** 2.4 **19.** 7 **20.** 32

Convert each percent to a fraction and reduce the fraction to its lowest terms.

21. 43% **22.** 71% **23.** 6% **24.** 8.4% **25.** 6.1%

26. 5.45% **27.** 0.1% **28.** $5\frac{1}{4}\%$ **29.** $12\frac{1}{2}\%$ **30.** $\frac{5}{8}\%$

Convert each fraction or mixed number to a percent. If the result is not a whole percent, round the result to the nearest tenth of a percent.

31. $\frac{14}{25}$ **32.** $\frac{37}{50}$ **33.** $\frac{13}{20}$ **34.** $\frac{7}{10}$ **35.** $\frac{3}{4}$

36. $\frac{13}{17}$ **37.** $\frac{2}{19}$ **38.** $\frac{24}{51}$ **39.** $\frac{5}{43}$ **40.** $5\frac{1}{4}$

Applications

41. The cost of living is 173% of the 1968 base year. Write this percent as a decimal.

42. A car dealer is offering a 0.069 interest rate on its model M15 Sports Car. What is the percent for this decimal?

43. A company estimates $8\frac{3}{4}\%$ of its salary for benefits. What is the fraction for this percent?

44. A department store is offering a one-fifth discount for its end-of-season sale. Change this fraction to a percent.

For Some Extra Thought

45. Eight students out of 4000 in a community college are given awards for special service to the college. What percent of the students do the eight represent?

46. The state sales tax of 6.2% is added to the county tax of 0.3% to find the total sales tax. Write the total sales tax as a fraction.

Review Problems for Maintaining Your Skills

47. Round 1,609,726 to the nearest ten thousand.

48. Find the quotient and remainder for $102 \div 37$.

In 49 through 51, reduce each fraction to its lowest terms and express any result that is an improper fraction as a mixed number.

49. $3\frac{2}{3} + 5\frac{3}{4} =$ **50.** $\frac{14}{15} - \frac{1}{12} =$ **51.** $\frac{5}{6} \div \frac{3}{4} =$

52. Round 3.4128 to the nearest tenth.

53. Round $2.108733 to the nearest cent.

54. $12 + 5.02 + 6.7 =$ **55.** $87.63 - 7.1 =$

56. $2.13 \times 5.1 =$ **57.** $534.6 \div 100 =$

58. Divide 57.4 by 1.3 and round the quotient to the nearest tenth.

59. $1.2 \times 3.4 + 4.8 =$ **60.** $12.5 - 3.1 \times 2 =$

4.2 FINDING THE AMOUNT

Preview Test

1. Find 8% of \$960.
2. Find 2.6% of 700.
3. Find 17.25% of 36,000.
4. Find 300% of 5.
5. Find $4\frac{1}{2}\%$ of \$600.

In many cities and states a sales tax is added to the cost of a purchase. The sales tax is generally quoted as a percent. To find the tax amount, you multiply the price by the sales tax rate. For example, if the price of a stereo is \$900 and the sales tax rate is 6%, the tax amount is found by multiplying \$900 by 6%. This is done by converting 6% to a decimal and carrying out the multiplication.

$$\text{Sales Tax} = \$900 \times 6\% = \$900 \times 0.06 = \$54$$

When you find a percent of a number the result is called the **amount**, the percent is called the **rate**, and the number you are finding the percent of is called the **base**. In the above example

\$900 is the base
6% is the rate
\$54 is the amount

To find the amount (that is, the percent of a number) use the formula

$$\text{AMOUNT} = \text{BASE} \times \text{RATE}$$

EXAMPLE 1: Find each amount.

a) 14.8% of \$86,000 **b)** 250% of 4. **c)** $7\frac{1}{2}\%$ of \$240

Solution:

a) BASE = \$86,000 and RATE = 14.8%

$$\begin{aligned}\text{AMOUNT} &= \text{BASE} \times \text{RATE} \\ &= \$86{,}000 \times 14.8\% \\ &= \$86{,}000 \times 0.148 \\ &= \$12{,}728\end{aligned}$$

b) BASE = 4 and RATE = 250%

$$\begin{aligned}\text{AMOUNT} &= \text{BASE} \times \text{RATE} \\ &= 4 \times 250\% \\ &= 4 \times 2.50 \\ &= 10\end{aligned}$$

ANSWERS FOR PREVIEW TEST

1. \$76.80 **2.** 18.2 **3.** 6,210 **4.** 15 **5.** \$27

c) BASE = \$240 and RATE = $7\frac{1}{2}\%$

$$\begin{aligned} \text{AMOUNT} &= \text{BASE} \times \text{RATE} \\ &= \$240 \times 7\tfrac{1}{2}\% \\ &= \$240 \times 0.075 \\ &= \$18 \end{aligned}$$

PRACTICE PROBLEM 1: Find each amount.

a) 8.4% of \$700. **b)** 400% of 7 **c)** $8\frac{3}{4}\%$ of \$660.

Mental Calculation of an Amount

Finding an amount is probably the most frequently used mental calculation. For example, it is not uncommon to mentally calculate a tip (15% of a dinner check) or an increase in salary (8% of a base salary). We will outline some helpful techniques for calculating an amount mentally.

1. To find 1% of a number, move the decimal point in the number two places to the left. For example,

$$1\% \text{ of } 4{,}752 = 47.52$$

2. To find 2%, 3%, . . . , or 9% of a number, first find 1% of the number. Then multiply the result by 2, 3, . . . , or 9. For example,

$$3\% \text{ of } 6{,}800 = 3 \times 68 = 204$$

3. To find 10% of a number, move the decimal point in the number one place to the left. For example,

$$10\% \text{ of } 3{,}679 = 367.9$$

4. To find 20, 30%, . . . , or 90% of a number, first find 10% of the number. Then multiply the result by 2, 3, . . . , or 9. For example,

$$70\% \text{ of } 230 = 7 \times 23 = 161$$

5. To find 25% of a number, divide the number by 4. For example,

$$25\% \text{ of } 1{,}700 = \frac{1{,}700}{4} = 425$$

6. To find 50% of a number, divide the number by 2. For example,

$$50\% \text{ of } 12{,}300 = \frac{12{,}300}{2} = 6{,}150$$

7. To find 15% of a number add 10% of the number to one-half of 10% of the number. For example,

$$15\% \text{ of } 70 = 7 + \frac{1}{2} \times 7 = 10.5$$

8. To find 35%, 45%, 55%, 65%, 75%, 85%, or 95% of a number, add one-half of 10% of the number to 30%, 40%, 50%, 60%, 70%, 80%, or 90% of the number. For example,

$$35\% \text{ of } 60 = 3 \times 6 + \frac{1}{2} \times 6 = 21$$

9. To find $33\frac{1}{3}\%$ or $66\frac{2}{3}\%$ of a number, multiply the number by $\frac{1}{3}$ or $\frac{2}{3}$. For example,

$$33\frac{1}{3}\% \text{ of } 60 = \frac{1}{3} \times 60 = 20 \text{ and } 66\frac{2}{3}\% \text{ of } 60 = \frac{2}{3} \times 60 = 40$$

EXAMPLE 2: Find each amount mentally.

a) 10% of 1,656 **b)** 75% of 180 **c)** 65% of 1,700 **d)** $33\frac{1}{3}\%$ of 450

Solution:

a) 10% of 1,656 = 165.6

b) $75\% \text{ of } 180 = 7 \times 18 + \frac{1}{2} \times 18 = 135$

c) $65\% \text{ of } 1{,}700 = 6 \times 170 + \frac{1}{2} \times 170 = 1{,}105$

d) $33\frac{1}{3}\% \text{ of } 450 = \frac{1}{3} \times 450 = 150$

PRACTICE PROBLEM 2: Find each amount mentally.

a) 1% of 145 **b)** 4% of 55 **c)** 25% of 960 **d)** $66\frac{2}{3}\%$ of 600

ANSWERS FOR PRACTICE PROBLEMS

1. **a)** $58.80 **b)** 28 **c)** $57.75 **2.** **a)** 1.45 **b)** 2.2 **c)** 240 **d)** 400

EXERCISE SET 4.2A

Skill Development

1. Find 13% of $15,600.
2. Find 4% of $740.
3. 1.4% of 800,000 equals what number?
4. 6.2% of 2,600 equals what number?
5. What number is 65% of $450?
6. What number is 24% of $6,700?
7. Find 0.2% of $6,200.
8. Find 0.04% of $9,000,000.
9. 140% of 28 equals what number?
10. 250% of 40 equals what number?
11. What number is $6\frac{1}{2}\%$ of $5,600?
12. What number is $12\frac{3}{4}\%$ of $8,000?
13. 500% of 2 is what number?
14. 2,000% of 1 is what number?
15. Find $1\frac{1}{8}\%$ of $45,000.
16. Find $2\frac{1}{4}\%$ of $6,800.
17. What number is $\frac{1}{2}\%$ of $870?
18. What number is $\frac{3}{4}\%$ of $90,000?
19. Find 1,200% of 0.4.
20. Find 800% of 0.06.

Applications

21. A store offers a 20% discount on a sports jacket whose list price is $120. Find the amount of the discount.

22. A car is purchased in a state where the sales tax is 7.5%. Find the sales tax on a car that costs $17,500.

23. A bank pays yearly interest of 8.37%. Find the interest on a $20,000 one-year certificate of deposit.

24. During a major snowstorm 65% of the 340 employees at the MGD show factory were absent. How many employees were absent?

For Some Extra Thought

25. Find $0.2\frac{1}{2}\%$ of \$50,000.

26. If the base and amount are equal, what is the rate?

Mental Calculations

Find each amount mentally.

27. 2% of 4,600.

28. 25% of \$560.

29. 65% of 80.

30. $66\frac{2}{3}\%$ of 240.

Review Problems for Maintaining Your Skills

31. Write 4,060,017 in words.

32. $364 \times 807 =$

In 33 through 35, reduce each fraction to its lowest terms and express any result that is an improper fraction as a mixed number.

33. $\frac{5}{6} + \frac{1}{2} =$ **34.** $4\frac{3}{4} \times 3 =$ **35.** $\frac{7}{8} \div \frac{3}{4} =$

36. Round 2.4068 to the nearest hundredth.

37. Round \$3.937814 to the nearest cent.

38. $6.52 - 1.059 =$ **39.** $2.1467 \times 10{,}000 =$

40. Divide 56 by 2.03 and round the quotient to the nearest thousandth.

41. $74 - 3.6 \times 8.7 + 1.2 \times 1.2 =$

42. Convert 7.3% to a decimal.

43. Convert 0.078 to a percent.

44. Convert 7.2% to a fraction and reduce the fraction to its lowest terms.

45. Convert $\frac{19}{20}$ to a percent.

EXERCISE SET 4.2B

Skill Development

1. Find 17% of $3,000.
2. Find 5% of $870.
3. 1.6% of 70,000 equals what number?
4. 3.7% of 4,100 equals what number?
5. What number is 45% of $390?
6. What number is 18% of $8,650?
7. Find 0.7% of $40,000.
8. Find 0.02% of $6,000,000.
9. 130% of 34 equals what number?
10. 450% of 20 equals what number?
11. What number is $4\frac{1}{2}\%$ of $4,000?
12. What number is $7\frac{3}{8}\%$ of $12,000?
13. 200% of 3 is what number?
14. 1,000% of 1 is what number?
15. Find $2\frac{1}{2}\%$ of $60,000.
16. Find $5\frac{1}{2}\%$ of $9,000.
17. What number is $\frac{1}{4}\%$ of $540?
18. What number is $\frac{7}{8}\%$ of $40,000?
19. Find 1,500% of 0.7.
20. Find 600% of 0.09.

Applications

21. If the city income tax rate is 1.5%, find the tax on income of $800.

22. A seed company claims that 83.3% of the grass seed in a package will germinate. If a package contains 20,000 seeds, what number does the company claim will germinate?

23. One year ago Merkin Buick sold 80 cars in one month. In the same month this year, the total sales was 150% of last year's. Find the number of cars sold in the month this year.

24. Tom Brett's present salary is $40,000. On December 31, he learned that he will be given an 11.5% increase for next year. Find Tom Brett's new salary.

For Some Extra Thought

25. Find $0.04\frac{1}{2}\%$ of \$30,000.

26. If the base is twice the amount, what is the rate?

Mental Calculations

Find each amount mentally.

27. 6% of 6,100.

28. 50% of 49.

29. 45% of \$600.

30. $33\frac{1}{3}\%$ of 180.

Review Problems for Maintaining Your Skills

31. Write 2,704,000,000 in words.

32. $31 \times 100{,}000 =$

In 33 through 35, reduce each fraction to its lowest terms and express any result that is an improper fraction as a mixed number.

33. $2\frac{4}{5} + 3\frac{2}{3} =$ **34.** $\frac{11}{12} \times \frac{3}{5} =$ **35.** $3 \div 5\frac{2}{7} =$

36. Round 3.07562 to the nearest ten-thousandth.

37. Round \$6.31488 to the nearest cent.

38. $34.12 - 23.8 =$ **39.** $64.38 \times 2.4 =$

40. Divide 64.1 by 1.05 and round the quotient to the nearest hundredth.

41. $100 - (3.2 \times 6.7 - 4.8) =$

42. Convert 5.6% to a decimal.

43. Convert 0.483 to a percent.

44. Convert 6.82% to a fraction and reduce the fraction to its lowest terms.

45. Convert $\frac{17}{25}$ to a percent.

4.3 FINDING THE RATE

Preview Test

1. 17 is what percent of 20?
2. What percent of 60 is 90?
3. 0.19 is what percent of 0.25?
4. To the nearest tenth of a percent, 19 is what percent of 30?
5. To the nearest hundredth of a percent, 2 is what percent of 3?

Suppose that 7 of the 20 students in a math class earned an A for the year. The part of the class that earned an A can be expressed as a fraction, which can then be converted to a percent.

$$\text{Part of the class that earned an A} = \frac{7}{20} = 0.35 = 35\%$$

From the above, you can say that the 7 students represent 35% of the class of 20. So this also answers the question, "7 is what percent of 20?" In this case

7 is the amount
20 is the base
35% is the rate

To find a rate, divide the amount by the base.

$$\text{RATE} = \frac{\text{AMOUNT}}{\text{BASE}}$$

EXAMPLE 1: Find each rate.

a) 23 is what percent of 40?
b) 14 is what percent of 19? Round to the nearest tenth of a percent.
c) 15 is what percent of 12?

Solution:

a) AMOUNT = 23 and BASE = 40

$$\text{RATE} = \frac{\text{AMOUNT}}{\text{BASE}} = \frac{23}{40} = 0.575 = 57.5\%$$

b) AMOUNT = 14 and BASE = 19

$$\text{RATE} = \frac{\text{AMOUNT}}{\text{BASE}} = \frac{14}{19} = 0.7368 = 73.7\%$$

c) AMOUNT = 15 and BASE = 12

$$\text{RATE} = \frac{\text{AMOUNT}}{\text{BASE}} = \frac{15}{12} = 1.25 = 125\%$$

ANSWERS FOR PREVIEW TEST

1. 85% **2.** 150% **3.** 76% **4.** 63.3% **5.** 66.67%

PRACTICE PROBLEM 1: Find each rate.

a) 13 is what percent of 50?

b) 7 is what percent of 11? Round to the nearest hundredth of a percent.

c) 8 is what percent of 2?

Mental Calculation of the Rate

When we are finding the rate by dividing the amount by the base we are in effect converting a fraction to a decimal and then converting the decimal to a percent. It is helpful to know these common conversions from fractions to decimals.

$$\frac{1}{2} = 0.5 \quad \frac{1}{3} = 0.333\ldots \quad \frac{2}{3} = 0.666\ldots \quad \frac{1}{4} = 0.25 \quad \frac{3}{4} = 0.75$$

$$\frac{1}{5} = 0.2 \quad \frac{2}{5} = 0.4 \quad \frac{3}{5} = 0.6 \quad \frac{4}{5} = 0.8$$

$$\frac{1}{8} = 0.125 \quad \frac{3}{8} = 0.375 \quad \frac{5}{8} = 0.625 \quad \frac{7}{8} = 0.875$$

For fractions with denominators of 20, 25, and 50, you can use the following rules for converting to decimals.

1. To convert a fraction with a denominator of 20 to a decimal, multiply the numerator by 5 and move the decimal point 2 places to the left. For example, convert $\frac{17}{20}$ to a decimal.
   Multiply: $5 \times 17 = 85$
   Moving the decimal point 2 places to the left = 0.85.
2. To convert a fraction with a denominator of 25 to a decimal, multiply the numerator by 4 and move the decimal point 2 places to the left. For example, convert $\frac{13}{25}$ to a decimal.
   Multiply: $4 \times 13 = 52$
   Moving the decimal point 2 places to the left = 0.52.
3. To convert a fraction with a denominator of 50 to a decimal, multiply the numerator by 2 and move the decimal point 2 places to the left. For example, convert $\frac{7}{50}$ to a decimal.
   Multiply: $2 \times 7 = 14$
   Moving the decimal point 2 places to the left = 0.14.

EXAMPLE 2: Find the rate mentally.

a) 3 is what percent of 8?

b) What percent of 50 is 19?

Solution:

a) $\text{RATE} = \frac{\text{AMOUNT}}{\text{BASE}} = \frac{3}{8} = 0.375 = 37.5\%$

b) $\text{RATE} = \frac{\text{AMOUNT}}{\text{BASE}} = \frac{19}{50} = 0.38 = 38\%$

PRACTICE PROBLEM 2: Find the rate mentally.

a) What percent of 5 is 2?

b) 7 is what percent of 20?

ANSWERS FOR PRACTICE PROBLEMS

1. **a)** 26% **b)** 63.64% **c)** 400% **2.** **a)** 40% **b)** 35%

EXERCISE SET 4.3A

Skill Development

1. 7 is what percent of 10?
2. 78 is what percent of 200?
3. 23 is what percent of 50?
4. 1 is what percent of 20?
5. What percent of 40 is 30?
6. What percent of 25 is 17?
7. What percent of 6 is 30?
8. What percent of 13 is 13?
9. 0.3 is what percent of 0.12?
10. 0.27 is what percent of 0.5?
11. 87.5 is what percent of 100?
12. 900 is what percent of 2,000?

Round each result to the nearest tenth of a percent.

13. What percent of 13 is 9?
14. What percent of 33 is 23?
15. What percent of 80 is 49?
16. What percent of 124 is 91?
17. 64 is what percent of 300?
18. 1 is what percent of 6?
19. 0.21 is what percent of 0.8?
20. 1.56 is what percent of 2.3?

Applications

21. Ruth Seaver received a \$3,000 discount on a car that had a list price of \$15,000. Find the discount rate as a percent.

22. The sales tax on a \$7.50 VCR tape is \$0.39. Find the sales tax as a percent.

23. On the 23 Model DXI Hondas sold last week 17 were beige. To the nearest tenth of a percent, what percent of the cars were beige?

24. The property tax on a home with an assessed value of \$200,000 is \$3,000. Find the tax rate as a percent.

For Some Extra Thought

25. What percent of $3\frac{1}{5}$ is $4\frac{4}{5}$?

26. If the rate is 200% and the amount is 4, what is the base?

Mental Calculations

Find the rate mentally.

27. 5 is what percent of 8?

28. 23 is what percent of 50?

29. What percent of 20 is 19?

30. What percent of 10 is 1?

Review Problems for Maintaining Your Skills

31. Round 3,552,712 to the nearest hundred thousand.

32. Find the quotient and remainder for $45 \div 11$.

In 33 through 35, reduce each fraction to its lowest terms and express any result that is an improper fraction as a mixed number.

33. $1\frac{5}{6} + 3\frac{1}{2} =$ **34.** $\frac{17}{25} - \frac{3}{50} =$ **35.** $4\frac{2}{5} \times 1\frac{1}{3} =$

36. Round 0.005178 to the nearest thousandth.

37. Round \$12.3992 to the nearest cent.

38. $52.1 - 43.86 =$ **39.** $2.3 \div 0.001 =$

40. Divide 4.02 by 0.372 and round the quotient to the nearest hundredth.

41. Convert 5.62% to a decimal.

42. Convert 0.1438 to a percent.

43. Convert 1.6% to a fraction and reduce the fraction to its lowest terms.

44. Convert $\frac{40}{64}$ to a percent.

45. Find 5.2% of \$640.

46. Find $7\frac{1}{2}\%$ of \$800.

47. 400% of 6 is what number?

48. What number is 0.3% of \$50,000?

EXERCISE SET 4.3B

Skill Development

1. 17 is what percent of 68?
2. 50 is what percent of 40?
3. 1 is what percent of 100?
4. 35 is what percent of 70?
5. What percent of 80 is 20?
6. What percent of 18 is 36?
7. What percent of 200 is 106?
8. What percent of 260 is 13?
9. 1.8 is what percent of 0.6?
10. 0.13 is what percent of 0.2?
11. 63.7 is what percent of 100?
12. 800 is what percent of 4,000?

Round each result to the nearest hundredth of a percent.

13. What percent of 17 is 7?
14. What percent of 51 is 43?
15. What percent of 149 is 37?
16. What percent of 23 is 39?
17. 83 is what percent of 300?
18. 5 is what percent of 6?
19. 0.65 is what percent of 0.14?
20. 3.47 is what percent of 12.8?

Applications

21. A $5,000 bond pays interest of $425 per year. Find the bond interest rate as a percent.

22. Fifteen out of twenty-eight families living on Brewster Street have more than two children. To the nearest hundredth of a percent, what percent does this represent?

23. The unpaid balance on a loan of $1,400 is $650. To the nearest tenth of a percent, what percent of the loan is still unpaid?

24. Mary Harkim works a 40-hour week at a job. After 18 hours, what percent of Mary's work week has passed?

For Some Extra Thought

25. What percent of $6\frac{1}{2}$ is $1\frac{3}{10}$?

26. If the rate is 400% and the amount is 20, what is the base?

Mental Calculations

Find the rate mentally.

27. 11 is what percent of 25?

28. 47 is what percent of 50?

29. 7 is what percent of 8?

30. 3 is what percent of 5?

Review Problems for Maintaining Your Skills

31. Round 4,683,211 to the nearest million.

32. Find the quotient and remainder for $174 \div 173$.

In 33 through 35, reduce each fraction to its lowest terms and express any result that is an improper fraction as a mixed number.

33. $\frac{3}{8} + \frac{4}{9} =$ **34.** $3 - 1\frac{2}{7} =$ **35.** $3\frac{2}{5} \div 1\frac{1}{3} =$

36. Round 0.104688 to the nearest ten-thousandth.

37. Round \$2.407826 to the nearest cent.

38. $34.21 - 24.6 =$ **39.** $5{,}763 \div 100 =$

40. Divide 6.8 by 1.34 and round the quotient to the nearest thousandth.

41. Convert 17.6% to a decimal.

42. Convert 0.341 to a percent.

43. Convert 3.14% to a fraction and reduce the fraction to its lowest terms.

44. Convert $\frac{45}{60}$ to a percent.

45. Find 12.5% of \$6,000.

46. Find $3\frac{1}{4}\%$ of 400.

47. 150% of 80 is what number?

48. What number is 0.7% of \$300?

4.4 FINDING THE BASE

Preview Test

1. 8 is 2% of what number?
2. \$56 is 7% of how many dollars?
3. 1.6 is 5% of what number?
4. 3.25% of how many dollars equals \$29.25?
5. $\frac{1}{4}$% of how many dollars equals \$36?

Suppose the \$2,460 allocated to the women's tennis team represents 2% of the total budget for team sports in a high school. To find the total budget we are asking,

2% of what budget total equals \$2,460?

Put another way, we need to know what budget total times 0.02 equals \$2,460?

> When you know the product and one of the factors in a multiplication problem, you divide the product by the known factor to find the second factor.

Here you should divide \$2,640 by 0.02 to find the total sports budget.

$$\text{Total Sports Budget} = \frac{\$2{,}460}{0.02} = \$123{,}000$$

To find a base, divide the amount by the rate.

$$\text{BASE} = \frac{\text{AMOUNT}}{\text{RATE}}$$

EXAMPLE 1: Find each base.

a) 4% of how many dollars equals \$63?

b) 53.28 is 7.2% of what number?

c) $\frac{3}{4}$% of what number equals 15?

Solution:

a) RATE = 4% and AMOUNT = \$63

$$\text{BASE} = \frac{\text{AMOUNT}}{\text{RATE}} = \frac{\$63}{0.04} = \$1{,}575$$

b) RATE = 7.2% and AMOUNT = 53.28

$$\text{BASE} = \frac{\text{AMOUNT}}{\text{RATE}} = \frac{53.28}{0.072} = 740$$

ANSWERS FOR PREVIEW TEST

1. 400 2. \$800 3. 32 4. \$900 5. \$14,400

c) RATE $= \frac{3}{4}\%$ and AMOUNT $= 15$

$$\text{BASE} = \frac{\text{AMOUNT}}{\text{RATE}} = \frac{15}{0.0075} = 2{,}000$$

PRACTICE PROBLEM 1: Find each base.

a) 5% of how many dollars equals \$36?

b) 4.2 is 200% of what number?

c) $4\frac{1}{2}\%$ of how many dollars equals \$40.50?

Mental Calculation of the Base

To determine the base mentally, use the rules for dividing by 0.1 and 0.01.

1. If the rate equals 10%, divide the amount by 0.1. Move the decimal point in the amount 1 place to the right.
2. If the rate equals 1%, divide the amount by 0.01. Move the decimal point in the amount 2 places to the right.
3. If the rate equals 20%, 30%, . . . , or 90%, first divide by 0.1. Then divide this result by 2, 3, . . . , or 9.
4. If the rate equals 2%, 3%, . . . , or 9%, first divide by 0.01. Then divide this result by 2, 3, . . . , or 9.

EXAMPLE 2: Find the base mentally. Round to the nearest dollar or unit.

a) \$13 is 10% of what number?

b) 8 is 3% of what number?

Solution:

a) Divide \$13 by 0.1.

$$\frac{\$13}{0.1} = \$130$$

b) Divide 8 by 0.01 and then divide by 3.

$$\frac{8}{0.01} = 800 \text{ and } \frac{800}{3} = 267 \text{ to the nearest unit.}$$

PRACTICE PROBLEM 2: Find the base mentally. Round to the nearest dollar or unit.

a) \$6.23 is 1% of how many dollars?

b) 42 is 80% of what number?

ANSWERS FOR PRACTICE PROBLEMS

1. a) \$720 **b)** 2.1 **c)** \$900 **2. a)** \$623 **b)** 53

EXERCISE SET 4.4A

Skill Development

1. 4 is 1% of what number?
2. 18 is 6% of what number?
3. \$23 is 2% of how many dollars?
4. \$600 is 15% of how many dollars?
5. 2.5% of what number equals 75?
6. 37.5% of how many dollars equal \$96?
7. 4% of what number equals 24.8?
8. 80% of how many dollars equals \$484?
9. 59.5 is 8.5% of what number?
10. 960 is 16% of what number?
11. \$400 is 400% of how many dollars?
12. \$24 is 150% of how many dollars?
13. 2 is 0.4% of what number?
14. 3.6 is 0.25% of what number?
15. $\frac{1}{2}\%$ of how many dollars equals \$60?
16. $6\frac{1}{4}\%$ of how many dollars equals \$52.50?

Round each result to the nearest cent.

17. \$5 is 1.3% of how many dollars and cents?
18. \$1.60 is 12% of how many dollars and cents?
19. \$17 is 3.5% of how many dollars and cents?
20. \$2.60 is 7% of how many dollars and cents?

Applications

21. Joan Hiller has been saving to buy a new car. She makes a 20% down payment of \$3,700. What is the total price of the car?

22. The \$24,000 allocated to the data processing department for equipment represents 15% of the budget for new instructional equipment. Find the total budget for new instructional equipment.

23. The 42,000 fans that attended the baseball game represents 1.6% of the total home attendance for the year. Find the total attendance for the year.

24. In one month Greene's Department Store collected \$243,750 in sales tax from its customers. If the sales tax rate is 7.5%, find the sales for the month.

For Some Extra Thought

25. If 5% of the price of an item equals \$10, what is 6% of the price?

26. Under what condition will the base be less than the amount?

Mental Calculation

Find the base mentally. Round to the nearest dollar or unit.

27. \$2.40 is 2% of how many dollars?

28. \$7.30 is 10% of how many dollars?

29. 62 is 30% of what number?

30. 2 is 7% of what number?

Review Problems for Maintaining Your Skills

31. Write 17,506 in words.

32. Find the quotient and remainder for $571 \div 39$.

In 33 through 35, reduce each fraction to its lowest terms and express any result that is an improper fraction as a mixed number.

33. $1\frac{2}{5} + 3\frac{1}{4} =$ **34.** $\frac{19}{20} - \frac{1}{25} =$ **35.** $2\frac{1}{4} \div 5\frac{3}{7} =$

36. Round 2.607 to the nearest hundredth.

37. Round \$1.48293 to the nearest cent.

38. Divide 45.3 by 1.47 and round the quotient to the nearest hundredth.

39. Convert 39.7% to a decimal.

40. Convert 5 to a percent.

41. Find 17.5% of 400.

42. $3\frac{1}{4}\%$ of \$8,600 equals what amount?

43. 6 is what percent of 25?

44. 8 is what percent of 2?

45. To the nearest tenth of a percent, 23 is what percent of 41?

EXERCISE SET 4.4B

Skill Development

1. 14 is 2% of what number?
2. 42 is 6% of what number?
3. \$38 is 5% of how many dollars?
4. \$840 is 12% of what amount?
5. 1.5% of what number equals 9?
6. 62.5% of what number equals 125?
7. 2% of how many dollars equals \$15.60?
8. 42% of how many dollars equals \$3,612?
9. 57 is 9.5% of what number?
10. 65 is 13% of what number?
11. \$356 is 200% of how many dollars?
12. 46 is 250% of what number?
13. 3 is 0.6% of what number?
14. \$1.46 is 0.25% of how many dollars?
15. $\frac{1}{4}\%$ of how many dollars equals \$28?
16. $\frac{1}{2}\%$ of what number equals 0.3?

Round each result to the nearest cent.

17. \$7 is 1.4% of how many dollars and cents?
18. \$1.84 is 14% of how many dollars and cents?
19. \$23 is 4.5% of how many dollars and cents?
20. \$1.45 is 8.3% of how many dollars and cents?

Applications

21. The minimum passing grade on a driver's test is 80%. Daniel Keller passed with a minimum score, getting 24 questions correct. How many questions were on the test?

22. Of the amount contributed to Butler Hospital, 40% comes from corporations. Last year corporations contributed \$560,000 to the hospital. Find the total dollar figure contributed to the hospital last year.

23. The \$7.50 tip left by the Russell family represents 15% of the restaurant check. Find the total for the restaurant check.

24. The property tax rate in Harwood county is 0.5% of assessed value. Determine how much assessed property value is needed to raise $490,000 in property tax.

For Some Extra Thought

25. If 25% of the price of an item equals $60, what is 30% of the price?

26. Under what condition will the base equal the amount?

Mental Calculations

Find the base mentally. Round to the nearest dollar or unit.

27. $25.84 is 4% of how many dollars?

28. $0.47 is 1% of how many dollars?

29. 72 is 60% of what number?

30. 6.4 is 9% of what number?

Review Problems for Maintaining Your Skills

31. Write 1,000,642 in words.

32. Find the quotient and remainder for 267 ÷ 100.

In 33 through 35, reduce each fraction to its lowest terms and express any result that is an improper fraction as a mixed number.

33. $5\frac{3}{4} + 4\frac{2}{3} =$ **34.** $\frac{4}{15} - \frac{1}{12} =$ **35.** $5\frac{2}{5} \div 1\frac{3}{5} =$

36. Round 12.60874 to the nearest thousandth.

37. Round $14.6807 to the nearest cent.

38. Divide 38.41 by 1.7 and round the quotient to the nearest thousandth.

39. Convert 6.43% to a decimal.

40. Convert 0.2493 to a percent.

41. Find 3.4% of 50.

42. Find $6\frac{3}{4}$% of $2,000.

43. 17 is what percent of 50?

44. 63 is what percent of 42?

45. To the nearest tenth of a percent, 37 is what percent of 80?

4.5 FINDING THE AMOUNT, RATE, OR BASE

Preview Test

1. 18 is what percent of 200?
2. Find 5.9% of \$700.
3. 4% of what number equals 12?
4. What number is 200% of 9?
5. To the nearest tenth of a percent, 14 is what percent of 39?

In this section you will not have the luxury of knowing from the title what type of percent problem you are solving. However, if you remember the base is the number that is multiplied by the rate to produce the amount, you will not have difficulty identifying the amount, rate, or base.

EXAMPLE 1: Identify the base, rate, and amount. Do not solve.

a) What number is 17% of 28,000?
b) 14 is what percent of 40?
c) 54 is 2% of what number?

Solution:

a) The rate is 17% and the base is 28,000. The number to be found represents the amount.
b) The amount is 14 and the base is 40. The number to be found represents the rate.
c) The amount is 54 and the rate is 2%. The number to be found represents the base.

PRACTICE PROBLEM 1: Identify the base, rate, and amount. Do not solve.

a) What percent of 34 is 23?
b) 5% of what number is 10?
c) Find 17% of 500.

Once you have identified which quantity (amount, rate or base) is to be found, you can use the following aid instead of memorizing the three formulas:

$$\text{AMOUNT} = \text{BASE} \times \text{RATE}$$

$$\text{RATE} = \frac{\text{AMOUNT}}{\text{BASE}}$$

$$\text{BASE} = \frac{\text{AMOUNT}}{\text{RATE}}$$

ANSWERS FOR PREVIEW TEST

1. 9% **2.** 41.30 **3.** 300 **4.** 18 **5.** 35.9%

Cover the word that represents the quantity to be found. Then you can "see" the operation that must be performed to find the quantity.

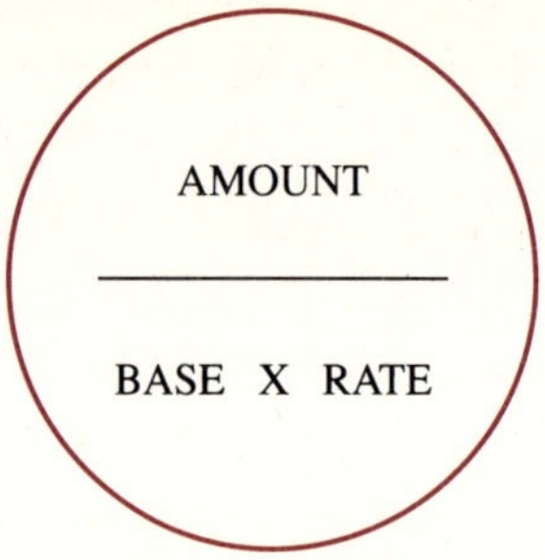

The **AMOUNT** is to be found.

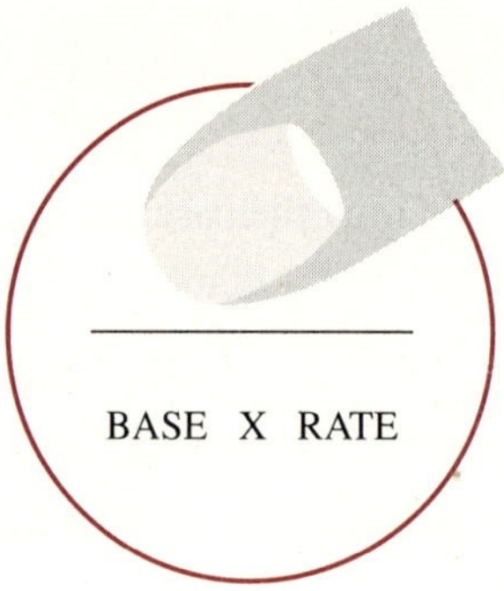

The **RATE** is to be found.

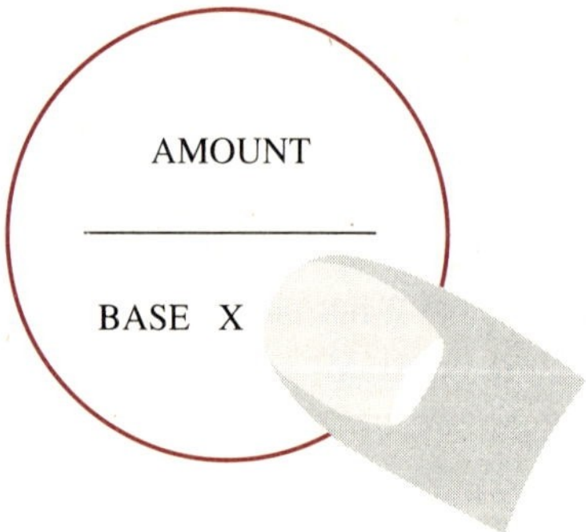

The **BASE** is to be found.

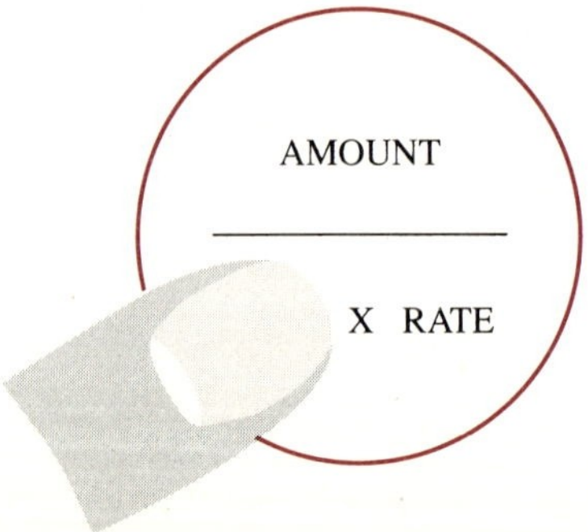

EXAMPLE 2: Find the indicated amount, rate, or base.

a) 15 is what percent of 40?

b) What number is 36% of 600?

c) 2% of what number equals 5?

Solution:

a) The amount is 15 and the base is 40.

$$\text{Rate} = \frac{15}{40} = 0.375 = 37.5\%$$

b) The rate is 36% and the base is 600.
Amount $= 600 \times 0.36 = 216$

c) The rate is 2% and the amount is 5.

$$\text{Base} = \frac{5}{0.02} = 250$$

PRACTICE PROBLEM 2: Find the indicated amount, rate, or base.

a) Find 45% of 8,000.

b) 20 is 8% of what number?

c) What percent of 6 is 12?

ANSWERS FOR PRACTICE PROBLEMS

1. **a)** 34 is the base and 23 is the amount. The rate is to be found.
   b) 5% is the rate and 10 is the amount. The base is to be found.
   c) 17% is the rate and 500 is the base. The amount is to be found.
2. **a)** 3,600 **b)** 250 **c)** 200%

EXERCISE SET 4.5A

Skill Development

1. Find 18% of 2,600.
2. 5% of what number equals 14?
3. 7.4% of 6,700 equals what number?
4. 8 is 1/2% of what number?
5. What number is 170% of 50?
6. To the nearest tenth of a percent, 19 is what percent of 24?
7. What number is 40% of 876?
8. 4% of what number equals 72?
9. 7 is what percent of 2?
10. 0.25% of what number equals 1?

Applications

11. Of the 6,300,000 cars sold in the northeast during a five-year period, 22.5% were imported. Find the number of imported cars.

12. Of the 124,000 tires sold by the Hill Tire Company, 3,267 were defective. Find the percent of defective tires to the nearest hundredth of a percent.

13. The 16 murders that took place in Smith County last year represent 40% of the murders in the state. Find the number of murders in the state.

14. As of the year 2001 it is expected that 18 out of every 40 persons who are alive at age 90 will survive to age 100. What percent of the ninety-year-old group will live to agc 100?

15. Of the 20,000 students at Whitney State University, 17% of their families have an income under $12,500 per year. Find the number of families in the under-$12,500 income group.

16. On a 250-question entrance examination 45 questions are multiple choice. What percent of the questions are multiple choice?

17. In 1991, Greene's Department Store had sales of $3,250,000. The state sales tax was 7.5%. Find the amount of sales tax that Greene's Department store collected from its customers in 1991.

18. The school tax in Williams County is 2.5% of the assessed property value. Find the assessed property value for a home that is sent a school tax bill of $2,050.

For Some Extra Thought

19. Do you ever solve a percent problem by multiplying by the amount?

20. If the amount is larger than the base, can the rate equal 80%?

Review Problems for Maintaining Your Skills

21. Round 51,807 to the nearest ten thousand.

22. $56 \times 8{,}094 =$

In 23 through 25, reduce each fraction to its lowest terms and express any result that is an improper fraction as a mixed number.

23. $\frac{2}{5} + \frac{3}{8} =$ **24.** $12 - 5\frac{3}{4} =$ **25.** $3\frac{2}{5} \times 1\frac{2}{3} =$

26. Round 0.057429 to the nearest thousandth.

27. Round $2.1382 to the nearest cent.

28. Divide 4.5 by 0.037 and round the quotient to the nearest hundredth.

29. Convert 6.05% to a decimal.

30. Convert 0.175 to a percent.

31. Find 300% of 17.

32. 2.5% of what number equals 200?

33. 13 is what percent of 20?

34. To the nearest tenth of a percent, 1 is what percent of 17?

35. 5% of what number equals 20.

EXERCISE SET 4.5B

Skill Development

1. Find 6.2% of 5,700.
2. 560 is what percent of 4,000?
3. What number is 300% of 4?
4. 5% of what number equals 24?
5. 1.5 is what percent of 50?
6. 8 is what percent of 2?
7. 0.5% of what number equals 3?
8. What number is 23.5% of 8,000?
9. 30 is 3/4% of what number?
10. Find 6.71% of 1,000.

Applications

11. The $8.50 tip left by the Russell family represents 17% of the restaurant check. Find the amount of the restaurant check.

12. In a class of 43 students, 12 are left-handed. Find to the nearest tenth of a percent the percent of the class that is left-handed.

13. Of the 1,300 houses in Walden County, 19% have a VCR. Find the number of houses that have a VCR.

14. Of the 32,000 gallons of gas sold by Mike's Gas Station, 23,100 were regular unleaded. Find the percent of regular unleaded gas sold by Mike's Gas Station and round to the nearest tenth of a percent.

15. The property tax rate in Harwood County is 0.5% of the assessed property value. Determine how much assessed property value is required to raise $490,000 in property tax.

16. At a motor vehicle inspection station, 42 out of 379 cars inspected during July needed brakes. Find to the nearest hundredth of a percent the percent of cars needing brakes.

17. A new car is purchased for \$12,600. The sales tax rate is 6.25%. Find the amount of the sales tax that the buyer of the car pays.

18. The Clarkstown College football team won 9 of its 12 games. What percent of their games did Clarkstown win?

For Some Extra Thought

19. Do you ever solve a percent problem by dividing by the amount?

20. If the base is larger than the amount, can the rate equal 80%?

Review Problems for Maintaining Your Skills

21. Round 1,506,872 to the nearest hundred thousand.

22. $23 \times 100{,}000 =$

In 23 through 25, reduce each fraction to its lowest terms and express any result that is an improper fraction as a mixed number.

23. $\frac{3}{8} + \frac{1}{4} =$ 24. $\frac{5}{6} - \frac{1}{5} =$ 25. $4\frac{1}{5} \div 3\frac{3}{4} =$

26. Round 1.6877 to the nearest tenth.

27. Round \$3.671 to the nearest cent.

28. Divide 3.17 by 1.9 and round the quotient to the nearest thousandth.

29. Convert 325% to a decimal.

30. Convert 0.0793 to a percent.

31. Find 26.5% of 8,600.

32. 4% of what number equals 9.2?

33. 24 is what percent of 16?

34. To the nearest hundredth of a percent, 3 is what percent of 7?

35. 1% of what number equals 0.4?

4.6 PERCENT INCREASE AND DECREASE

Preview Test

Find each percent increase.

1. A price increases from \$20 to \$24
2. A price increases from \$600 to \$1,200
3. A price increases from \$12 to \$17 (round to the nearest tenth of a percent)

Find each percent decrease.

4. A price decreases from \$45 to \$27
5. A price decreases from \$1,000 to \$700
6. A price decreases from \$41 to \$30 (round to the nearest hundredth of a percent)

When a number changes over a period of time, the change can be expressed as a percent increase or percent decrease. For example, if the ticket price to a movie increases from \$5.00 to \$5.25, the amount of the price increase is

$$\$5.25 - \$5.00 = \$0.25$$

If you divide the \$0.25 price increase by the \$5.00 old price and convert the decimal from this division to a percent you have found the percent increase in the ticket price.

$$\text{Percent Increase in the Ticket Price} = \frac{\$0.25}{\$5.00} = 0.05 = 5\%$$

If a number (such as a price) changes, the old number is referred to as the **old base** and the new number is referred to as the **new base**. If the new base is larger than the old base, the number has increased and the percent increase is given by

$$\text{PERCENT INCREASE} = \frac{\text{NEW BASE} - \text{OLD BASE}}{\text{OLD BASE}}$$

If the new base is smaller than the old base, the number has decreased and the percent decrease is given by

$$\text{PERCENT DECREASE} = \frac{\text{OLD BASE} - \text{NEW BASE}}{\text{OLD BASE}}$$

EXAMPLE 1: Find each percent increase or percent decrease. If necessary, round to the nearest tenth of a percent.

a) A price increases from \$40 to \$46

b) A price decreases from \$470 to \$410

ANSWERS FOR PREVIEW TEST

1. 20% **2.** 100% **3.** 41.7% **4.** 40% **5.** 30%
6. 26.83%

Solution:

a) OLD BASE = \$40 and NEW BASE = \$46

$$\text{PERCENT INCREASE} = \frac{\text{NEW BASE} - \text{OLD BASE}}{\text{OLD BASE}}$$

$$= \frac{\$46 - \$40}{\$40} = \frac{\$\ 6}{\$40} = 0.15 = 15\%$$

b) OLD BASE = \$470 and NEW BASE = \$410

$$\text{PERCENT DECREASE} = \frac{\text{OLD BASE} - \text{NEW BASE}}{\text{OLD BASE}}$$

$$= \frac{\$470 - \$410}{\$470} = \frac{\$\ 60}{\$470} = 0.1276\% = 12.8\%$$

PRACTICE PROBLEM 1: Find each percent increase or percent decrease. If necessary, round to the nearest tenth of a percent.

a) A price increases from \$50 to \$54

b) A price decreases from \$19 to \$17

ANSWERS FOR PRACTICE PROBLEM

1. a) 8% **b)** 10.5%

EXERCISE SET 4.6A

Skill Development

Find each percent increase.

1. A price increases from \$25 to \$28.
2. A price increases from \$500 to \$580.
3. A price increases from \$0.80 to \$0.84.
4. A price increases from \$20 to \$29.
5. An SAT math score increases from 480 to 552.
6. An SAT verbal score increases from 560 to 616.

Round each percent increase to the nearest tenth of a percent.

7. The number of units sold increases from 52 to 59.
8. The number of units sold increases from 240 to 273.

Find each percent decrease.

9. A price decreases from \$200 to \$192.
10. A price decreases from \$0.25 to \$0.19.
11. A price decreases from \$1,000 to \$930.
12. A price decreases from \$2.40 to \$2.16.
13. An SAT math score decreases from 600 to 576.
14. An SAT verbal score decreases from 500 to 495.

Round each percent decrease to the nearest hundredth of a percent.

15. The temperature decreases from 78 degrees to 71 degrees.
16. The temperature decreases from 45 degrees to 37 degrees.

Applications

Round each percent increase or percent decrease to the nearest tenth of a percent.

17. The number of new accounts opened at Bay State Bank during January and February were 56 and 172, respectively. Find the percent increase in the number of new accounts from January to February.

18. In 1951 the average price of a 12-inch black and white TV set was \$143. In 1991 the average price of the same set dropped to \$87. Find the percent decrease in the price.

19. Over the past 5 years, the average car insurance premium for an adult male under age 25 has increased from \$1,060 per year to \$2,897 per year. Find the percent increase.

20. Over a one-year period the annual cost of energy for the Brinkman family home increased from \$2,360 to \$2,534. Find the percent increase.

For Some Extra Thought

21. If the price triples, what is the percent increase?

22. Can a price decrease by more than 100%?

Review Problems for Maintaining Your Skills

23. Write 1,205,700 in words.

24. Find the quotient and remainder for 721 divided by 37.

In 25 throught 27, reduce each fraction to its lowest terms and express any result that is an improper fraction as a mixed number.

25. $2\frac{2}{5} + 1\frac{3}{4} =$ **26.** $\frac{13}{15} - \frac{7}{20} =$ **27.** $\frac{5}{8} \times \frac{2}{5} =$

28. Round 2.30478 to the nearest thousandth.

29. Round \$14.5637 to the nearest cent.

30. Divide 23 by 1.7 and round the quotient to the nearest tenth.

31. Convert 17.56% to a decimal.

32. Convert 37.5% to a fraction and reduce the fraction to its lowest terms.

33. Convert 0.007 to a percent.

34. 19 is what percent of 50?

35. Find 250% of 3.

36. 4% of what amount equals \$62.50?

EXERCISE SET 4.6B

Skill Development

Find each percent increase.

1. A price increases from \$20 to \$27.
2. A price increases from \$12.50 to \$17.
3. A price increases from \$0.30 to \$0.51.
4. A price increases from \$300 to \$324.
5. An SAT math score increases from 520 to 598.
6. An SAT verbal score increases from 620 to 651.

Round each percent increase to the nearest hundredth of a percent.

7. The number of units sold increases from 341 to 386.
8. The number of units sold increases from 23 to 27.

Find each percent decrease.

9. A price decreases from \$400 to \$300.
10. A price decreases from \$25 to \$22.
11. A price decreases from \$500 to \$432.
12. A price decreases from \$3.60 to \$3.24.
13. An SAT math score decreases from 480 to 456.
14. An SAT verbal score decreases from 500 to 482.

Round each percent decrease to the nearest tenth of a percent.

15. The temperature decreases from 58 degrees to 49 degrees.
16. The temperature decreases from 92 degrees to 81 degrees.

Applications

Round each percent increase or decrease to the nearest tenth of a percent.

17. With the installation of a traffic light at the corner of Main Street and Cook Avenue, the number of accidents decreased from an average of 7 per month to an average of 2 per month. Find the percent decrease.

18. In 1955 the average price for a reserved seat at a major league baseball game was \$2.85. In 1989 the average price was \$9.70. Find the percent increase.

19. After hiring a production manager, the number of complete units produced by the Wright Typewriter Company increased from an average of 120 to 144 per day. Find the percent increase.

20. In 1990 the Katona College women's basketball team won 23 games and in 1991 the team won 29 games. Find the percent increase in the number of wins.

For Some Extra Thought

21. If a price is increased by 100%, what percent decrease is needed to reduce it back to its original amount?

22. Can a price increase by more than 100%?

Review Problems for Maintaining Your Skills

23. Write 2,000,053,000 in words.

24. Find the quotient and remainder for 85 divided by 9.

In 25 through 27, reduce each fraction to its lowest terms and express any result that is an improper fraction as a mixed number.

25. $3\frac{3}{4} + 2\frac{1}{2} =$ **26.** $14 - \frac{5}{7} =$ **27.** $\frac{6}{7} \times \frac{4}{7} =$

28. Round 0.041863 to the nearest hundredth.

29. Round \$3.5444 to the nearest cent.

30. Divide 4.9 by 0.45 and round the quotient to the nearest hundredth.

31. Convert 2.87% to a decimal.

32. Convert 12.5% to a fraction and reduce the fraction to its lowest terms.

33. Convert 0.003 to a percent.

34. 13 is what percent of 25?

35. Find 200% of 9.

36. 2% of what amount equals \$3.64?

PRACTICE TEST 4A

1. Convert 47.8% to a decimal.
2. Convert 0.0425 to a percent.
3. Convert 24% to a fraction reduced to its lowest terms.
4. Convert $\frac{14}{19}$ to the nearest tenth of a percent.
5. 14 is 2% of what amount?
6. Find 250% of \$40.
7. 9 is what percent of 25?
8. Out of a class of 32 students, 7 earned an A. To the nearest whole percent find the percent of students who earned an A.
9. The sales tax rate in Alomar County is 4.5%. Find the sales tax for a car whose price is \$15,600.
10. The 8.4 inches of rain for April represents 5% of the rainfall for the year. Find the rainfall for the year.
11. Find the percent decrease if a price decreases from \$8.80 to \$7.48.
12. Over a five-year period the tuition cost at a state university has increased from \$3,600 per year to \$5,150. Find the percent increase to the nearest tenth of a percent.

PRACTICE TEST 4B

1. Convert 450% to a decimal.
2. Convert 0.173 to a percent.
3. Convert 2.5% to a fraction reduced to its lowest terms.
4. Convert $\frac{125}{400}$ to the nearest tenth of a percent.
5. 17 is what percent of 50?
6. \$1.24 is 4% of what amount?
7. Find 6.75% of \$6,000.
8. The \$2,000 for the debate club represents 8% of the total activities budget. Find the amount of the activities budget.
9. Out of the \$3,500,000 of sales for the Sports Shed, \$1,860,000 took place in the fourth quarter of the year. To the nearest tenth of a percent, find the percent of sales for the fourth quarter.
10. Twenty-three percent of the Lyman's \$42,000 yearly income is spent on food. How much do the Lymans spend on food in one year?
11. Find the percent increase to the nearest hundredth of a percent if a price increases from \$10.56 to \$12.37?
12. The price of the TRX500 computer has dropped from \$900 to \$666. Find the percent decrease.

CUMULATIVE REVIEW EXERCISES FOR CHAPTERS 1 THROUGH 4

1. Write 42,657,023 in words.

2. Round 1,269,140 to the nearest ten thousand.

3. Find the quotient and remainder for $178 \div 51$.

4. Reduce $\frac{64}{96}$ to its lowest terms.

5. Convert $\frac{230}{100}$ to a mixed number.

6. Convert $3\frac{7}{9}$ to an improper fraction.

In 7 through 10, reduce each fraction to its lowest terms and express any result that is an improper fraction as a mixed number.

7. $\frac{10}{13} \times \frac{2}{5} =$

8. $5\frac{3}{4} \div \frac{1}{4} =$

9. $6\frac{2}{9} + 1\frac{1}{6} =$

10. $\frac{17}{18} - \frac{5}{12} =$

11. Write 2.009 in words.

12. Write the number for five and twenty-three hundred-thousandths.

13. Round 0.0054545 to the nearest millionth.

14. Round $45.8047 to the nearest cent.

15. $12.066 - 1.007 =$

16. $1.23 \times 100{,}000 =$

17. Divide 12.7 by 1.06 and round the quotient to the nearest hundredth.

18. $4.2 \times 10 + 5.3 \times 10 + 6.7 \times 10 =$

19. Convert 17.34% to a decimal.

20. Convert 0.003 to a percent.

21. Convert 62% to a fraction reduced to its lowest terms.

22. Convert $\frac{41}{53}$ to the nearest hundredth of a percent.

23. Find 62.5% of $8,000.

24. 15.2 is 4% of what number?

25. 23 is what percent of 200?

26. Fifteen percent of the freshman class at Prescott University comes from Ohio. If the size of the freshman class is 2,580, how many of these students come from Ohio?

27. The \$12,000 yearly rent represents 15% of the total yearly expenses for the ABC Stationery Store. Find the yearly expenses.

28. In a math textbook 125 of the 200 pages have a geometric illustration. What percent of the pages have a geometric illustration?

29. Over a ten-year period the average cost for a computer disk decreased from \$2.31 to \$1.12. Find the percent decrease to the nearest tenth of a percent.

C H A P T E R 5

RATIO AND PROPORTION

5.1 WRITING RATIOS

5.2 CHECKING PROPORTIONS

5.3 SOLVING PROPORTIONS

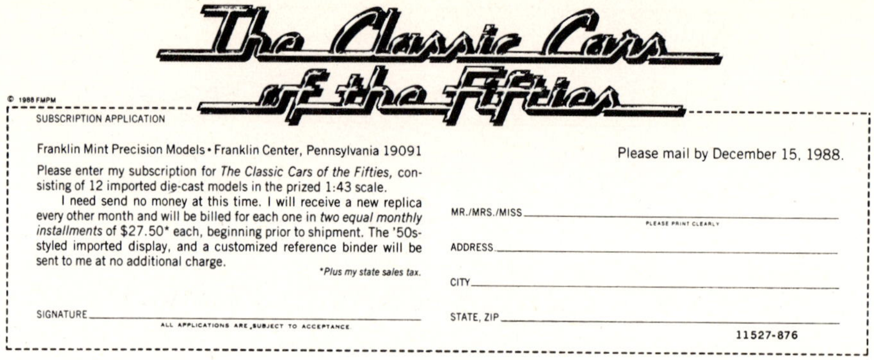
The Classic Cars of the Fifties

© 1988 FMPM

SUBSCRIPTION APPLICATION

Franklin Mint Precision Models • Franklin Center, Pennsylvania 19091

Please mail by December 15, 1988.

Please enter my subscription for *The Classic Cars of the Fifties*, consisting of 12 imported die-cast models in the prized 1:43 scale.

I need send no money at this time. I will receive a new replica every other month and will be billed for each one in *two equal monthly installments* of $27.50* each, beginning prior to shipment. The '50s-styled imported display, and a customized reference binder will be sent to me at no additional charge.

**Plus my state sales tax.*

MR./MRS./MISS ________

PLEASE PRINT CLEARLY

ADDRESS ________

CITY ________

SIGNATURE ________

ALL APPLICATIONS ARE SUBJECT TO ACCEPTANCE.

STATE, ZIP ________

11527-876

In the order coupon shown above, the classic cars are sold in a 1:43 scale. What does 1:43 mean? The notation 1:43 is a ratio. It means that the actual 1950s classic car was 43 times as large as the model that Franklin Mint is selling. Their model of the 1957 Corvette measures 4 inches. This means that the car was 43×4 inches or 172 inches. Dividing 172 inches by 12 we can calculate the length of the car in feet and inches as 14 feet, 4 inches.

As you can see, ratios are used to compare two amounts, such as the size of a die-cast model of a car to the actual size of the car. Ratios have many other applications. For example, if you travel to Canada you may find that the ratio of Canadian dollars to United States dollars is $1.21:$1.00. This means that for each United States dollar you will receive $1.21 in Canadian dollars.

In this chapter you will learn how to work with ratios and with proportions, which are equations involving ratios.

5.1 WRITING RATIOS

Preview Test

1. The National Provident Bank has two assistant managers and nine tellers.

a) Write the ratio of assistant managers to tellers in verbal form.

b) Write the ratio of tellers to assistant managers in fraction form.

2. A work shift at the Monks Point Smelting Plant is 8 hours. Ann Lyman has completed 3 hours and 45 minutes of her shift. Express each ratio in minutes and reduce to its lowest terms.

a) Write the ratio of the time completed to total shift time for Ann Lyman using colon format.

b) Write the ratio of time completed to time not completed for Ann Lyman using fraction format.

3. A cashier has 27 nickels and 19 dimes in his cash drawer.

a) Write the ratio of dimes to nickels in verbal format.

b) Convert each coin amount to cents and write the ratio of nickels to dimes in colon format.

A **ratio** is used to compare two amounts. For example, if there are 6 licensed drivers in a family that owns 4 cars, then the ratio of licensed drivers to cars owned is 6 to 4. When the ratio is written as 6 to 4, it is expressed in verbal form. Two other ways of writing ratios are

6:4	Colon Form
$\frac{6}{4}$	Fraction Form

Ratios can be reduced by dividing by a common divisor. In the example above, since 2 is a common divisor of 6 and 4, the ratio can be written as

3 to 2 in verbal form,

3:2 in colon form,

or $\frac{3}{2}$ in fraction form.

EXAMPLE 1: Eric Wright has $141 in his checking account and $1,239 in his savings account.

a) Use colon form to write the ratio of money in his checking account to money in his savings account.

b) Use fraction form to write the ratio of money in his checking account to the total amount in both accounts.

Solution:

a) The ratio is 141:1,239

b) The total amount in both accounts is

$$\$141 + \$1{,}239 = \$1{,}380$$

Therefore, the ratio of money in his checking account to the total in both accounts is $\frac{141}{1{,}380}$

ANSWERS FOR PREVIEW TEST

1. a) 2 to 9 **b)** $\frac{9}{2}$ **2. a)** 15:32 **b)** $\frac{15}{17}$

3. a) 19 to 27 **b)** 135:190

PRACTICE PROBLEM 1: On Wednesday the RPG gas station sold 782 gallons of regular gas and 514 gallons of super unleaded gas. Reduce each ratio to its lowest terms.

a) Use verbal form to write the ratio of gallons of super unleaded gas to gallons of regular unleaded.

b) Use colon form to write the ratio of gallons of regular unleaded to total gallons.

Converting to Common Units

If you are comparing two amounts where different units of measure are used, it sometimes makes sense to convert to a common unit of measure. For example, a ratio comparing two time periods should be expressed in the same units, such as seconds, minutes, hours, days, and so on.

EXAMPLE 2: It takes Cathy Collins 1 hour and 12 minutes to drive from home to her law office. If she has been driving for 48 minutes, find

a) The ratio of the time driven to the total time for her trip in fraction format;

b) The ratio of the time driven to the time remaining for the trip in colon format.

Solution:

a) To express the ratio in minutes, convert 1 hour and 12 minutes to $60 + 12 = 72$ minutes. Then the ratio of time driven to total time is $\frac{48}{72}$, which can be reduced to $\frac{2}{3}$.

b) The time remaining for the trip is $72 - 48 = 24$ minutes. The ratio of time driven to time remaining is 48:24, which reduces to 2:1.

PRACTICE PROBLEM 2: In a bank teller's cash drawer there are 100 twenty-dollar bills and 85 ten-dollar bills. Reduce each ratio to its lowest terms.

a) What is the ratio of twenty-dollar bills to ten-dollar bills in colon format?

b) Convert each amount to dollars and find the ratio of the dollar amount in twenty-dollar bills to the dollar amount in ten-dollar bills, using fraction format.

ANSWERS FOR PRACTICE PROBLEMS

1. a) 257 to 391 **b)** 391:648 **2. a)** 20:17 **b)** $\frac{40}{17}$

EXERCISE SET 5.1A

Skill Development

Reduce each ratio to its lowest terms.

1. 20 to 16

2. 32:20

3. 70:90

4. $\frac{36}{54}$

5. $\frac{51}{34}$

6. $\frac{1{,}200}{2{,}500}$

Convert each ratio to both colon and fraction form.

7. 5 to 2

8. 1 to 6

9. 23 to 3

10. 2 to 7

11. 1 to 1

12. 13 to 19

Convert each ratio to both verbal and fraction form.

13. 3:5

14. 17:50

15. 6:5

16. 4:1

17. 5:83

18. 100:1

Applications

Write each ratio in reduced form.

19. For the first 20 trading days last year, the stock market was over 2,500 for 16 days.

a) Use colon form to write the ratio of the days the average was over 2,500 to the number of trading days.

b) Use fraction form to write the ratio of days the average was not over 2,500 to days the average was over 2,500.

20. At the Regis Hotel Corporation there are 14 managers and 24 assistant managers.

a) Use verbal form to write the ratio of managers to assistant managers.

b) Use colon form to write the ratio of assistant managers to managers.

21. It is recommended that a student spend 2 hours in study outside of class for every hour spent in class. Use verbal form to write the ratio of the time spent in class to the time spent studying outside class.

22. A three-part sales presentation runs for three hours. The first part takes 43 minutes. Use colon form to write the ratio of the time for the first part to the time for the whole presentation, in minutes.

For Some Extra Thought

23. The ratio of passing plays to running plays in a football game is 3 to 2. If there are 45 passing plays, find the number of running plays in the game.

24. Can the ratio 5 to 2 be reduced to 2.5 to 1?

Review Problems for Maintaining Your Skills

25. Round 4,652 to the nearest hundred.

26. $245 \times 78 =$

In 27 and 28, reduce each fraction to its lowest terms and express any result that is an improper fraction as a mixed number.

27. $\frac{3}{8} + \frac{1}{5} =$ **28.** $4\frac{1}{3} - 2 =$

29. Round $4.2608 to the nearest cent.

30. Divide 40 by 1.03 and round the quotient to the nearest hundredth.

31. Convert 3.45% to a decimal.

32. Convert 2 to a percent.

33. 14 is what percent of 40?

34. 2.5% of what amount equals $50?

35. Find 16.5% of $4,000.

36. Find the percent increase if a price increases from $50 to $56.

37. Find the percent decrease to the nearest tenth of a percent if the stock market index drops from 2560 to 2480?

EXERCISE SET 5.1B

Skill Development

Reduce each ratio to its lowest terms.

1. 25 to 15 **2.** 24 to 60 **3.** 100 to 300

4. $\frac{13}{39}$ **5.** $\frac{2}{12}$ **6.** $\frac{55}{45}$

Convert each ratio to first colon, then fraction form.

7. 4 to 7 **8.** 16 to 23 **9.** 13 to 16

10. 1 to 9 **11.** 42 to 39 **12.** 3 to 2

Convert each ratio to first verbal, then fraction form.

13. 4:13 **14.** 1:1 **15.** 16:27

16. 2:5 **17.** 125:102 **18.** 33:37

Applications

Write each ratio in reduced form.

19. The Moraga Springs Shopping Mall has 32 stores. Twenty-two of the stores have yearly sales of $500,000 or more.

a) Use fraction form to write the ratio of stores with yearly sales under $500,000 to stores with yearly sales of $500,000 or more.

b) Use verbal form to write the ratio of stores with yearly sales of $500,000 or more to the number of stores in the mall.

20. Nancy Reynolds has 5 ten-dollar bills and 7 five-dollar bills in her wallet.

a) In fraction form, what is the ratio of five-dollar bills to ten-dollar bills?

b) In verbal form, what is the ratio of the value of the ten-dollar bills to the value of the five-dollar bills?

21. On a test, the ratio of true-false questions to multiple choice questions is 25 to 20. In fraction form, what is the ratio of true-false questions to total questions on the test?

22. It is recommended that 12 ounces of dry gas be added to every 10 gallons of car gas. If each gallon contains 128 ounces, what is the recommended ratio of dry gas to car gas in colon format?

For Some Extra Thought

23. If the ratio of men to women that work at the Schindler Insurance Company is 3 to 1 and there are 48 men, how many women work for Schindler?

24. The ratio of sales of 90-minute cassette tapes to 60-minute cassette tapes is 2 to 5 at Parkin Drug Store. Use colon form to state the ratio of sales of 90-minute tapes to the combined sales of 90- and 60-minute tapes?

Review Problems for Maintaining Your Skills

25. Round 41,687,900 to the nearest hundred thousand.

26. $10{,}000 \times 26 =$

In 27 and 28, reduce each fraction to its lowest terms and express any result that is an improper fraction as a mixed number.

27. $13\frac{1}{2} - 4\frac{3}{4} =$ **28.** $\frac{10}{16} \times \frac{4}{5} =$

29. Round 4.016278 to the nearest thousandth.

30. Divide 3.4 by 0.7 and round the quotient to the nearest tenth.

31. Convert 45.3% to a decimal.

32. Convert 0.1785 to a percent.

33. To the nearest tenth of a percent, 13 is what percent of 31?

34. 5% of what number equals 72?

35. Find 14% of \$9,000.

36. Find the percent decrease if a price decreases from \$25 to \$18.75.

37. Find the percent increase to the nearest hundredth of a percent if an SAT score increases from 514 to 579.

5.2 CHECKING PROPORTIONS

Preview Test

Use cross multiplication to determine if each proportion is true.

1. $\frac{10}{18} = \frac{14}{21}$ **2.** $\frac{20}{25} = \frac{12}{20}$ **3.** $\frac{90}{75} = \frac{30}{25}$ **4.** $\frac{1.2}{1.8} = \frac{0.2}{0.3}$

When one ratio is equal to another ratio, the statement is called a **proportion**. For example, since the ratios $\frac{2}{3}$ and $\frac{4}{6}$ are equal, $\frac{2}{3} = \frac{4}{6}$ is a proportion.

You can check the truth of a proportion by a method known as **cross multiplication**. To cross multiply, you must have two ratios in fraction form. First multiply the numerator of the first fraction by the denominator of the second fraction. Then multiply the denominator of the first fraction by the numerator of the second fraction. If the two products are equal, the proportion is true. If the products are not equal, the proportion is false.

Cross-Multiplication Rule for Checking Proportions

$$\frac{\text{NUMERATOR 1}}{\text{DENOMINATOR 1}} = \frac{\text{NUMERATOR 2}}{\text{DENOMINATOR 2}}$$

The proportion is true if

$$\text{NUMERATOR 1} \times \text{DENOMINATOR 2} = \text{DENOMINATOR 1} \times \text{NUMERATOR 2}$$

EXAMPLE 1: Use cross-multiplication to determine if $\frac{5}{8} = \frac{35}{56}$ is true or false.

Solution:

$5 \times 56 = 280$

$8 \times 35 = 280$

Since $5 \times 56 = 8 \times 35$, the proportion is true.

PRACTICE PROBLEM 1: Use cross-multiplication to determine if each proportion is true or false.

a) $\frac{2}{3} = \frac{12}{20}$ **b)** $\frac{6}{7} = \frac{54}{63}$

ANSWERS FOR PRACTICE PROBLEM

1. a) False **b)** True

ANSWERS FOR PREVIEW TEST

1. No **2.** No **3.** Yes **4.** Yes

EXERCISE SET 5.2A

Skill Development

Use cross-multiplication to determine whether each proportion is true.

1. $\frac{2}{3} = \frac{20}{30}$

2. $\frac{5}{6} = \frac{20}{25}$

3. $\frac{12}{20} = \frac{18}{30}$

4. $\frac{6}{7} = \frac{14}{16}$

5. $\frac{1}{3} = \frac{12}{42}$

6. $\frac{5}{6} = \frac{50}{60}$

7. $\frac{30}{45} = \frac{120}{180}$

8. $\frac{2}{1} = \frac{16}{8}$

9. $\frac{4}{3} = \frac{24}{18}$

10. $\frac{3}{7} = \frac{27}{62}$

11. $\frac{24}{22} = \frac{72}{64}$

12. $\frac{7}{2} = \frac{35}{10}$

Applications

13. Bernhard's sells a table lamp for $15 that was originally priced at $24. The store also sells a floor lamp for $55 that originally sold for $86. Do the ratios of the current selling price to the original price for the table lamp and floor lamp form a true proportion?

14. Gyroscope Inc. has located 10 defective ball bearings in one lot of 2,500 ball bearings. In another lot of 3,000 ball bearings, the company located 12 defective bearings. Do the ratios of defective ball bearings to total ball bearings in the two lots form a true proportion?

15. Save-more Office Supplies offers an $11.74 discount on a $58.70 filing cabinet. The Budget Office Store advertises $8 off on a $40 filing cabinet. Do the ratios of the discounts to the original prices for the two stores form a true proportion?

16. After checking 2,400 printed circuit boards one day Gary Laverna rejected 12 boards. The next day he rejected 17 out of 3,400 circuit boards. Do the ratios of rejected boards to total boards for the two days form a true proportion?

For Some Extra Thought

17. Jolene Scrivener receives a \$7 commission for every \$140 of sales she makes. What commission does Jolene receive for \$100 of sales?

18. A county charges \$487.50 property tax for a home with a \$65,000 assessed value. If the ratio of tax to assessed value does not change, how much will the tax increase if the assessed value increases by \$10,000?

Review Problems for Maintaining Your Skills

19. Write 1,956,422 in words.

20. Find the quotient and remainder for 200 divided by 17.

In 21 and 22, reduce each fraction to its lowest terms and express any result that is an improper fraction as a mixed number.

21. $5\frac{2}{5} + 7\frac{1}{2}$ **22.** $\frac{12}{13} - \frac{2}{39} =$

23. Round 0.045267 to the nearest ten-thousandth.

24. Divide 4.56 by 0.7 and round the quotient to the nearest hundredth.

25. Convert 250% to a decimal.

26. Convert $\frac{23}{50}$ to a percent.

27. Find 7.5% of \$8,950.

28. 24 is what percent of 25?

29. 2% of what amount equals \$1.30?

30. Find the percent decrease if a price decreases from \$40 to \$34.

31. Convert the ratio 1 to 4 to colon form.

32. Reduce the ratio 72 to 90 to its lowest terms.

EXERCISE SET 5.2B

Skill Development

Use cross-multiplication to determine whether each proportion is true.

1. $\frac{5}{6} = \frac{65}{78}$

2. $\frac{3}{5} = \frac{12}{20}$

3. $\frac{1}{4} = \frac{6}{27}$

4. $\frac{3}{7} = \frac{17}{42}$

5. $\frac{2}{3} = \frac{14}{21}$

6. $\frac{50}{35} = \frac{20}{14}$

7. $\frac{1}{4} = \frac{10}{40}$

8. $\frac{100}{125} = \frac{100}{125}$

9. $\frac{2}{7} = \frac{18}{28}$

10. $\frac{6}{9} = \frac{48}{63}$

11. $\frac{1}{2} = \frac{1}{3}$

12. $\frac{3}{4} = \frac{27}{36}$

Applications

13. Teknika Interior Design's head office listed 13 absent employees from its staff of 1,335 last week. One of its branch stores listed 4 absent employees out of 285 employees last week. Are the ratios of absent employees to total employees equal for the head office and the branch store?

14. One batter had 150 hits in 600 times at bat. Another batter had 190 hits in 760 times at bat. Are the ratios of hits to times at bat equal for the two batters?

15. In a class of 32 students, 8 earned an A. In another class of 36 students, 9 earned an A. Do the ratios of students who earned an A to total students form a true proportion?

16. One cereal company has determined that it needs a box that weighs 2 ounces to contain 15.5 ounces of cereal. Another company uses a box that weighs 2.5 ounces for 18.6 ounces of cereal. Do the ratios of box weight to cereal weight form a true proportion?

For Some Extra Thought

17. A \$742.50 personal property tax is charged for an \$82,500 house. What tax is charged for each \$100 of house value?

18. Of 225,000 registered voters in a state, 90,000 are registered Republicans. How many registered voters that are not Republicans would be expected if the number of registered voters increased to 300,000?

Review Problems for Maintaining Your Skills

19. Write 23,897,462 in words.

20. Find the quotient and remainder for 146 divided by 41.

In 21 and 22, reduce each fraction to its lowest terms and express any result that is an improper fraction as a mixed number.

21. $\frac{12}{13} - \frac{1}{2} =$ 22. $3\frac{1}{4} \times 5\frac{2}{3} =$

23. Round 2.30487 to the nearest tenth.

24. Divide 6.7 by 23 and round the quotient to the nearest thousandth.

25. Convert 4.56% to a decimal.

26. Convert $\frac{21}{25}$ to a percent.

27. Find 3.4% of \$700.

28. To the nearest tenth of a percent, 13 is what percent of 17?

29. 5% of what amount equals \$65?

30. Find the percent increase if a price increases from \$200 to \$212.

31. Convert the ratio 6:5 to fraction form.

32. Reduce the ratio 60 to 75 to its lowest terms.

5.3 SOLVING PROPORTIONS

Preview Test

Solve each proportion using cross-multiplication.

1. $\frac{N}{4} = \frac{7}{2}$ 2. $\frac{6}{9} = \frac{W}{6}$ 3. $\frac{0.5}{A} = \frac{40}{1.6}$ 4. $\frac{12}{1} = \frac{R}{\frac{3}{4}}$

5. A map is drawn with a legend that 2 inches equals 8 miles. How many miles apart are two cities that are drawn 5 inches apart?

When you solve proportion problems, one of the four terms is not known. This term is represented by a letter. You have to find the value of the letter that makes the proportion true.

EXAMPLE 1: Solve $\frac{N}{12} = \frac{2}{3}$

Solution:
To find the number for N that makes the proportion true, begin by cross-multiplying.

$$N \times 3 = 12 \times 2$$
$$N \times 3 = 24$$

Now divide each side by the number multiplying the letter, which in this case is 3.

$$\frac{N \times 3}{3} = \frac{24}{3}$$

Finally, carry out the division on the left and right side of the equals sign. On the left side 3 divided by 3 equals 1. On the right side 24 divided by 3 equals 8.

$$\frac{N \times \overset{1}{\cancel{3}}}{\cancel{3}} = \frac{\overset{8}{\cancel{24}}}{\cancel{3}}$$

Therefore, $N \times 1 = 8$. But, since 1 times a number represented by N equals N,

$$N \times 1 = 8$$

can be written as

$$N = 8$$

So, 8 is the number that makes the proportion $\frac{N}{12} = \frac{2}{3}$ true.

ANSWERS FOR PREVIEW TEST

1. 14 2. 4 3. 0.02 4. 9 5. 20 miles

PRACTICE PROBLEM 1: Solve the proportion $\frac{2}{5} = \frac{T}{40}$.

The numerator or denominator of a proportion can include decimals or fractions. Although the calculations may be a bit more difficult, the method of solution is the same as illustrated in Example 1.

EXAMPLE 2: Solve each proportion.

a) $\frac{10}{2.5} = \frac{0.4}{T}$ **b)** $\frac{4}{7} = \frac{y}{\frac{1}{2}}$

Solution:

a) First cross-multiply.

$$10 \times T = 2.5 \times 0.4$$
$$10 \times T = 1$$

Now divide each side by 10.

$$\frac{10 \times T}{10} = \frac{1}{10}$$

Therefore, $T = \frac{1}{10}$ or 0.1

b) $4 \times \frac{1}{2} = 7 \times y$

$2 = 7 \times y$

$\frac{2}{7} = \frac{7 \times y}{7}$

Therefore, $y = \frac{2}{7}$

Note: Since $\frac{2}{7}$ does not convert to an exact decimal, the answer is left in fraction form.

PRACTICE PROBLEM 2: Solve each proportion.

a) $\frac{0.5}{D} = \frac{1.5}{1.2}$ **b)** $\frac{A}{2} = \frac{6}{\frac{3}{8}}$

Applied proportion problems involve setting up two ratios for different quantities. The quantities might be miles and hours, commissions and sales, time and production, and so on. The key to setting up the ratios for the proportions is to keep like quantities in the numerators and denominators of the ratios. Consider the following example.

EXAMPLE 3: A salesperson receives \$11 commission for \$55 worth of sales. What commission would the salesperson receive for \$72 worth of sales?

Solution:

One way to set up the proportion so like quantities appear in the numerator and denominator is:

$$\frac{\text{commission}}{\text{sales}} = \frac{\text{commission}}{\text{sales}}$$

Notice that "commission" is the numerator term in both ratios, and "sales" is the denominator in both ratios. Another way to set up the proportion is:

$$\frac{\text{sales}}{\text{commission}} = \frac{\text{sales}}{\text{commission}}$$

Both ratios are still similar, because "sales" appears as the numerator term and "commission" appears as the denominator term in both ratios. However, the proportion

$$\frac{\text{sales}}{\text{commission}} = \frac{\text{commission}}{\text{sales}}$$

is incorrect because "sales" and "commission" do not hold the same positions in the ratios. "Sales" is the numerator term in one ratio and the denominator term in the other ratio. The same statement can be made about the placement of "commission." Let's solve the proportion using the first correct proportion. In substituting values, it's important to get the commission and sales numbers together. Since there is an \$11 commission for \$55 of sales, these two numbers are placed together in the same ratio. Use the letter C for the commission associated with sales of \$72. Then

$$\frac{\$11}{\$55} = \frac{C}{\$72}$$

Cross-multiplying and solving for C we have

$$11 \times 72 = 55 \times C$$

$$\frac{11 \times 72}{55} = \frac{55 \times C}{55}$$

$$14.4 = C \text{ or } C = 14.4$$

Therefore, there will be \$14.40 commission for \$72 in sales. This answer can be checked by substituting 14.4 for C in the proportion and checking the cross-products.

PRACTICE PROBLEM 3: The annual tax on \$25,000 of income is \$625 in a certain state. At this same rate, what income would result in a \$687.50 tax?

Be Careful About This

You may have noticed that for some proportions you can reduce before cross-multiplying. Make sure that when you reduce, you divide out common factors for fractions on the same side of the equals symbol. For example, you can do the following reduction.

Divide 20 and 12 by the common divisor 4.

$$\frac{1}{P} = \frac{\overset{5}{\cancel{20}}}{\underset{3}{\cancel{12}}}$$

In this proportion 20 and 12 are both on the right side of the equals symbol. However, you cannot do the following reduction.

$$\frac{\overset{5}{\cancel{20}}}{P} = \frac{1}{\underset{3}{\cancel{12}}}$$

Here, the 20 is on the left side and the 12 is on the right side of the equals symbol.

ANSWERS FOR PRACTICE PROBLEMS

1. 16 **2. a)** 0.4 **b)** 32 **3.** $27,500

EXERCISE SET 5.3A

Skill Development

Solve each proportion using cross-multiplication.

1. $\frac{N}{2} = \frac{6}{4}$

2. $\frac{B}{3} = \frac{5}{1}$

3. $\frac{2}{A} = \frac{15}{15}$

4. $\frac{2}{Z} = \frac{8}{1}$

5. $\frac{M}{4} = \frac{2.5}{8}$

6. $\frac{1.4}{0.1} = \frac{7}{X}$

7. $\frac{5}{3} = \frac{K}{\frac{1}{5}}$

8. $\frac{40}{P} = \frac{16}{\frac{2}{5}}$

9. $\frac{S}{2\frac{1}{2}} = \frac{24}{10}$

Applications

10. A map is drawn with a scale of 2 inches for 5 miles distance. If 2 cities are 24 miles apart, how many inches apart would they be drawn on the map?

11. A saline solution contains 3.5 cubic centimeters of salt for every 10 cubic centimeters of solution. How many cubic centimeters of solution can be made with 14 cubic centimeters of salt?

12. A salesperson receives $12.50 commission for every $100 of sales. If she receives $80 in commissions, how much in sales does she generate?

13. The Zelaya family earns $35,500 and pays $2,485 in state income tax. At the same rate, what state income tax would the family pay for $40,000 of income?

14. A chemical process requires 5 minutes reaction time for every 250 grams of substance. A technician finds the reaction time for one batch of substance to be 64 minutes. How many grams of substance does the batch contain?

15. A lot of 1,000 locking mechanisms contains 3 defective pieces. How many defective pieces can be expected in a lot of 6,000 locking mechanisms?

16. According to Labor Department statistics several years ago, a woman earns 55 cents for every dollar earned by a man. In a certain state the average yearly income for a man is $32,000. If the 55 cents to one dollar ratio holds in this state, find the average yearly income for a woman.

17. At Central Banking a home buyer can expect to pay $690 in points for a $46,000 mortgage loan. If the ratio of points to loan amount remains the same, how much would a home buyer pay in points for a $55,000 loan?

For Some Extra Thought

18. A music store offers a discount sale of one dollar off for every $9 of list price. What is the discounted price of a car stereo listed at $180?

19. A floor supervisor estimates 4 defective cables for every 1,500 cables made. How many nondefective cables can be expected for a lot of 9,000 cables?

Review Problems for Maintaining Your Skills

20. Round 563,987,324 to the nearest ten million.

21. $10{,}000 \times 213 =$

22. Convert $5\frac{3}{11}$ to an improper fraction.

23. Convert $\frac{100}{23}$ to a mixed number.

24. Round $12.06397 to the nearest cent.

25. Divide 145.6 by 12.7 and round the quotient to the nearest hundredth.

26. Convert 23.75% to a decimal.

27. Convert 0.024 to a percent.

28. Find 7.29% of $2,500.

29. To the nearest tenth of a percent, 17 is what percent of 30?

30. 5% of what amount equals \$84?

31. Reduce the ratio 54:36 to its lowest terms.

32. Convert the ratio 29 to 37 to fraction form.

In 33 and 34, use cross-multiplication to determine whether each proportion is true.

33. $\frac{12}{13} = \frac{5}{6}$ **34.** $\frac{2}{3} = \frac{10}{15}$

EXERCISE SET 5.3B

Skill Development

Solve each proportion using cross-multiplication.

1. $\frac{H}{5} = \frac{10}{25}$

2. $\frac{2}{1} = \frac{J}{8}$

3. $\frac{25}{T} = \frac{40}{12}$

4. $\frac{3}{2} = \frac{B}{6}$

5. $\frac{1.6}{4} = \frac{K}{7}$

6. $\frac{2.5}{10} = \frac{3.6}{Z}$

7. $\frac{4}{1} = \frac{J}{\frac{1}{4}}$

8. $\frac{200}{2\frac{1}{2}} = \frac{X}{5}$

9. $\frac{2}{5} = \frac{H}{\frac{5}{2}}$

Applications

10. The Johnsons are charged \$792 personal property tax for their \$72,000 home. At the same tax rate, what tax would Anne Lynn pay for her \$75,000 home?

11. The Lyman family drove to New England, averaging 55 miles per hour. If the trip takes 6.5 hours, how far did they drive?

12. A company can process 200 circuit boards in 8 hours. How long will it take to process 850 boards?

13. A drawing is done with 1/8 inch representing 1 foot. What length would a 15-foot room be drawn?

14. At a certain time of day a 2.5-meter fence post casts a 7.5-meter shadow. What height pole will cast a 2.4-meter shadow?

15. A department store is selling its wares at a 33-cent discount for each dollar of list price. What would be the discount of a pocket-size AM/FM receiver listed at $15?

16. Of every 15,000 cars sold at a car dealership, 1,800 are white. For a year that 20,000 cars are sold, how many are white if the 15,000 to 1,800 ratio holds?

17. An American League baseball team has won 32 out of 56 games played by mid-June. If their ratio of wins to games played remains the same through 70 games, how many games should they have won after 70 games have been played?

For Some Extra Thought

18. The net proceeds is the amount of a purchase that goes to a store after the salesperson's commission has been deducted. What are the net proceeds for a $400 VCR if the salesperson receives a commission of $1.25 for every $10 purchase?

19. Historically, in School District M, 900 out of every 2,000 students have begun college following high school graduation. If the historical ratio holds, how many of the 1,200 students in the graduating class would not be going to college?

Review Problems for Maintaining Your Skills

20. Round 1,628 to the nearest ten.
21. $246 \times 47 =$
22. Convert $4\frac{2}{7}$ to an improper fraction.
23. Convert $\frac{73}{70}$ to a mixed number.
24. Round 1.4509 to the nearest hundredth.
25. Divide 12.78 by 64.3 and round the quotient to the nearest thousandth.

26. Convert 125% to a decimal.
27. Convert 0.0083 to a percent.
28. To the nearest hundredth of a percent, 45 is what percent of 56?

29. 2% of what amount equals $1.29?
30. Find 17.8% of 60,000.
31. Reduce the ratio 100 to 250 to its lowest terms.

32. Convert the ratio $\frac{31}{64}$ to colon form.

In 33 and 34, use cross-multiplication to determine whether each proportion is true.

33. $\frac{20}{60} = \frac{33}{99}$

34. $\frac{2}{5} = \frac{14}{34}$

PRACTICE TEST 5A

1. Reduce each ratio to its lowest terms.

a) 55 to 35 **b)** 72:72 **c)** $\frac{380}{760}$

2. During the first 15 days in March the temperature dropped below zero degrees Celsius on six days. Use colon form to write, in its lowest terms, the ratio of days that the temperature dropped below zero degrees Celsius to the days that the temperature did not drop below zero degrees Celsius.

3. Use cross-multiplication to determine whether each proportion is true.

a) $\frac{10}{15} = \frac{14}{20}$ **b)** $\frac{1.3}{100} = \frac{5.2}{400}$

4. In a sample of 800 tires, 6 were defective. In a sample of 2,000 tires, 15 were defective. Do the ratios of defective tires to the sample size for the two samples form a true proportion?

5. Solve each proportion using cross-multiplication.

a) $\frac{N}{4} = \frac{25}{10}$ **b)** $\frac{1.2}{W} = \frac{2}{9}$

6. A map has a scale of 0.1 inch per 10 miles. If two cities are 1.56 inches apart on the map, what is the mileage between these cities?

PRACTICE TEST 5B

1. Reduce each ratio to its lowest terms.

   a) 124 to 31 **b)** 14:35 **c)** $\frac{240}{320}$

2. During the 21 trading days in June, the Dow Jones Industrial Average increased by over 25 points on 12 of the days. Use fraction form to write the ratio of the days the Dow increased by over 25 points to the total trading days in the month.

3. Use cross-multiplication to determine whether each proportion is true.

   a) $\frac{100}{250} = \frac{36}{80}$ **b)** $\frac{0.02}{0.03} = \frac{3.4}{5.1}$

4. In a history class of 24 students, 9 earned an A. In another section of history, 8 out of 22 students earned an A. Do the ratios of A's to class size for the two sections of history form a true proportion?

5. Solve each proportion using cross-multiplication.

   a) $\frac{R}{8} = \frac{3}{2}$ **b)** $\frac{8}{100} = \frac{10}{N}$

6. A company pays its sales personnel a \$7.50 commission for every \$50 of sales. How much commission does a salesperson earn for \$2,400 of sales?

CUMULATIVE REVIEW EXERCISES FOR CHAPTERS 1 THROUGH 5

1. Write 1,003,500,000 in words.

2. Round 23,894 to the nearest hundred.
3. Find the quotient and remainder for $189 \div 12$.
4. Reduce $\frac{120}{220}$ to its lowest terms.
5. Convert $\frac{100}{9}$ to a mixed number.
6. Convert $4\frac{5}{7}$ to an improper fraction.

In 7 through 10, reduce each fraction to its lowest terms and express any result that is an improper fraction as a mixed number.

7. $4\frac{1}{3} \times 5\frac{2}{3} =$
8. $\frac{13}{16} \div \frac{1}{12} =$
9. $\frac{14}{15} + \frac{3}{5} =$
10. $15\frac{1}{2} - 7\frac{3}{4} =$
11. Write 1.0587 in words.

12. Write the number for ninety-three ten-thousandths.

13. Round 12.48371 to the nearest thousandth.
14. Round \$3.07499 to the nearest cent.
15. $1.23 + 4.006 + 0.3 =$
16. $4.01 \times 5.3 =$
17. Divide 45.6 by 21.44 and round the quotient to the nearest tenth.

18. $12.6 - 2.1 \times 1.6 =$
19. Convert 2.25% to a decimal.
20. Convert 3.5 to a percent.
21. Convert 45% to a fraction reduced to its lowest terms.
22. Convert $\frac{17}{30}$ to the nearest tenth of a percent.
23. 26 is what percent of 200?
24. 19.4 is 2% of what number?
25. Find 7.5% of \$16,800.
26. If the sales tax rate is 8%, find the sales tax on a car that is purchased for \$14,900.

27. The \$600 spent for the band represents 15% of the cost of a wedding reception. Find the total cost of the reception.

28. During April it rained on 12 of the 30 days. Find the percent of the days it rained.

29. Jane Simmon's grade point average increased from 2.59 to 3.14. Find the percent increase to the nearest tenth of a percent.

30. Reduce each ratio to its lowest terms.

a) 144:180 **b)** 45 to 54

31. Use cross-multiplication to determine whether each proportion is true.

a) $\frac{12}{25} = \frac{96}{200}$ **b)** $\frac{1.4}{4.2} = \frac{1}{4}$

32. Solve each proportion using cross-multiplication.

a) $\frac{R}{5} = \frac{13}{10}$ **b)** $\frac{1.6}{3.2} = \frac{20}{T}$

33. On the first page of an 80-question multiple choice test Mike Tessler answers 13 of 16 questions correctly. If the ratio of correct answers to questions is maintained for the entire 80-question test, how many questions will Mike answer correctly?

CHAPTER 6

SIGNED NUMBERS

6.1 AN INTRODUCTION TO SIGNED NUMBERS

6.2 ADDING SIGNED NUMBERS

6.3 SUBTRACTING SIGNED NUMBERS

6.4 MULTIPLYING SIGNED NUMBERS

6.5 DIVIDING SIGNED NUMBERS

6.6 MIXED OPERATIONS

In the first five chapters of this book you worked with different kinds of numbers to answer questions such as

> In a class of 24 students, 2/3 have grade point averages in excess of 2.75. How many students in the class have grade point averages below 2.75?

Recall that 24 is a *whole number*, 2/3 is a *fraction*, and 2.75 is a *decimal*. A property shared by these three numbers is that they are all larger than zero. Numbers larger than zero are called **positive numbers**. We have not used the modifier "positive" in previous chapters because, with the exception of zero, all the numbers in the examples and exercises were positive.

There are numbers that are less than zero, and these numbers are called **negative numbers**. It is very likely that you have seen negative numbers used to describe some familiar situations. For example,

- $-17°$C (negative seventeen degrees Celsius) for a temperature reading that is 17 degrees below 0°C celsius.
- $-\$3$ (negative three dollars) for a decrease in the price of a corporation's stock of \$3 per share.
- -10 (negative ten) for a ten-point deduction on a test question.
- -5 (negative 5) to show that a team is five games behind the first-place team in the baseball standings.

Notice that for each of these numbers a negative sign ($-$) is placed in front of the number and the name for the number includes the word "negative." Negative numbers, such as -17, -3, -10, and -5, must include the negative sign. A positive sign "+" can, but does not have to, precede a positive number. So, you can write 24 or $+24$ for positive twenty-four.

For the remainder of this chapter we will refer to the set of negative and positive numbers as **signed numbers**, (that is, numbers with positive and negative signs).

6.1 AN INTRODUCTION TO SIGNED NUMBERS

Preview Test

In questions 1 through 4, insert the appropriate symbol, $<$ or $>$, in the blank.

1. -8 ____ -5 **2.** 0 ____ -4

3. 15 ____ -13 **4.** -89 ____ -88

In questions 5 through 8, find the indicated absolute value.

5. $|-3|$ **6.** $|12|$ **7.** $|0|$ **8.** $|-3.45|$

When rounding was discussed in earlier chapters of this book, numbers were positioned on a horizontal number line to help you understand the "Which number is

ANSWERS FOR PREVIEW TEST

1. $<$ **2.** $>$ **3.** $>$ **4.** $<$ **5.** 3 **6.** 12 **7.** 0
8. 3.45

closer?" method of rounding. (See Figures 1.2.1 and 3.2.1 in Chapters 1 and 3.) The number line can be extended to include negative numbers. Notice in the number line illustrated in Figure 6.1.1 that the negative numbers -4, -3, -2, and -1 lie to the left of zero. Also, the arrows at each end of the line indicate that the line and associated numbers extend indefinitely to the left and right.

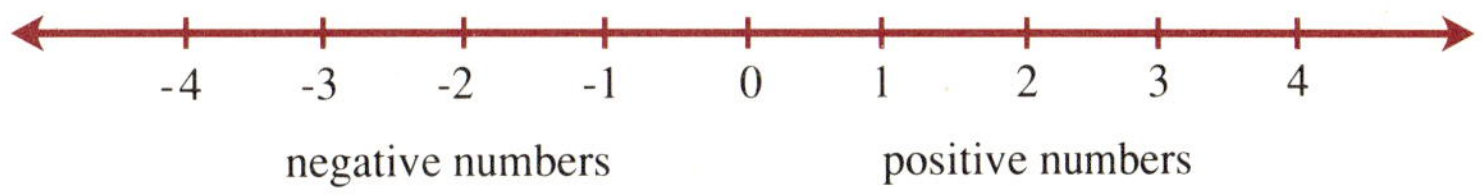

Figure 6.1.1

As we move from left to right on the number line the numbers get larger. To help you understand this property, think of the number line as a temperature scale. As we move to the right on the scale the temperature is getting warmer (it is rising) and the warmer the temperature the larger the number (see Figure 6.1.2).

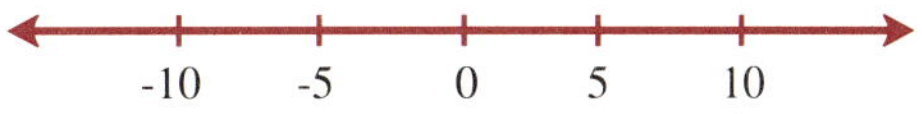

left to right ⟶ the direction of warming temperatures

left to right ⟶ the direction of increasing numerical values

Figure 6.1.2

Comparing Signed Numbers

We are now ready to compare signed numbers using the symbols "$<$" (less than) and "$>$" (greater than). For example, to show that 5 is less than 24 we can write

$$5 < 24$$

and to show that 14 is greater than 9 we can write

$$14 > 9$$

EXAMPLE 1: Insert the appropriate symbol, $<$ or $>$, in the blank.

a) -13 ____ -17 **b)** -4 ____ 2

Solution:

a) $-13 > -17$, because -13 lies to the right of -17 on the number line (see Figure 6.1.3(a)). Also, -13 degrees is warmer than -17 degrees.

Figure 6.1.3(a)

b) $-4 < 2$, because -4 lies to the left of 2 on the number line (see Figure 6.1.3(b)). Also, note that any negative number is less than any positive number.

Figure 6.1.3(b)

PRACTICE PROBLEM 1: Insert the appropriate symbol, $<$ or $>$, in the blank.

a) -19 ____ -17 **b)** -9 ____ 0

Absolute Value

Sometimes we are interested in the value of a number independent of its sign. For example, a sudden big change in blood pressure (up *or* down) is potentially dangerous. That is, whether the change is from 90 to 120 (positive thirty) or from 90 to 60 (negative thirty), the blood pressure has changed by 30 points and a health problem may exist.

When we are reporting the value of a number independent of its sign, we are using its **absolute value**. The symbol for absolute value is | |. So,

$|+30| = 30$ The absolute value of $+30$ equals 30.

$|-30| = 30$ The absolute value of -30 equals 30.

The absolute value of a number can be defined as its distance from zero on the number line. Notice that both -30 and $+30$ lie 30 units from zero (see Figure 6.1.4).

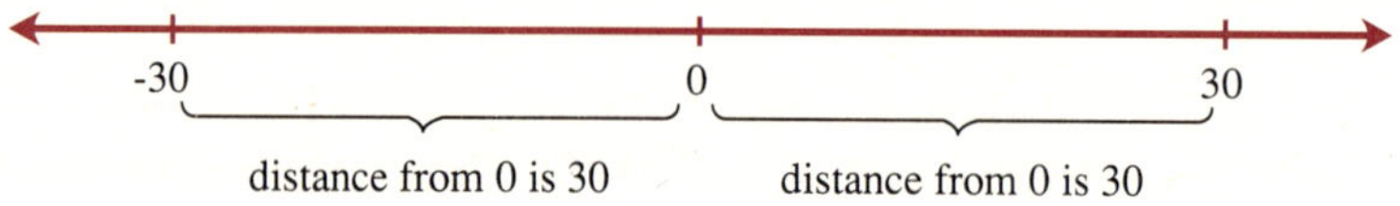

Figure 6.1.4

EXAMPLE 2: Find the indicated absolute value.

a) $|-5|$ **b)** $|+4.1|$ **c)** $|0|$

Solution:

a) $|-5| = 5$ because -5 lies 5 units from 0.

b) $|+4.1| = 4.1$ because $+4.1$ lies 4.1 units from 0.

c) $|0| = 0$ because 0 is 0 units from 0.

PRACTICE PROBLEM 2: Find the indicated absolute value.

a) $|-7.3|$ **b)** $|+2|$ **c)** $|5|$

ANSWERS FOR PRACTICE PROBLEMS

1. a) $<$ **b)** $<$ **2. a)** 7.3 **b)** 2 **c)** 5

EXERCISE SET 6.1A

Skill Development

In questions 1 through 12, insert the appropriate symbol, $<$ or $>$, in the blank.

1. $4 \_\_\_\_ -3$	**2.** $-2 \_\_\_\_ -5$	**3.** $0 \_\_\_\_ 12$
4. $-8 \_\_\_\_ 5$	**5.** $19 \_\_\_\_ -13$	**6.** $-45 \_\_\_\_ -43$
7. $-10 \_\_\_\_ -20$	**8.** $-2.5 \_\_\_\_ -2.6$	**9.** $-3.45 \_\_\_\_ -3.44$
10. $-\frac{3}{8} \_\_\_\_ -\frac{5}{8}$	**11.** $\frac{1}{7} \_\_\_\_ -\frac{3}{7}$	**12.** $-\frac{24}{25} \_\_\_\_ -\frac{23}{25}$

In questions 13 through 20, find the indicated absolute value.

13. $\lvert -5 \rvert$	**14.** $\lvert +17 \rvert$	**15.** $\lvert -12 \rvert$	**16.** $\lvert 0 \rvert$
17. $\lvert 2.5 \rvert$	**18.** $\lvert -3.5 \rvert$	**19.** $\lvert +1.6 \rvert$	**20.** $\lvert -22.5 \rvert$

Applications

21. Over a three-hour period the temperature increased from -3 degrees to 5 degrees. By how many degrees did the temperature increase?

22. The change in the price of gold from its opening to its closing price was $+\$6.32$ per ounce. If the opening price was \$415.85 per ounce, find the closing price.

23. If the temperature moves from -17.45 to -17.35 is it rising or falling?

24. The change in the price of a stock is $-\$0.75$. What is the absolute value of this price change?

For Some Extra Thought

In questions 25 and 26, insert the appropriate symbol, $<$ or $>$, in the blank.

25. $-\frac{8}{9} \_\_\_\_ -\frac{7}{8}$ **26.** $-2.01 \_\_\_\_ -2.005$

27. Does the point for $-\frac{6}{7}$ lie to the left or right of the point for -0.8 on the number line?

Review Problems for Maintaining Your Skills

28. Round 28,672 to the nearest thousand.

29. Divide 584 by 19 and find the quotient and remainder.

In questions 30 and 31, reduce each fraction to its lowest terms and express any result that is an improper fraction as a mixed number.

30. $\frac{13}{15} - \frac{1}{2} =$ **31.** $3\frac{1}{4} \times 2 =$

32. Divide 23.4 by 1.7 and round the quotient to the nearest hundredth.

33. Find 2.75% of \$8,400.

34. To the nearest tenth of a percent, 17 is what percent of 30?

35. \$2.50 is 2% of what amount?

36. Find the percent decrease if a price decreases from \$25 to \$21.

37. Reduce the ratio 48 to 56 to its lowest terms.

In 38 and 39, use cross-multiplication to determine whether each proportion is true.

38. $\frac{5}{6} = \frac{8}{9}$ **39.** $\frac{2}{5} = \frac{12}{30}$

In 40 and 41, use cross-multiplication to solve each proportion.

40. $\frac{R}{8} = \frac{1}{4}$ **41.** $\frac{2.5}{N} = \frac{10}{4}$

EXERCISE SET 6.1B

Skill Development

In questions 1 through 12, insert the appropriate symbol, $<$ or $>$, in the blank.

1. -2 ____ -6 **2.** 5 ____ -2 **3.** -9 ____ -10

4. 14 ____ -1 **5.** -10 ____ -12 **6.** -200 ____ -150

7. -23 ____ -17 **8.** -2.5 ____ -2.4 **9.** -5.3 ____ -5.7

10. $-\frac{1}{9}$ ____ $-\frac{2}{9}$ **11.** $-\frac{47}{60}$ ____ $-\frac{49}{60}$ **12.** $-\frac{3}{5}$ ____ $-\frac{2}{5}$

In questions 13 through 20, find the indicated absolute value.

13. $|-7|$ **14.** $|-16|$ **15.** $|+13|$ **16.** $|-33|$

17. $|-1.3|$ **18.** $|-26.1|$ **19.** $|+3.45|$ **20.** $|-3.75|$

Applications

21. Over a five-hour period the temperature decreased from 6 degrees to -7 degrees. By how many degrees did the temperature fall?

22. In relation to the first-place team, a baseball team went from -12 games to -7 games. By how many games did they improve their position?

23. If the temperature moves from -3.65 to -3.8 is it rising or falling?

24. An automobile radio is protected to -40 degrees Fahrenheit. What is the absolute value of this temperature protection?

For Some Extra Thought

In questions 25 and 26, insert the appropriate symbol, $<$ or $>$, in the blank.

25. $-\frac{6}{7}$ ____ $-\frac{5}{6}$ **26.** -3.0054 ____ -3.054

27. Does the point for $-\frac{5}{6}$ lie to the left or right of the point for -0.9 on the number line?

Review Problems for Maintaining Your Skills

28. Round 1,026 to the nearest hundred.

29. Divide 89 by 14 and find the quotient and remainder.

In questions 30 and 31, reduce each fraction to its lowest terms and express any result that is an improper fraction as a mixed number.

30. $5\frac{2}{3} + 1\frac{1}{4} =$ **31.** $\frac{3}{5} - \frac{1}{5} =$

32. Divide 143.9 by 1.7 and round the quotient to the nearest tenth.

33. To the nearest hundredth of a percent, 13 is what percent of 18?

34. Find 150% of 6.

35. 2% of what amount equals $3.50?

36. Find the percent increase if a price increases from $40 to $48.

37. Reduce the ratio 20 to 45 to its lowest terms.

In 38 and 39, use cross-multiplication to determine whether each proportion is true.

38. $\frac{10}{15} = \frac{80}{120}$ **39.** $\frac{5}{7} = \frac{60}{84}$

In 40 and 41, use cross-multiplication to solve each proportion.

40. $\frac{W}{5} = \frac{2}{5}$ **41.** $\frac{12}{P} = \frac{2}{3}$

6.2 ADDING SIGNED NUMBERS

Preview Test

In questions 1 through 8, find each sum.

1. $(-8)+(-5)=$ **2.** $(+17)+(-26)=$

3. $(-3)+(+12)=$ **4.** $-6+(+10)=$

5. $-2.3+5.6=$ **6.** $-7+10-12=$

7. $\left(-\frac{3}{4}\right)+\left(-\frac{1}{2}\right)=$ **8.** $-\frac{7}{8}+\frac{1}{4}-\frac{3}{4}-\frac{5}{8}=$

To help you understand how the addition process works for signed numbers, let's use the number line to add $+3$ and $+4$. Starting from zero, move to the right three units (for $+3$) and then move another four units (for $+4$). The finishing point after these two movements is $+7$ (see Figure 6.2.1).

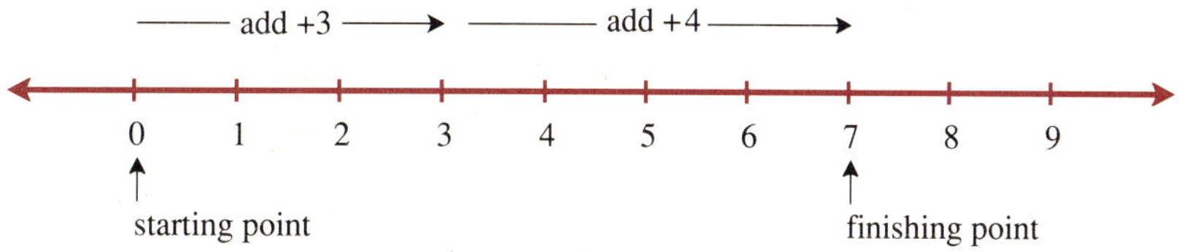

Figure 6.2.1

So, $(+3)+(+4)=+7$

You may believe that using a number line to add two positive numbers like $+3$ and $+4$ is like using an atomic bomb to kill an ant. However, this process (which we admit is really not needed to find the sum of two positive numbers) is very useful for understanding the rules for addition involving negative numbers. Let's consider the following problem.

$$(-2)+(-4)=$$

To add a negative number, move to the left the appropriate distance. In this case, starting from zero, move two units to the left (for -2) and then move another four units to the left (for -4). The finishing point after these two movements is -6 (see Figure 6.2.2). So, $(-2)+(-4)=-6$

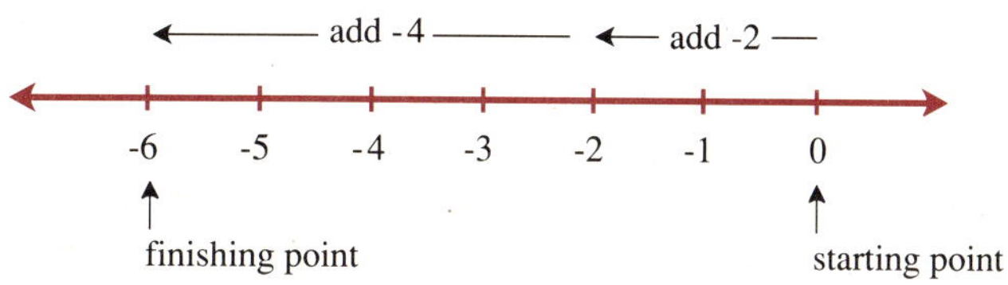

Figure 6.2.2

ANSWERS FOR PREVIEW TEST

1. -13 **2.** -9 **3.** $+9$ **4.** $+4$ **5.** $+3.3$ **6.** -9

7. $-1\frac{1}{4}$ **8.** -2

Instead of depending on the number line for adding two positive numbers or two negative numbers, we can observe the following.

1. The sum of two positive numbers must be positive. Since we are making two consecutive movements to the right of zero, the finishing point must be to the right of zero.
2. The sum of two negative numbers must be negative. Since we are making two consecutive movements to the left of zero, the finishing point must be to the left of zero.

These observations can be translated into the following rule.

To Add Two Numbers With Like Signs

Step 1: Add their absolute values.
Step 2: Give the sum the common sign.

For $(+3) + (+4)$

Step 1: $|+3| + |+4| = 3 + 4 = 7$
Step 2: Assign a plus sign to 7.

So, $(+3) + (+4) = +7$
For $(-4) + (-2)$

Step 1: $|-4| + |-2| = 4 + 2 = 6$
Step 2: Assign a negative sign to 6.

So, $(-4) + (-2) = -6$

EXAMPLE 1: $(-7) + (-5) =$

Solution:
Step 1: $(-7) + (-5) = |-7| + |-5| = 7 + 5 = 12$
Step 2: Assign a negative sign to 12.

So, $(-7) + (-5) = -12$

PRACTICE PROBLEM 1: $(+8) + (+7) =$

We will now go back to the number line to discuss the sum of two numbers with unlike signs. Let's start with $(+3) + (-8)$. Starting from zero, move three units to the right (for $+3$) and then move eight units to the left (for -8). The finishing point after these two movements is -5 (see Figure 6.2.3). So, $(+3) + (-8) = -5$

Notice that the sum for $(+3) + (-8)$ is the negative of the difference in the absolute values of $(+3)$ and (-8). The negative sign is the result of moving more units

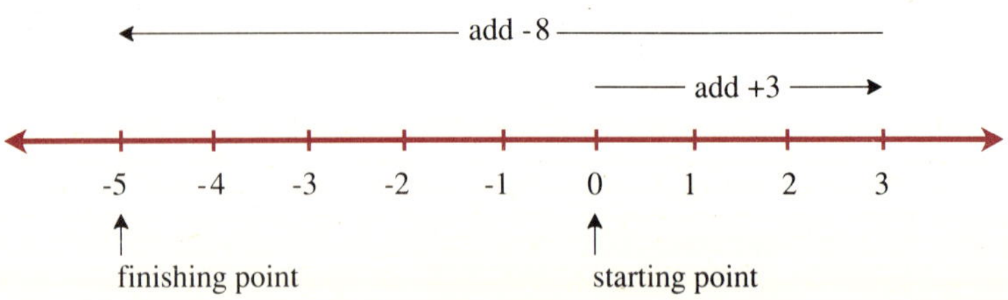

Figure 6.2.3

to the left (8 units) than to the right (3 units). In general, if we start from zero and the left movement is longer than the right movement, the finishing point must be on the left side (negative side) of zero.

The reason the sum for $(+3)$ and (-8) involves the difference in the absolute values is because the movements are in opposite directions (3 to the right followed by 8 to the left). The net effect is 5 units to the left. The rule can now be stated.

To Add Two Numbers With Unlike Signs

Step 1: Find the difference between the absolute values.
Step 2: Give the sum the sign of the number with the larger absolute value.

For $(+3) + (-8)$

Step 1: $|-8| - |+3| = 8 - 3 = 5$
Step 2: Since $|-8|$ is larger than $|+3|$, we assign a negative sign to 5. So, $(+3) + (-8) = -5$

EXAMPLE 2: Find each sum.

a) $(-4) + (+7) =$ **b)** $(+8) + (-17) =$

Solution:

a) **Step 1:** $|+7| - |-4| = 7 - 4 = 3$
Step 2: Since $|+7|$ is larger than $|-4|$, we assign a positive sign to 3. So, $(-4) + (+7) = +3$

b) **Step 1:** $|-17| - |+8| = 17 - 8 = 9$
Step 2: Since $|-17|$ is larger than $|+8|$, we assign a negative sign to 9. So, $(+8) + (-17) = -9$

PRACTICE PROBLEM 2: Find each sum.

a) $(+19) + (-13) =$ **b)** $(-7) + (+1) =$

Be Careful About This

Remember that the rule for adding signed number with unlike signs requires a subtraction operation. If you are wondering why you are subtracting when you are finding a sum, remember that in long division you have to multiply and subtract. To carry out an operation such as division, or addition of signed numbers, other operations have to be performed.

To add more than two signed numbers, add two numbers at a time using the rules stated in this section, and carry forward the sum. Let's consider the sum for $(-3) + (+5) + (-10)$

$$\underbrace{(-3) + (+5)}_{\text{add}} + (-10)$$

$$= \underbrace{(+2) \quad + (-10)}_{\text{add}}$$

$$= -8$$

This same sum could be found by adding from right to left, or adding the two negative numbers first. That is

$$(-3) + \underbrace{(+5) + (-10)}_{\text{add}}$$

$$= \underbrace{(-3) + \qquad (-5)}_{\text{add}}$$

$$= \qquad -8$$

or

$$(-3) + (+5) + (-10)$$

$$\text{add } (-3) \text{ and } (-10)$$

$$= \qquad \underbrace{(-13) + (+5)}_{\text{add}}$$

$$= \qquad -8$$

EXAMPLE 3: $(-9) + (+3) + (+2) + (-12) =$

Solution:
We will add from left to right.

$$\underbrace{(-9) + (+3)}_{\text{add}} + (+2) + (-12)$$

$$= \quad \underbrace{(-6) \qquad + (+2)}_{\text{add}} + (-12)$$

$$= \quad \underbrace{(-4) \qquad + (-12)}_{\text{add}}$$

$$= \qquad -16$$

PRACTICE PROBLEM 3: $(-4) + (-5) + (-2) + (+7) =$

EXAMPLE 4: $(-3.92) + (+2.07) =$

Solution:
The sum of signed decimal numbers is handled the same as the sum of signed whole numbers.

Step 1: $|-3.92| - |+2.07| = 3.92 - 2.07 = 1.85$
Step 2: Since $|-3.92|$ is larger than $|+2.07|$ we assign a negative sign to 1.85. So, $(-3.92) + (+2.07) = -1.85$

PRACTICE PROBLEM 4: $(-2.91) + (-5.32) =$

EXAMPLE 5: $\left(-\frac{3}{8}\right) + \left(-\frac{1}{4}\right) =$

Solution:

Since the LCD is 8, we will convert $-\frac{1}{4}$ to the equivalent fraction $-\frac{2}{8}$. Then

$$\left(-\frac{3}{8}\right)+\left(-\frac{1}{4}\right)=\left(-\frac{3}{8}\right)+\left(-\frac{2}{8}\right)=-\frac{5}{8}$$

PRACTICE PROBLEM 5: $\left(-\frac{3}{5}\right)+\left(\frac{1}{2}\right)=$

So far all the examples and practice problems in this section were stated using parentheses, such as $(+7)+(-10)=-3$. The problem $(+7)+(-10)$ can be restated with the omission of the first parentheses or both parentheses. In each case the sum is still -3.

With the parentheses around $+7$ omitted we have

$$+7+(-10)=-3$$

With both parentheses omitted we have

$$+7-10=-3$$

Notice that when both parentheses are omitted, the plus sign that appeared between parentheses is also omitted. When an addition problem appears without parentheses each sign is "attached" to the number on its right. For $-9+3-1+10$ we have

$$-9 \text{ plus } +3 \text{ plus } -1 \text{ plus } +10$$

which equals $+3$.

EXAMPLE 6: $-4+2+7-1=$

Solution:

$$-4 \text{ plus } +2 \text{ plus } +7 \text{ plus } -1=+4$$

PRACTICE PROBLEM 6: $+9-12+3-3-6=$

ANSWERS FOR PRACTICE PROBLEMS

1. $+15$ **2. a)** $+6$ **b)** -6 **3.** -4 **4.** -8.23

5. $-\frac{1}{10}$ **6.** -9

EXERCISE SET 6.2A

Skill Development

In questions 1 through 18, find each sum.

1. $(-5) + (-9) =$
2. $(+12) + (-8) =$
3. $(-35) + (+48) =$
4. $(-5) + (-7) + (+12) =$
5. $(+7) + (+15) + (-30) =$
6. $(-7) + (-4) + (-1) =$
7. $-10 + (-14) =$
8. $17 + (-23) =$
9. $(-7.3) + (-2.6) =$
10. $(-4.87) + (+2.33) =$
11. $-3 + 9 + 6 - 10 =$
12. $12 - 8 - 3 - 2 - 5 =$
13. $-5.63 + 7.22 - 2.87 =$
14. $5.6 - 3.3 - 2.8 - 9.1 =$
15. $+\frac{1}{4} - \frac{3}{4} =$
16. $+\frac{2}{3} - \frac{1}{3} - \frac{1}{6} =$
17. $\left(+\frac{3}{5}\right) + \left(-\frac{1}{10}\right) =$
18. $-\frac{2}{3} + \left(-\frac{1}{4}\right) =$

Applications

19. On a driver's test you pass if you do not lose more than 20 points. For various driving errors Eric Jones received scores of -3, -7, -2, -6, and -1. Did Eric pass his driver's test?

20. The closing price of a stock on Monday was \$28 per share. On the next four trading days the price changes were: $-\$2$, $+\$4.50$, $-\$1$, and $-\$3.75$. Find the closing price after the fourth day.

21. At 4 A.M. the temperature was -6 degrees Celsius. Over the next six hours the temperature increased by 15 degrees. What was the temperature at 10 A.M.?

22. At 3 A.M. the temperature was -9 degrees Celsius. Over the next five hours the temperature increased by 7 degrees. What was the Celsius temperature at 8 A.M.?

For Some Extra Thought

23. What must be added to -2.3 for the sum to be -6.5?

24. If the sum of three numbers is -15 and one of the numbers is $+50$, must the other two numbers be negative? Explain.

Review Problems for Maintaining Your Skills

25. Write 2,358,674 in words.

26. Reduce $\frac{175}{450}$ to its lowest terms.

27. Convert $9\frac{7}{8}$ to an improper fraction.

28. Round 17.0648 to the nearest hundredth.

29. $6.74 \times 10{,}000 =$

30. Find 4.75% of \$8,600.

31. 10% of what amount equals \$17?

32. 45 is what percent of 60?

In 33 and 34, use cross-multiplication to solve each proportion.

33. $\frac{M}{2} = \frac{3}{4}$ **34.** $\frac{1.2}{R} = \frac{10}{25}$

In 35 and 36, insert the appropriate symbol, $<$ or $>$, in the blank.

35. -10 ____ -6 **36.** 0 ____ -19

In 37 and 38, find the indicated absolute value.

37. $|-1.65|$ **38.** $|+75|$

EXERCISE SET 6.2B

Skill Development

In questions 1 through 18, find each sum.

1. $(+7)+(-15)=$

2. $(-4)+(-9)=$

3. $(-23)+(-15)=$

4. $(-3)+(+10)+(-7)=$

5. $(-5)+(-14)+(+40)=$

6. $(-1)+(-5)+(-8)+(-1)=$

7. $18+(-13)=$

8. $-17+(-6)+(+8)=$

9. $(-12.3)+(-5.3)=$

10. $(+5.63)+(-4.17)=$

11. $-9-8-12+14=$

12. $11-8-15+13=$

13. $-9.87+7.33-1.05=$

14. $+7.1-4.15-6=$

15. $+\frac{2}{5}-\frac{1}{5}=$

16. $-\frac{1}{3}+\frac{2}{9}+\frac{1}{3}=$

17. $\left(-\frac{5}{6}\right)+\left(\frac{1}{6}\right)=$

18. $\left(+\frac{3}{4}\right)+\left(-\frac{1}{8}\right)=$

Applications

19. What was the year number 75 years after 80 BC?

20. What was the year number 52 years after 125 BC?

21. On a hundred-point history test the point deductions on the five essay questions are -6, -2, -1, -3, and -1. Find the grade on the test.

22. The opening Dow Jones Industrial Average is 2,856.31 and the change in the average for the day is -5.63. Find the closing average.

For Some Extra Thought

23. What must be added to -5.6 for the sum to be $+8.9$?

24. Find four numbers whose sum is -2 if two are positive and two are negative.

Review Problems for Maintaining Your Skills

25. Write 268,974,056 in words.

26. Reduce $\frac{72}{84}$ to its lowest terms.

27. Convert $\frac{117}{25}$ to a mixed number.

28. Round 0.0078046 to the nearest ten-thousandth.

29. $2.78 \times 1.3 =$

30. Find 250% of \$30.

31. 2% of what amount equals \$28?

32. 33 is what percent of 1000?

In 33 and 34, use cross-multiplication to solve each proportion.

33. $\frac{W}{2} = \frac{25}{10}$ **34.** $\frac{200}{300} = \frac{1.8}{H}$

In 35 and 36, insert the appropriate symbol, < or >, in the blank.

35. -4 ____ 2 **36.** -17.6 ____ -17.7

In 37 and 38, find the indicated absolute value.

37. $|-5.68|$ **38.** $|0|$

6.3 SUBTRACTING SIGNED NUMBERS

Preview Test

In questions 1 through 8, find each difference.

1. $(-9)-(+7)=$ **2.** $(+15)-(-2)=$

3. $(-10)-(+3)=$ **4.** $(-2)-(-1)=$

5. $(-3.6)-(-1.7)=$ **6.** $(5.1)-(-7)=$

7. $\left(-\frac{1}{4}\right)-\left(-\frac{1}{8}\right)=$ **8.** $\left(+\frac{3}{5}\right)-\left(-\frac{4}{5}\right)=$

If we subtract $+3$ from $+7$ the result is $+4$. That is,

$$(+7)-(+3)=+4$$

Here we are carrying out subtraction with two positive numbers where the subtrahend is smaller than the minuend. If we were always subtracting a positive number from a larger positive number, there would be no need for this section. But such is not always the case. What happens when the minuend or subtrahend is negative or the minuend is smaller than the subtrahend?

To develop a method to handle these situations let's return to our original example $(+7)-(+3)=+4$. Instead of subtracting $+3$ from $+7$, let's add -3 to $+7$.

$$(+7)+(-3)=+4$$

Notice that we get the same result. So adding -3 to $+7$ is equivalent to subtracting $+3$ from $+7$.

$$(+7)-(+3)=(+7)+(-3)=+4$$

In order to perform the subtraction, the sign of the subtrahend is changed and the rules for addition are used.

$$(+7)-(+3)$$

Change the operation to addition. Change the sign of the subtrahend.

$$(+7)+(-3)=+4$$

Think of it this way—taking away something positive is like adding something negative and taking away something negative is like adding something positive.

THE PRICE OF A STOCK OPENS
AT $28 AND DROPS $4

Taking away something positive.	Adding something negative.
$(+28)-(+4)=+24$	$(+28)+(-4)=+24$

ANSWERS FOR PREVIEW TEST

1. -16 **2.** 17 **3.** -13 **4.** -1 **5.** -1.9 **6.** 12.1

7. $-\frac{1}{8}$ **8.** $1\frac{2}{5}$

A 20-POINT ESSAY QUESTION IS SCORED 16 AFTER A 4-POINT DEDUCTION. AFTER SOME CAREFUL CONSIDERATION, THE 4-POINT DEDUCTION IS REINSTATED.

Taking away something negative.	Adding something positive.
$(+16) - (-4) = +20$	$(+16) + (+4) = +20$

EXAMPLE 1: $(-7) - (-9) =$

Solution:
Change the sign of the subtrahend and change the operation to addition.

$$(-7) - (-9) = (-7) + (+9) = +2$$

PRACTICE PROBLEM 1: $(+8) - (-3) =$

EXAMPLE 2: $(-3.92) - (+9.17) =$

Solution:
Change the sign of the subtrahend and change the operation to addition.

$$(-3.92) - (+9.17) = (-3.92) + (-9.17) = -13.09$$

PRACTICE PROBLEM 2: $(+4.8) - (+7.6) =$

EXAMPLE 3: $\left(-\frac{7}{15}\right) - \left(-\frac{1}{5}\right) =$

Solution:
The LCD is 15. Convert $-\frac{1}{5}$ to $-\frac{3}{15}$. Then change the sign of the subtrahend and change the operation to addition.

$$\left(-\frac{7}{15}\right) - \left(-\frac{3}{15}\right) = \left(-\frac{7}{15}\right) + \left(+\frac{3}{15}\right) = -\frac{4}{15}$$

PRACTICE PROBLEM 3: $\left(-\frac{3}{8}\right) - \left(+\frac{1}{2}\right) =$

ANSWERS FOR PRACTICE PROBLEMS

1. $+11$ **2.** -2.8 **3.** $-\frac{7}{8}$

EXERCISE SET 6.3A

Skill Development

In questions 1 through 16, find each difference.

1. $(-9) - (+7) =$
2. $(+13) - (-14) =$
3. $(-4) - (-5) =$
4. $(15) - (-6) =$
5. $(-3) - (22) =$
6. $(-33) - (+72) =$
7. $(-8) - (-8) =$
8. $(+23) - (+62) =$
9. $(-2.4) - (-7.8) =$
10. $(+4.25) - (-2.65) =$
11. $(+1.4) - (-3.68) =$
12. $(-5.9) - (-8.72) =$
13. $\left(-\frac{4}{5}\right) - \left(+\frac{3}{5}\right) =$
14. $\left(+\frac{1}{2}\right) - \left(+\frac{7}{8}\right) =$
15. $\left(-1\frac{4}{7}\right) - \left(-2\frac{3}{7}\right) =$
16. $(-3) - \left(-4\frac{1}{6}\right) =$

Applications

17. If the temperature drops from 5 degrees Celsius to -12 degrees Celsius what is the change in temperature?

18. If the temperature drops from -8 degrees Celsius to -18 degrees Celsius, what is the change in temperature?

19. If a person was born in the year 107 BC and died in the year 39 BC, how many years did the person live?

20. In a 4-week period the Cincinnati Reds went from 5 games behind the first-place team to 7.5 games in front of the second-place team. By how many games did they improve their position?

For Some Extra Thought

21. What number must be subtracted from -7 for the result to be $+2$?

22. Can the difference of two negative numbers be positive? Explain.

Review Problems for Maintaining Your Skills

23. Round 12,687 to the nearest hundred.
24. Divide 145 by 17 and find the quotient and remainder.

In questions 25 and 26, reduce each fraction to its lowest terms and express any result that is an improper fraction as a mixed number.

25. $3\frac{5}{6} - 1\frac{1}{2} =$
26. $\frac{10}{11} \times \frac{22}{25} =$

27. Divide 4.6 by 17 and round the quotient to the nearest hundredth.

28. 8% of what amount equals $10?

29. 84 is what percent of 200?

30. Find 9.75% of $12,000.

In 31 and 32, use cross-multiplication to determine whether each proportion is true.

31. $\frac{20}{50} = \frac{3}{8}$

32. $\frac{1.6}{200} = \frac{2.4}{300}$

In 33 and 34, find each sum.

33. $(-8) + (+12) + (-19) =$

34. $-4.92 - 12.33 =$

EXERCISE SET 6.3B

Skill Development

In questions 1 through 16, find each difference.

1. $(+7) - (-10) =$
2. $(-12) - (-15) =$
3. $(-21) - (13) =$
4. $(16) - (24) =$
5. $(-54) - (-19) =$
6. $(+100) - (+700) =$
7. $(+32) - (-65) =$
8. $(-52) - (+14) =$
9. $(-5.6) - (-3.4) =$
10. $(+7.16) - (-1.04) =$
11. $(+3.4) - (-20) =$
12. $(-5.41) - (+6.8) =$
13. $\left(-\frac{7}{8}\right) - \left(+\frac{3}{8}\right) =$
14. $\left(+\frac{2}{3}\right) - \left(-\frac{1}{6}\right) =$
15. $\left(-3\frac{1}{4}\right) - \left(+2\frac{3}{4}\right) =$
16. $(+12) - \left(-5\frac{4}{9}\right) =$

Applications

17. On a tour near the Dead Sea, Marie Collins climbed from 20.1 feet below sea level to 37.6 feet above sea level. By how many feet did she increase her elevation relative to sea level?

18. The ABC Packing Company lost \$2,695,000 last year and made a profit of \$3,240,000 this year. By how much did the company improve with respect to their profit from last year to this year?

19. At 11 A.M. the change in the Dow Jones Industrial Average from the previous day was -5.73. By 3 P.M. the change read $+11.61$. Find the change from 11 A.M. to 3 P.M.

20. In a football game, a running back's total of 127 yards gained includes a 13-yard loss. Without the 13-yard loss, what would have been his total yards for the game?

For Some Extra Thought

21. What number must be subtracted from -12 for the result to be -4?

22. If the difference of two negative numbers is zero, what conclusion can we make about the numbers?

Review Problems for Maintaining Your Skills

23. Round 2,497,355 to the nearest million.
24. Divide 37 by 31 and find the quotient and remainder.

In questions 25 and 26, reduce each fraction to its lowest terms and express any result that is an improper fraction as a mixed number.

25. $4\frac{2}{5} + 5\frac{3}{4} =$ **26.** $\frac{15}{16} - \frac{1}{4} =$

27. Divide 6.37 by 1.4 and round the quotient to the nearest tenth.

28. 10% of what amount equals $2.37?

29. 4 is what percent of $400?

30. Find 250% of $60.

In 31 and 32, use cross-multiplication to determine whether each proportion is true.

31. $\frac{2}{5} = \frac{16}{80}$ **32.** $\frac{2.5}{100} = \frac{3.5}{140}$

In 33 and 34, find each sum.

33. $(+18) + (-24) + (-1) =$

34. $-24.09 + 1.3 - 4.6 =$

6.4 MULTIPLYING SIGNED NUMBERS

Preview Test

In questions 1 through 8, find each product.

1. $(-3) \times (+6) =$ **2.** $(-7)(-3) =$

3. $(+5) \times (-2) \times (-4) =$ **4.** $(-6)(-1)(-8) =$

5. $(-3.4)(+2.1) =$ **6.** $(-3.24) \times (10.5) =$

7. $\left(+\frac{3}{5}\right) \times \left(-\frac{1}{7}\right) =$ **8.** $\left(+2\frac{2}{3}\right) \times \left(-2\frac{1}{2}\right) =$

The rule for multiplying two signed numbers is

Step 1: Multiply the absolute values.
Step 2: Find the sign of the product, which is
positive if the signs of the factors are alike,
negative if the signs of the factors are not alike.

Using this rule we can see that

with like signs

POSITIVE × POSITIVE = POSITIVE $(+3) \times (+6) = +18$

NEGATIVE × NEGATIVE = POSITIVE $(-3) \times (-6) = +18$

and with unlike signs

POSITIVE × NEGATIVE = NEGATIVE $(+3) \times (-6) = -18$

NEGATIVE × POSITIVE = NEGATIVE $(-3) \times (+6) = -18$

To help you understand the rule, let's imagine that we are taking some action scenes with a videocamera. Scene I shows a person walking forward from point A to point B at 4 miles per hour. Scene II shows a person walking backward from point B to point A at -3 miles per hour (see Figure 6.4.1).

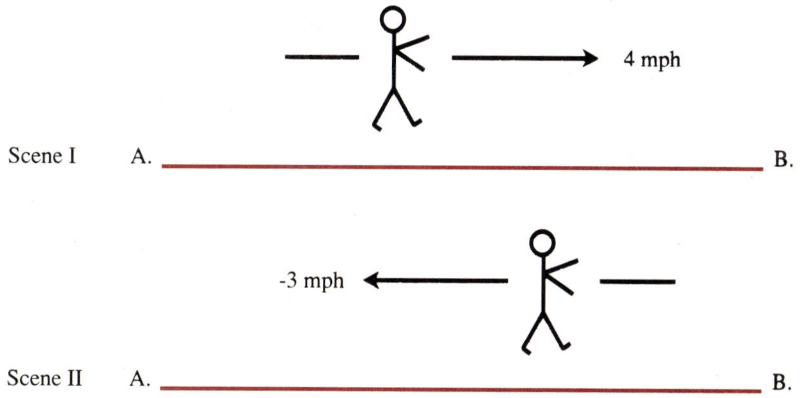

Figure 6.4.1

ANSWERS FOR PREVIEW TEST

1. -18 **2.** 21 **3.** 40 **4.** -48 **5.** -7.14

6. -34.02 **7.** $-\frac{3}{35}$ **8.** $-6\frac{2}{3}$

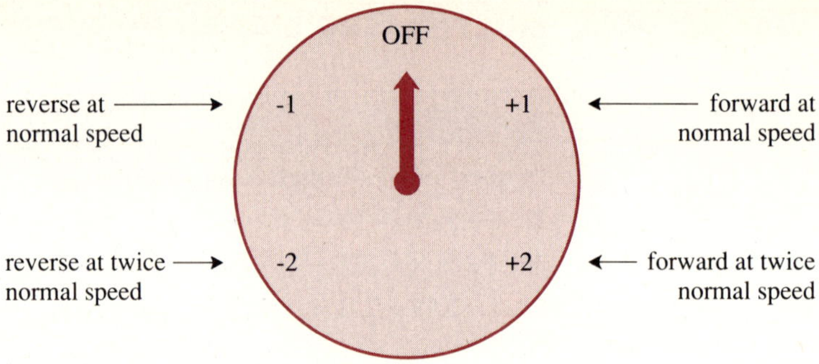

Figure 6.4.2

The video player has the capability of running the tape forward or running it in reverse, at normal or twice normal speed. The dial illustrated in Figure 6.4.2 shows the two forward and two reverse settings.

When the tape is played, what we view on the TV screen is given by the formula

$$\text{SETTING ON DIAL} \times \text{ACTUAL SPEED OF WALKER} = \text{SPEED OF WALKER ON SCREEN}$$

Let's set the dial at $+2$ and view Scene I.

$$\text{Dial Setting} \times \text{Actual Speed} = \text{Speed On Screen}$$
$$(+2) \quad \times \quad (+4) \quad = \quad +8$$

So we see the person walking at $+8$ miles per hour. Here, the positive dial setting times the actual speed gives us a positive projected speed.

POSITIVE × POSITIVE = POSITIVE

Let's now set the dial at -2 and view Scene I from its end to its beginning. This will take the person from point B (the end of the walk) to point A (the beginning of the walk) at twice the original speed.

$$\text{Dial Setting} \times \text{Actual Speed} = \text{Speed On Screen}$$
$$(-2) \quad \times \quad (+4) \quad = \quad -8$$

So we see the person walking in reverse at -8 miles per hour. Here, the negative dial setting times the actual speed gives us a negative projected speed.

NEGATIVE × POSITIVE = NEGATIVE

Now let's set the dial at $+2$ and view Scene II.

$$\text{Dial Setting} \times \text{Actual Speed} = \text{Speed on Screen}$$
$$(+2) \quad \times \quad (-3) \quad = \quad -6$$

We see the person walking from point B to point A at -6 miles per hour. The positive dial setting times the negative actual speed gives us a negative projected speed.

POSITIVE × NEGATIVE = NEGATIVE

Finally, set the dial at -2 and view Scene II from its end to its beginning. This will take the person from point A (the end of the backward walk) to point B (the beginning of the backward walk). So the person will appear to be moving forward on the TV screen.

$$\text{Dial Setting} \times \text{Actual Speed} = \text{Speed On Screen}$$
$$(-2) \quad \times \quad (-3) \quad = \quad +6$$

So we see the person walking forward at $+6$ miles per hour. That is, the negative dial setting times the negative actual speed gives us a positive projected speed.

$$\text{NEGATIVE} \times \text{NEGATIVE} = \text{POSITIVE}$$

Be Careful About This

There is a possibility that you found the video story that was used to explain the rules for multiplying signed numbers hard to follow. Don't let this distract you from memorizing the rules. Understanding rules should always be your goal, but do not feel guilty about memorizing first and then trying to understand later what you have memorized. Also, remember that the rules for multiplying signed numbers are not the same as the rules for adding signed numbers. The *sum* of two negative numbers is negative but the *product* of two negative numbers is positive. The *sum* of a positive and negative number can be positive or negative, but the *product* is always negative.

There are five ways that four times nine can be written numerically. They are

$$4 \times 9 = 36$$
$$(4) \times (9) = 36$$
$$4 \cdot 9 = 36$$
$$4(9) = 36$$
$$(4)(9) = 36$$

In this section we will use parentheses with and without the symbol "$\times$" for multiplication. That is, for -5 times $+6$ we will write $(-5) \times (+6) = -30$ or $(-5)(+6) = -30$.

EXAMPLE 1: $(-3) \times (+7) =$

Solution:
Since the product of two numbers with unlike signs is negative, $(-3) \times (+7) = -21$.

PRACTICE PROBLEM 1: $(-4)(-9) =$

If there are more than two factors, multiply two and carry the product to the next factor. This is illustrated in the next example.

EXAMPLE 2: $(+3) \times (-4) \times (-2) =$

Solution:

$$\underbrace{(+3) \times (-4)}_{\text{multiply}} \times (-2)$$
$$= \underbrace{(-12) \quad \times (-2)}_{\text{multiply}}$$
$$= +24$$

PRACTICE PROBLEM 2 $(-5) \times (-2) \times (-1) \times (+3) =$

EXAMPLE 3: $(-2.3)(-4.5) =$

Solution:
Since the signs are alike, the product is positive. So,

$$(-2.3)(-4.5) = +10.35$$

PRACTICE PROBLEM 3: $(-1.4) \times (+6.3) \times (+0.6) =$

EXAMPLE 4: $\left(-1\frac{1}{2}\right) \times \left(+3\frac{2}{5}\right) =$

Solution:
First change the mixed numbers to improper fractions. Then multiply and convert the product to a mixed number. So,

$$\left(-1\frac{1}{2}\right) \times \left(+3\frac{2}{5}\right) = \left(-\frac{3}{2}\right) \times \left(+\frac{17}{5}\right) = -\frac{51}{10} = -5\frac{1}{10}$$

PRACTICE PROBLEM 4: $\left(-\frac{3}{5}\right)\left(-\frac{1}{2}\right)\left(-\frac{2}{7}\right) =$

ANSWERS FOR PRACTICE PROBLEMS

1. $+36$ **2.** -30 **3.** -5.292 **4.** $-\frac{3}{35}$

EXERCISE SET 6.4A

Skill Development

In questions 1 through 16, find each product.

1. $(-6) \times (+8) =$	**2.** $(-2) \times (-7) =$
3. $(-9)(-4) =$	**4.** $(+12)(-15) =$
5. $(-5) \times (-3) \times (+7) =$	**6.** $(+6) \times (-9) \times (+3) =$
7. $(-6)(-4)(-3) =$	**8.** $(-10)(-10)(+10) =$
9. $(-4.5) \times (-2.6) =$	**10.** $(+6.1) \times (-3.08) =$
11. $(9.3)(-2.4)(-1.5) =$	**12.** $(-3.4)(2.4)(10) =$
13. $\left(-\frac{1}{5}\right) \times \left(-\frac{5}{12}\right) =$	**14.** $\left(+\frac{2}{3}\right)\left(-\frac{3}{4}\right) =$
15. $\left(+4\frac{1}{4}\right)\left(-1\frac{2}{3}\right) =$	**16.** $\left(-1\frac{4}{7}\right) \times \left(-5\frac{1}{2}\right) =$

For Some Extra Thought

17. How do we know without multiplying that the product of 80 negative factors is positive?

18. Negative seven times what number equals one?

Review Problems for Maintaining Your Skills

19. Write 1,605,000,000 in words.

20. Reduce $\frac{74}{94}$ to its lowest terms.

21. Divide 56.87 by 3.1 and round the quotient to the nearest hundredth.

22. Find 1.5% of \$86.

23. 400% of what amount equals \$280?

24. To the nearest hundredth of a percent, 1 is what percent of 7?

25. Find the percent decrease if a price decreases from \$200 to \$186.

26. Use cross-multiplication to solve $\frac{10}{35} = \frac{60}{W}$

In 27 and 28, insert the appropriate symbol, < or >, in the blank.

27. 14.5 ____ 14.43 **28.** −156 ____ −163

In 29 and 30, find each sum.

29. $-4 + 13 + 21 - 17 =$

30. $(-1.2) + (-6.9) + (24) =$

In 31 and 32, find each difference.

31. $(-27) - (-29) =$

32. $(+4.8) - (-6.75) =$

EXERCISE SET 6.4B

Skill Development

1. $(+6) \times (-12) =$
2. $(-8) \times (-9) =$
3. $(-4)(-14) =$
4. $(+7)(-20) =$
5. $(-3) \times (-7) \times (-5) =$
6. $(+8) \times (-13) \times (-1) =$
7. $(20)(-6)(-2)(-3) =$
8. $(-5)(4)(-2)(-1)(10) =$
9. $(-3.64) \times (+7.1) =$
10. $(-8.1) \times (+3.45) =$
11. $(-10.1)(-5.3)(-8) =$
12. $(20)(-2.25)(+9.1) =$
13. $\left(+\frac{4}{5}\right) \times \left(-\frac{1}{2}\right) =$
14. $\left(-\frac{3}{5}\right)\left(-\frac{2}{3}\right) =$
15. $\left(+1\frac{1}{2}\right)\left(-5\frac{3}{4}\right) =$
16. $\left(-3\frac{2}{5}\right) \times \left(-5\frac{1}{4}\right) =$

For Some Extra Thought

17. How do we know without multiplying that the product of twenty-one negative numbers is negative?

18. Negative five times what number equals five-tenths?

Review Problems for Maintaining Your Skills

19. Write 13,678,304 in words.

20. Reduce $\frac{480}{720}$ to its lowest terms.

21. Divide 21.6 by 1.9 and round the quotient to the nearest tenth.

22. Find 8.56% of $20,000.

23. 25% of what amount equals $56?

24. To the nearest hundredth of a percent, 18 is what percent of 43?

25. Find the percent increase if a price increases from $50 to $53.

26. Use cross-multiplication to solve $\frac{N}{3} = \frac{1.8}{1.2}$

In 27 and 28, insert the appropriate symbol, $<$ or $>$, in the blank.

27. -15 ____ -18
28. -78.3 ____ -78.29

In 29 and 30, find each sum.

29. $-5 - 9 + 8 - 12 =$

30. $(-17) + (+14) + (-3) =$

In 31 and 32, find each difference.

31. $(+4) - (-9) =$

32. $(-2.54) - (-1.68) =$

6.5 DIVIDING SIGNED NUMBERS

Preview Test

In questions 1 through 8, find each quotient.

1. $(-24) \div (+6) =$
2. $(+42) \div (-14) =$
3. $(-3) \div (-3) =$
4. $(-120) \div (15) =$
5. $(+20) \div (-0.5) =$
6. $(-1) \div (-0.002) =$
7. $\left(-\frac{4}{5}\right) \div \left(-\frac{1}{2}\right) =$
8. $\left(-2\frac{3}{4}\right) \div \left(+1\frac{5}{6}\right) =$

The rule for dividing two signed numbers is

Step 1: Divide the absolute value of the dividend by the absolute value of the divisor.
Step 2: Find the sign of the quotient, which is
positive if the signs of the dividend and divisor are alike,
negative if the signs of the dividend and divisor are not alike.

Using this rule we can see that

with like signs

POSITIVE ÷ POSITIVE = POSITIVE	$(+32) \div (+8) = +4$
NEGATIVE ÷ NEGATIVE = POSITIVE	$(-32) \div (-8) = +4$

and with unlike signs

POSITIVE ÷ NEGATIVE = NEGATIVE	$(+32) \div (-8) = -4$
NEGATIVE ÷ POSITIVE = NEGATIVE	$(-32) \div (+8) = -4$

To help you understand the rule of signs for division, use the fact that a quotient can be checked by multiplication. That is,

24	÷	6	=	4	because	4	×	6	=	24
DIVIDEND	÷	DIVISOR	=	QUOTIENT		QUOTIENT	×	DIVISOR	=	DIVIDEND

So,

DIVIDEND	÷	DIVISOR	=	QUOTIENT	because	QUOTIENT	×	DIVISOR	=	DIVIDEND
POSITIVE	÷	POSITIVE	=	POSITIVE	because	POSITIVE	×	POSITIVE	=	POSITIVE
NEGATIVE	÷	NEGATIVE	=	POSITIVE	because	POSITIVE	×	NEGATIVE	=	NEGATIVE
POSITIVE	÷	NEGATIVE	=	NEGATIVE	because	NEGATIVE	×	NEGATIVE	=	POSITIVE
NEGATIVE	÷	POSITIVE	=	NEGATIVE	because	NEGATIVE	×	POSITIVE	=	NEGATIVE

Let's now look at some examples involving division of signed numbers.

EXAMPLE 1: $(-48) \div (+6) =$

Solution:
Since the quotient of two numbers with unlike signs is negative, $(-48) \div (+6) = -8$.

ANSWERS FOR PREVIEW TEST

1. -4 2. -3 3. 1 4. -8 5. -40 6. 500
7. $1\frac{3}{5}$ 8. $-1\frac{1}{2}$

PRACTICE PROBLEM 1: $(-18) \div (-3) =$

EXAMPLE 2: $(-12.4) \div (-0.04) =$

Solution:
Since the signs are alike, the quotient is positive. So, $(-12.4) \div (-0.04) = +310$

PRACTICE PROBLEM 2: $(+0.5) \div (-0.25) =$

EXAMPLE 3: $\left(-\frac{2}{5}\right) \div \left(-\frac{1}{2}\right) =$

Solution:
Remember, to divide we must invert and multiply. Also, since the signs are alike, the quotient is positive. Therefore, $\left(-\frac{2}{5}\right) \div \left(-\frac{1}{2}\right) = \left(-\frac{2}{5}\right) \times \left(-\frac{2}{1}\right) = +\frac{4}{5}$

PRACTICE PROBLEM 3: $\left(+\frac{7}{8}\right) \div \left(-\frac{1}{4}\right) =$

EXAMPLE 4: $\left(2\frac{1}{4}\right) \div (-3) =$

Solution:
First change the mixed number $2\frac{1}{4}$ to an improper fraction $\frac{9}{4}$ and write the whole number -3 as $-\frac{3}{1}$. Then invert and multiply. Therefore,

$$\left(2\frac{1}{4}\right) \div (-3) = \left(\frac{9}{4}\right) \div \left(-\frac{3}{1}\right) = \left(\frac{9}{4}\right) \times \left(-\frac{1}{3}\right) = -\frac{9}{12} = -\frac{3}{4}$$

PRACTICE PROBLEM 4: $\left(-4\frac{2}{3}\right) \div \left(-5\frac{1}{2}\right) =$

ANSWERS FOR PRACTICE PROBLEMS

1. $+6$ **2.** -2 **3.** $-3\frac{1}{2}$ **4.** $\frac{28}{33}$

EXERCISE SET 6.5A

Skill Development

In questions 1 through 16, find each quotient.

1. $(-42) \div (-7) =$
2. $(+30) \div (-2) =$
3. $(+48) \div (+16) =$
4. $(-12) \div (+12) =$
5. $(-36) \div (-9) =$
6. $(+40) \div (-8) =$
7. $(-200) \div (+25) =$
8. $(+160) \div (-20) =$
9. $(-30) \div (+1.5) =$
10. $(-0.24) \div (-10) =$
11. $(+1.44) \div (0.012) =$
12. $(-24.6) \div (+0.02) =$
13. $\left(-\frac{3}{5}\right) \div \left(+\frac{2}{5}\right) =$
14. $\left(-\frac{5}{8}\right) \div \left(-\frac{13}{14}\right) =$
15. $\left(+2\frac{1}{2}\right) \div (-4) =$
16. $\left(-4\frac{3}{8}\right) \div \left(+1\frac{1}{4}\right) =$

For Some Extra Thought

17. Does zero divided by negative two equal zero divided by positive two?

18. Will the sum of two negative numbers divided by the sum of two positive numbers always have a negative quotient?

Review Problems for Maintaining Your Skills

19. Round 21,809 to the nearest thousand.
20. Divide 407 by 25 and find the quotient and remainder.

21. $6\frac{1}{3} + 8\frac{1}{4} =$
22. Divide 56.8 by 1.06 and round the quotient to the nearest hundredth.

23. Find 6.25% of \$6,500.
24. Find the percent increase if a price increases from \$50 to \$57.

25. Use cross-multiplication to solve the proportion $\frac{N}{5} = \frac{10}{2}$

26. Insert the correct symbol (< or >) -25.6 ____ 25.57
27. $-9 + 8 + 17 - 13 =$
28. $(-125) - (-123) =$
29. $(-12) \times (+15) =$
30. $(-1.5)(-2.5) =$

EXERCISE SET 6.5B

Skill Development

In questions 1 through 16, find each quotient.

1. $(-56) \div (+8) =$
2. $(-60) \div (-20) =$
3. $(+15) \div (-15) =$
4. $(-100) \div (+25) =$
5. $(+72) \div (+36) =$
6. $(-400) \div (-80) =$
7. $(-31) \div (+1) =$
8. $(+160) \div (-32) =$
9. $(-50) \div (-0.5) =$
10. $(+20.15) \div (-100) =$
11. $(-2.64) \div (-0.132) =$
12. $(+0.36) \div (-1.2) =$
13. $\left(+\frac{1}{3}\right) \div \left(-\frac{2}{5}\right) =$
14. $\left(-\frac{20}{21}\right) \div \left(-\frac{5}{7}\right) =$
15. $(-3) \div \left(-2\frac{1}{3}\right) =$
16. $\left(+6\frac{1}{2}\right) \div \left(-2\frac{2}{3}\right) =$

For Some Extra Thought

17. What is the quotient for the absolute value of negative four divided by the absolute value of positive four?

18. Will the difference of two negative numbers divided by a positive number always have a negative quotient?

Review Problems for Maintaining Your Skills

19. Round 56,987,364 to the nearest hundred thousand.

20. Divide 55 by 17 and find the quotient and remainder.

21. $\frac{3}{7} \times \frac{2}{9} =$

22. Divide 1.9 by 20 and round the quotient to the nearest tenth.

23. Find 12.4% of \$23,000.

24. Find the percent decrease if a price decreases from \$40 to \$32.

25. Use cross-multiplication to solve the proportion $\frac{1.2}{R} = \frac{6}{7}$

26. Insert the correct symbol ($<$ or $>$) -1.75 ____ 0.204
27. $-1.2 - 3.6 - 4.5 =$
28. $(+48) - (-67) =$
29. $(55) \times (-100) =$
30. $(-4.8)(-0.2) =$

6.6 MIXED OPERATIONS

Preview Test

1. $(-8) - (-6) \times (+4) =$
2. $(-24) \div (+3) + (-2) \times (-2) =$
3. $(-4) \times (+3) - (16) \div (-16) =$
4. $[(-4) + (-6) \times (+3)] - [(-5) - (2)(-3)] =$
5. $-100 + [25 - ((-6) \times (-1) + (-4))] =$

The order in which operations should be performed is the same for signed numbers as it was for positive whole numbers (Section 1.7), positive fractions (Section 2.7), and positive decimals (Section 3.6). For your convenience, we repeat here the rules that were originally stated in Section 1.7.

If no parentheses ( ) or brackets [ ] appear, the order in which operations are performed is

1. multiplication and /or division in the order in which the operations occur from left to right
2. addition and/or subtraction in the order in which the operations occur from left to right.

If parentheses or brackets are included, the operations within the parentheses or brackets should be performed first in the correct order; that is, multiplication and/or division from left to right and then addition and/or subtraction from left to right.

EXAMPLE 1: $(-14) - (+6) \times (-8) =$

Solution:

$$\begin{aligned} &(-14) - \underbrace{(+6) \times (-8)}_{\text{multiply}} \\ &= \underbrace{(-14) - \quad (-48)}_{\text{subtract}} \\ &= (-14) + \quad (+48) \\ &= +34 \end{aligned}$$

PRACTICE PROBLEM 1: $(-40) \div (-8) - (+6) \times (-5) =$

ANSWERS FOR PREVIEW TEST

1. $+16$ **2.** -4 **3.** -11 **4.** -23 **5.** -77

EXAMPLE 2: $[-28 \div ((+4) + (-5) \times (-2))] - (7)(-1) =$

Solution:

$$[-28 \div ((+4) + \underbrace{(-5) \times (-2)}_{\text{multiply}})] - \underbrace{(7)(-1)}_{\text{multiply}}$$

$$= [-28 \div \underbrace{((+4) + (+10))}_{\text{add}}] - (-7)$$

$$= [\underbrace{-28 \div (+14)}_{\text{divide}}] - (-7)$$

$$= \underbrace{[-2] - (-7)}_{\text{subtract}}$$

$$= [-2] + (+7)$$

$$= +5$$

PRACTICE PROBLEM 2: $-100 + [-20 - ((-12) \div (-3) + (+5))] =$

ANSWERS FOR PRACTICE PROBLEMS

1. $+35$ **2.** -129

EXERCISE SET 6.6A

Skill Development

1. $(-7) \times (-6) - (+23) =$

2. $(+3) - (-7) \times (-8) =$

3. $-50 - (-5)(-4) + (-42) \div (-21) =$

4. $(-84) - (-4) \times (-1) + (+50) =$

5. $(-6) + (-28) - (+7) - (-3)(-3) =$

6. $[-5 - (-3) \times (+2)] - (-6) =$

7. $[(-3) + (+5)] \times [(+4) - (-6)] =$

8. $-72 - [-80 + (-20) \times (-2)] =$

9. $[(-3) - ((+5) - (-6) \times (-2))] =$

10. $(+56) - [(-5) \times (-4) + (+3)] - (-11) =$

For Some Extra Thought

11. $0 - [-6 - ((-8) \times (-8) - (-3) \times (-4))] =$

12. $(-0.1) \div (-0.1) \div (-0.1)$ plus what number equals 10?

Review Problems for Maintaining Your Skills

13. Write 237,908 in words.

14. Convert $9\frac{2}{3}$ to an improper fraction.

15. $5\frac{2}{3} - 3\frac{3}{4} =$

16. $34.8 \times 2.56 =$

17. To the nearest tenth of a percent, 8 is what percent of 35?

18. 5% of what amount equals $23.75?

19. Use cross-multiplication to solve $\frac{2}{3} = \frac{H}{5.4}$

20. Insert the correct symbol ($<$ or $>$) 5.067 ____ 5.0608

21. $(-9) + (26) + (-34) =$

22. $(-1.26) - (+3.7) =$

23. $(7) \times (-5) \times (+3) =$

24. $(0.005)(-0.04) =$

25. $(-240) \div (+0.3) =$

26. $(-5.4) \div (-0.9) =$

EXERCISE SET 6.6B

Skill Development

1. $(+8) - (-9) \times (+7) =$

2. $(-7) \times (+2) - (-8) \times (-1) =$

3. $(-24) + (-20) \div (-5) - (+6) \times (-5) =$

4. $-60 - (-4) \times (+6) - (-12) + (+14) =$

5. $(+42) - (-7) + (-20) \div (-20) =$

6. $[-12 - (-6) \times (+2)] - (+15) =$

7. $[(-4) - (-5)] - [(-7) \times (-3) - (+5)] =$

8. $[(-7) \times (-3) - ((+5) \times (-2) + (-14))] =$

9. $64 - [-72 - ((+6) + (-16) \div (-8))] =$

10. $[(-8) - (-7 + 9) \times (-4 - 3)] + (-22) =$

For Some Extra Thought

11. $[-6 + (2) \times (3)] - [-40 + 85] =$

12. $(-20) + (-30) \div (-30)$ plus what number equals 25?

Review Problems for Maintaining Your Skills

13. Write 6,708,005 in words.

14. Convert $5\frac{13}{17}$ to an improper fraction.

15. $14\frac{2}{5} - 10\frac{6}{7} =$

16. $10{,}000 \times 0.045 =$

17. To the nearest hundredth of a percent, 17 is what percent of 23?

18. 1.5% of what amount equals \$18.00?

19. Use cross-multiplication to solve $\frac{2.5}{M} = \frac{100}{2.4}$

20. Insert the correct symbol (< or >) -4.5 ____ -4.48

21. $-12 - 15 - 17 =$

22. $(+8.45) - (+12.678) =$

23. $(-5) \times (-8) \times (-10) =$

24. $(-7.2)(-4.6) =$

25. $(+360) \div (-4.5) =$

26. $(-2.478) \div (-1{,}000) =$

PRACTICE TEST 6A

1. Insert the appropriate symbol, < or >, in the blank.

   a) -9 ____ -14 b) -2.51 ____ -2.607

2. Find the indicated absolute value.

   a) $|6.7|$ b) $|-4.29|$

3. Find each sum.

   a) $(-17) + (-13) =$ b) $(-2.6) + (+9.5) =$

   c) $-7 + 9 - 12 =$ d) $\left(-\frac{1}{2}\right) + \left(-\frac{2}{5}\right) =$

4. Find each difference.

   a) $(-4) - (+19) =$ b) $(2.56) - (-6.7) =$

5. If the temperature increases from -5.6 degrees Celsius to -2.7 degrees Celsius, find the change in temperature.

6. Find each product.

   a) $(-7) \times (+9) =$ b) $(2.3)(-6)(-2) =$

   c) $\left(-2\frac{2}{3}\right) \times \left(+8\frac{1}{2}\right) =$ d) $\left(-\frac{7}{9}\right) \times \left(+\frac{3}{4}\right) =$

7. Find each quotient.

   a) $(-42) \div (+14) =$ b) $(+24.4) \div (-0.2) =$

   c) $\left(-\frac{4}{5}\right) \div \left(-\frac{2}{5}\right) =$ d) $(-8) \div \left(-2\frac{1}{2}\right) =$

8. $(-4) \times (+7) + (-6) \times (-2) =$

PRACTICE TEST 6B

1. Insert the appropriate symbol, < or >, in the blank.

a) $+7$ ____ -6 **b)** -6.75 ____ -10

2. Find the indicated absolute value.

a) $|6.89|$ **b)** $|-5.44|$

3. Find each sum.

a) $(+9)+(-25)=$ **b)** $(-4)+(-7)=$

c) $\left(-\frac{6}{7}\right)+\left(+\frac{1}{2}\right)=$ **d)** $-2\frac{3}{4}+3\frac{2}{5}=$

4. Find each difference.

a) $(-5)-(-12)=$ **b)** $(+52.4)-(+72.6)=$

5. If a person was born in 124 BC and died in 37 BC, for how many years did the person live?

6. Find each product.

a) $(-6)(-6)(-6)=$ **b)** $(-1.2)\times(+6.7)=$

c) $\left(-1\frac{2}{5}\right)\times\left(+1\frac{1}{2}\right)=$ **d)** $\left(+\frac{8}{9}\right)\times\left(+\frac{3}{4}\right)=$

7. Find each quotient.

a) $(+80)\div(-16)=$ **b)** $(-3.75)\div(-0.125)=$

c) $\left(-\frac{7}{8}\right)\div\left(+\frac{3}{4}\right)=$ **d)** $12\div\left(-\frac{1}{5}\right)=$

8. $20-(-5)\times(-6)+(-17)=$

CUMULATIVE REVIEW EXERCISES FOR CHAPTERS 1 THROUGH 6

1. Write 23,584,100 in words.

2. Round 41,230 to the nearest ten thousand.

In 3 through 6, reduce each fraction to its lowest terms and express any result that is an improper fraction as a mixed number.

3. $\frac{11}{12} \times \frac{6}{7} =$

4. $4\frac{2}{3} \div 1\frac{1}{4} =$

5. $7\frac{1}{2} + 2\frac{5}{8} =$

6. $\frac{13}{15} - \frac{1}{5} =$

7. Write 0.0023 in words.

8. Round 4.36087 to the nearest thousandth.

9. Round $12.60712 to the nearest cent.

10. $4.02 \times 1.3 =$

11. Divide 14.3 by 2.91 and round the quotient to the nearest hundredth.

12. $4.2 \times 3 + 5.2 \times 1.6 =$

13. Convert 4.2% to a decimal.

14. Convert 0.175 to a percent.

15. Convert 24% to a fraction reduced to its lowest terms.

16. Find 24.5% of $15,000.

17. 4% of what amount equals $6.30?

18. 13 is what percent of 25?

19. The income tax rate in Baron County is 2.5%. Find the income tax on $42,300.

20. The $3,000 budget for the school band represents 12% of the total activities budget. Find the amount of the activities budget.

21. The cost of a sports car increased from $16,000 to $17,120. Find the percent increase in the price.

22. Use cross-multiplication to determine whether the proportion is true.

$$\frac{12}{20} = \frac{15}{25}$$

23. Solve the proportion using cross-multiplication.

$$\frac{N}{2} = \frac{5}{4}$$

24. Insert the appropriate symbol, $<$ or $>$, in the blank.

a) -9 ____ -8 **b)** 2.34 ____ 2.057

25. Find the indicated absolute value.

a) $|-34|$ **b)** $|-4.09|$

26. Find each sum.

a) $-13 + 8 - 17 =$ **b)** $(-6) + (-7) + (10) =$

27. Find each difference.

a) $(-4.52) - (-7.5) =$ **b)** $(300) - (-500) =$

28. Find each product.

a) $(-5) \times (-7) =$ **b)** $\left(-6\frac{1}{2}\right) \times (10) =$

29. Find each quotient.

a) $(-56) \div (+1.4) =$ **b)** $(-100) \div (-0.05) =$

30. $(-6) \times (-10) + (2) \times (-9) =$

C H A P T E R 7

ALGEBRAIC EXPRESSIONS AND EQUATIONS

7.1 ALGEBRAIC EXPRESSIONS

7.2 EQUIVALENT EQUATIONS

7.3 SOLVING EQUATIONS BY ADDITION

7.4 SOLVING EQUATIONS BY MULTIPLICATION OR DIVISION

7.5 SOLVING EQUATIONS BY MULTIPLE OPERATIONS

Words, abbreviations, and single letters that are used to represent numbers are called **variables**. When a computer is programmed to calculate a quantity, such as sales tax, sales commission, or total tuition, a formula is entered in the program using variables to represent numerical quantities. For example, the formula for the calculation of sales tax is

$$\text{SALES TAX} = \text{PRICE} \times \text{SALES TAX RATE}$$

If the price of an item is \$56 and the sales tax rate is 4%, the sales tax is calculated by substituting these numbers in the formula.

$$\begin{aligned} \text{SALES TAX} &= \text{PRICE} \times \text{SALES TAX RATE} \\ \text{SALES TAX} &= \$56 \times 4\% \\ &= \$56 \times 0.04 \\ &= \$2.24 \end{aligned}$$

Instead of using words like SALES TAX as a variable, abbreviations or single letters can be used. Using TX for SALES TAX, PR for PRICE, and RT for SALES TAX RATES, the sales tax formula is

$$TX = PR \times RT$$

and using the letters T for SALES TAX, P for PRICE, and R for SALES TAX RATE, the sales tax formula is

$$T = P \times R$$

In this chapter variables will be used along with numbers to build algebraic expressions and equations. Then methods for solving algebraic equations will be discussed.

7.1 ALGEBRAIC EXPRESSIONS

Preview Test

1. If P represents the number of people attending a football game, write the expression for the difference between 50,000 and the attendance.

2. Write $4w + 3z$ in words.

3. How many terms are there in the expression $3x + 4y - 10z$?

4. For each term name the numerical coefficient.

   a) $-7w$ **b)** z

5. Add the like terms in each expression.
   a) $4x - 9x + 7w + w$
   b) $-8y + 2x + 7y$
6. Remove the parentheses and simplify by combining like terms.
   a) $3(4w - 6) + 9w$
   b) $8(3 - 4z) - (7z - 4)$

When variables are combined with numbers and other variables the result is called an **algebraic expression**. For example, if E represents a person's gross pay for the month, we can write the following algebraic expressions.

$E + \$200$ for the gross pay plus a \$200 raise.

$E - \$150$ for the gross pay less \$150 for withholding taxes.

$12E$ for the gross pay for 12 months.

Notice that $12E$ means "12 times E." In algebra when a number and a variable or a set of variables are written with no symbol between them, multiplication is implied. So, $4w$ means "4 times w" and pqr means "p times q times r."

EXAMPLE 1: If M and N represent the number of students in sections I and II of a math course, write the expressions for

a) the sum of the students in the two sections;
b) the number of students in section I multiplied by 2;
c) 15 added to the number of students in section II.

Solution:

a) The sum is $M + N$.
b) $2M$ represents the number of students in section I multiplied by 2.
c) $N + 15$ represents 15 added to the number of students in section II.

PRACTICE PROBLEM 1: If S represents the sales and E represents the expenses for a company, write the expressions for

a) the difference between the sales and expenses;
b) the expenses multiplied by 0.75;
c) the sales increased by \$40,000.

EXAMPLE 2: Write each expression in words.

a) $4y + 7z$ b) $w + 3x - 7y$

ANSWERS FOR PREVIEW TEST

1. $50{,}000 - P$ 2. four times w plus three times z 3. 3
4. a) -7 b) 1 5. a) $-5x + 8w$ b) $-y + 2x$
6. a) $21w - 18$ b) $-39z + 28$

Solution:

a) Four times y plus seven times z.

b) w plus three times x minus seven times y.

PRACTICE PROBLEM 2: Write each expression in words.

a) $8x - y$ **b)** $-2x + y - 6z$

There is a name for the parts of an expression such as $3w + 5y$. $3w$ and $5y$ are called **terms**. Terms are separated by plus or minus signs ($+$ or $-$) that are not contained in a numerator, denominator, or within parentheses. For example,

$3w + 5y$ has two terms,

$5x - 3y - 7z$ has three terms,

$7P$ has one term,

$\frac{3 + w}{4}$ has one term,

$\frac{3 + w}{4} + \frac{7z}{2}$ has two terms,

and $5x + (7P + 4)$ has two terms.

$\frac{3 + w}{4}$ is considered one term because the plus sign is in the numerator and $(7P + 4)$ is considered one term because the plus sign is contained within the parentheses.

EXAMPLE 3: How many terms are there in each expression?

a) $4w + 3$ **b)** $7x - 6w + 9z$

c) $(8M + 5) - (7M + 6)$ **d)** $\frac{7 + R}{3}$

Solution:

a) 2 **b)** 3 **c)** 2 **d)** 1

PRACTICE PROBLEM 3: How many terms are there in each expression?

a) $4x + 5y - z$ **b)** $(7x + 2) - 9x$

c) $(3 - w) + (7 + w)$ **d)** $\frac{x}{2} + \frac{y - z}{6}$

The number multiplying the variable in a term is called its **numerical coefficient**. For example, the numerical coefficient for the term $5w$ is 5. If a variable appears with no number, its numerical coefficient is

$+1$ if the variable is preceded by no sign or a plus sign

-1 if the variable is preceded by a minus sign.

That is, the numerical coefficient for $+N$ or N is $+1$ and the numerical coefficient for $-N$ is -1. In other words, N means "one times N" and $-N$ means "negative one times N."

EXERCISE SET 7.1A

Skill Development

In questions 1 through 4, determine the number of terms in each expression.

1. $7w - 3x + 5$ **2.** $3 - (6 + 7y)$

3. $\frac{x + 3}{4} - 17z$ **4.** $\frac{a + b + c}{a - b - c}$

In questions 5 through 8, find the numerical coefficient for each term.

5. $-8z$ **6.** M **7.** $0.25w$ **8.** $-x$

In questions 9 through 12, add the like terms in each expression.

9. $7x + 5x - 14x$

10. $w - 3x + 7w$

11. $-z + 3z - 7y - 2y$

12. $10M - 6K + 17K - M$

In questions 13 through 16, remove the parentheses and simplify by combining like terms.

13. $5(2x - 7) + 4x$

14. $-2(3 - 4x) + 15$

15. $6(7w - 2) - 3(5 - 7w)$

16. $20y - (3 - 2y) + 6(y - 1)$

In questions 17 through 20, write each expression in words.

17. $6x - 5$

18. $3w + 5y - 2z$

19. $A - 3B - 10$

20. $3x - 4y + 2z - 1$

Applications

21. If A represents the advertising charge for one minute of advertising, write the expression for the charge for

a) 2.5 minutes of advertising;

b) the per minute charge increased by \$20,000.

22. If x represents the amount of rain in January and y represents the amount of rain in February, write the expression for

a) the total rainfall for the two months;

b) the absolute value of the difference in the rainfall for the two months;

c) 2.6 inches more than the February rainfall.

23. If S represents a salesperson's monthly sales and B represents the salesperson's base salary, write the expression for

a) 0.04 times the sales;

b) the base salary plus 0.15 times the sales.

24. If G represents the gross pay, F represents the federal withholding tax, and S represents the state withholding tax, write the expression for

a) the sum of the federal and state withholding tax;

b) the gross pay less the sum of the federal and state withholding tax.

For Some Extra Thought

25. $-w - w - w - w - w - w - w - w$ plus what equals zero?

26. What is the numerical value of $5x - 17$ if $x = 3$?

Review Questions for Maintaining Your Skills

27. Round 7,925,000,000 to the nearest ten million.

28. Divide 189 by 73 and find the quotient and remainder.

29. $4\frac{2}{3} + 8\frac{1}{2} + 1\frac{4}{5} =$

30. Find 2.5% of \$840.

31. \$13 is what percent of \$200?

32. 8% of what amount equals \$56.24?

33. Use cross-multiplication to determine if the proportion $\frac{12}{25} = \frac{60}{125}$ is true.

34. Insert the correct symbol ($<$ or $>$) -0.034 ____ -0.0351

35. $-4 + 6 + 17 - 9 =$

36. $(-13.5) - (-5.7) =$

37. $(-8)(-7)(-10) =$

38. $(-240) \div (+1.2) =$

39. $(-8) + (-3) \times (+6) =$

40. $-100 - [4 \times 6 + (-5) \times (-2)] =$

EXERCISE SET 7.1B

Skill Development

In questions 1 through 4, determine the number of terms in each expression.

1. $8x - 9y$
2. $(5x - 2) + (3x + 5) - 4$
3. $\frac{x}{2} + \frac{y}{2} + \frac{z}{2}$
4. $\frac{x + y + z}{2}$

In questions 5 through 8, find the numerical coefficient for each term.

5. $6w$
6. $-R$
7. $2.25w$
8. $\frac{3}{4}z$

In questions 9 through 12, add the like terms in each expression.

9. $4x - 5x + 2x$
10. $3w - 7w + 5y - y$
11. $3P - P - 2R + R$
12. $16x + y + 10y - 16x$

In questions 13 through 16, remove the parentheses and simplify by combining like terms.

13. $4(5w - 3) + 2w$
14. $7(3 - 5x) + 2(6x - 7)$
15. $-8(6x - 5) - 3(4x + 5)$
16. $42z - (3 - 6z) - 18$

In questions 17 through 20, write each expression in words.

17. $7y - 3z$

18. $A + 7B - 5$

19. $2.5x - 3.5y$

20. $-u + 3v$

Applications

21. If B represents the number of points scored by the Chicago Bears and M represents the number of points scored by the Miami Dolphins, write the expression for
    a) the sum of the points scored by both teams;

    b) the Bear's points increased by 7;

    c) the absolute value of the difference in the teams' points.

22. If x represents Jim's grade on the first sociology test and y represents Jim's grade on the second sociology test, write the expression for

a) the sum of the two grades;

b) the sum of the two grades divided by two;

c) the sum of twice the first grade plus three times the second grade divided by five.

23. If H represents a worker's hourly wage and N represents the number of hours worked, write the expression for

a) the hourly wage times the number of hours worked;

b) the number of hours worked times the sum of the hourly wage plus 0.25.

24. If D represents the closing Dow Jones Industrial Average, write the expression for

a) the closing average increased by 6.5;

b) the sum of the closing average plus 2,000 divided by 2.

For Some Extra Thought

25. $-z + z - z + z - z + z - z$ plus what equals zero?

26. What is the numerical value of $0.05S + 100$ if $S = 1000$?

Review Questions for Maintaining Your Skills

27. Round 4,562 to the nearest ten.

28. Divide 47 by 17 and find the quotient and remainder.

29. $\frac{7}{8} + \frac{1}{4} + \frac{3}{10} =$

30. Find 8.75% of \$20,600.

31. \$5.50 is what percent of \$50?

32. 20% of what amount equals \$420?

33. Use cross-multiplication to determine if the proportion $\frac{2}{5} = \frac{3.1}{8.2}$ is true.

34. Insert the correct symbol (< or >) 704.2 ____ 704.09

35. $-6 + 7 - 7 - 12 =$

36. $(+47.8) - (+52.67) =$

37. $(-4.2) \times (100) =$

38. $(-20.48) \div (-0.004) =$

39. $-7 + (+9) \times (-14) =$

40. $[(-7) + (5)(-9)] - (-24) =$

7.2 EQUIVALENT EQUATIONS

Preview Test

1. Write an equation equivalent to $4w - 5 = 8$ by adding 5 to both sides.
2. Write an equation equivalent to $-6x + 7 = 5$ by adding -7 to both sides.
3. Write an equation equivalent to $2z - 8 = 10$ by dividing both sides by 2.
4. Write an equation equivalent to $\frac{m}{3} + 1 = -4$ by multiplying both sides by 3.
5. The solution for the equation $7x - 3 = 11$ is 2. Which of the following equations are equivalent to $7x - 3 = 11$?

   a) $7x = 14$ **b)** $7x + 3 = 17$ **c)** $14x = 14$

Let's consider a salesperson whose earnings are based on a base salary of \$400 per week plus a commission of 2% of his or her sales. With sales of \$7,000 the earnings for the week are

$$\begin{aligned} &2\% \times \$7{,}000 + \$400 \\ &= 0.02 \times \$7{,}000 + \$400 \\ &= \$140 + \$400 = \$540 \end{aligned}$$

If the sales are \$12,000 the earnings for the week are

$$\begin{aligned} &2\% \times \$12{,}000 + \$400 \\ &= 0.02 \times \$12{,}000 + \$400 \\ &= \$240 + \$400 = \$640 \end{aligned}$$

Suppose the salesperson wants to know how much he or she must sell to earn \$900 in one week. The question can be expressed by asking, "What number can we substitute for the variable S so that $0.02S + 400 = 900$?" The answer to this question is \$25,000 because $0.02(\$25{,}000) + \$400 = \$900$. We call $0.02S + 400 = 900$ an **equation**. Generally when we write an equation we are looking for the number we can substitute for the variable that will make the left side and right side equal. This number is called the **solution** for the equation. So, for $0.02S + 400 = 900$ the solution is \$25,000.

If we subtract the base salary (\$400) from the left side and right side of the equation $0.02S + 400 = 900$ we have

$$0.02S + 400 - 400 = 900 - 400$$

or

$$0.02S = 500$$

The solution for $0.02S = 500$ is \$25,000, because

$$0.02(\$25{,}000) = \$500$$

ANSWERS FOR PREVIEW TEST

1. $4w = 13$ **2.** $-6x = -2$ **3.** $z - 4 = 5$ **4.** $m + 3 = -12$
5. a) equivalent **b)** equivalent **c)** not equivalent

It is no coincidence that $0.02S + 400 = 900$ and $0.02S = 500$ have the same solution since the questions

2% times what sales figure plus $400 equals $900?

and

2% times what sales figure equals $500?

are equivalent. (That is, two ways of asking the same question).

In mathematics, instead of saying that the questions are equivalent we say that the equations are equivalent.

> Two equations are **equivalent** if they have the same solution.

So, the equations $0.02S + 400 = 900$ and $0.02S = 500$ are equivalent because both have the solution $25,000. Recall that $0.02S = 500$ evolved from $0.02S + 400 = 900$ by subtracting 400 from each side.

> In general, starting with an equation we can generate an equivalent equation by
>
> 1. adding the same number or variable term to the left and right sides;
> 2. subtracting the same number or variable term from the left and right sides;
> 3. multiplying both sides by the same number (except zero);
> 4. dividing both sides by the same number (except zero).

In the examples and exercises that follow you will be asked to carry out some operations that generate equivalent equations. At this point it may not be clear why these operations are being performed. In the sections that follow, the "Why are we doing this?" will be answered.

EXAMPLE 1: Write an equation equivalent to $3x - 4 = 7$ by adding 4 to both sides.

Solution:

$$3x - 4 = 7$$
$$3x - 4 + 4 = 7 + 4$$
$$3x + 0 = 11$$
$$3x = 11$$

PRACTICE PROBLEM 1: Write an equation equivalent to $5w + 6 = 19$ by adding -6 to both sides.

EXAMPLE 2: Write an equation equivalent to $4w - 8 = 12$ by dividing both sides by 4.

Solution:

$$4w - 8 = 12$$
$$\frac{4w}{4} - \frac{8}{4} = \frac{12}{4}$$
$$w - 2 = 3$$

PRACTICE PROBLEM 2: Write an equation equivalent to $5z + 10 = -20$ by dividing both sides by 5.

EXAMPLE 3: Write an equation equivalent to $\frac{y}{3} + \frac{1}{4} = 1$ by multiplying both sides by 12.

Solution:

$$12\left(\frac{y}{3}\right) + 12\left(\frac{1}{4}\right) = 12(1)$$

$$4y + 3 = 12$$

PRACTICE PROBLEM 3: Write an equation equivalent to $\frac{z}{6} + \frac{3}{8} = 2$ by multiplying both sides by 24.

EXAMPLE 4: The solution for the equation $5x - 4 = 11$ is 3. Which of the following equations are equivalent?

a) $5x = 20$ **b)** $5x = 15$ **c)** $5x + 10 = 25$

Solution:

a) $5x = 20$ is not equivalent because $5(3) = 20$ is not true.

b) $5x = 15$ is equivalent because $5(3) = 15$ is true.

c) $5x + 10 = 25$ is equivalent because $5(3) + 10 = 25$ is true.

PRACTICE PROBLEM 4: The solution for the equation $-2x = 8$ is -4. Which of the following equations are equivalent?

a) $-2x + 2 = 10$ **b)** $-2x = 12$ **c)** $2x = -8$

ANSWERS FOR PRACTICE PROBLEMS

1. $5w = 13$ **2.** $z + 2 = -4$ **3.** $4z + 9 = 48$

4. **a)** equivalent **b)** not equivalent **c)** equivalent

EXERCISE SET 7.2A

Skill Development

In questions 1 through 12, write an equivalent equation by performing the indicated operation.

1. $x - 7 = 12$; add 7 to both sides
2. $w + 9 = -3$; add -9 to both sides
3. $4x - 3 = 10$; add 3 to both sides
4. $-5z + 13 = 0$; add -13 to both sides
5. $0.2S + 100 = 500$; add -100 to both sides
6. $0.05S + 450 = 1000$; add -450 to both sides
7. $-4z = 8$; divide both sides by -4
8. $8x = 17$; divide both sides by 8
9. $4x + 8 = 12$; divide both sides by 4
10. $5y - 10 = 10$; divide both sides by 5
11. $\frac{w}{4} = 3$; multiply both sides by 4
12. $\frac{x}{-3} = 2$; multiply both sides by -3
13. The solution for the equation $x + 7 = 19$ is 12. Which of the following equations are equivalent to $x + 7 = 19$?
    a) $x - 12 = 0$ **b)** $2x + 10 = 22$ **c)** $-2x = -24$
14. The solution for the equation $3x - 1 = 5$ is 2. Which of the following equations are equivalent to $3x - 1 = 5$?
    a) $x - 3 = 2$ **b)** $x = 2$ **c)** $5x + 100 = 110$
15. The solution for the equation $\frac{z}{2} = -6$ is -12. Which of the following equations are equivalent to $\frac{z}{2} = -6$?
    a) $z + 9 = -12$ **b)** $z + 12 = 0$ **c)** $4z - 10 = -58$

For Some Extra Thought

16. Write an equation equivalent to $x = 3$.
17. Write an equation equivalent to $w + 5 = 15$.
18. Write an equation equivalent to $3y = 12$.

Review Questions for Maintaining Your Skills

19. Write 5,604,800 in words.

20. Divide 563 by 100 and find the quotient and remainder.

21. $\frac{3}{8} + \frac{11}{12} =$

22. $7\frac{1}{2} - 3\frac{2}{5} =$

23. Round 4.036874 to the nearest ten-thousandth.

24. $0.003879 \times 100{,}000 =$

25. To the nearest tenth of a percent, 67 is what percent of 91?

26. 2% of what amount equals $4.56?

27. Use cross-multiplication to solve $\frac{N}{4} = \frac{1.5}{3}$

28. $-9 + 12 - 17 - 2 =$

29. $(-12.5) - (-9.72) =$

30. $(-0.5)(-3.5) =$

31. $(-500) \div (+0.25) =$

32. $100 - [(4)(-3) + (-6)(0)] =$

33. Find the number of terms in the expression $-9w + 17x - 1$

34. Find the numerical coefficient for $-7N$

35. Add the like terms: $-5w + 7x - 9x + 13w$

36. Remove the parentheses and simplify: $-7(4r - 5) + 9r$

37. Write the expression $7N + 5P - 2.5$ in words.

EXERCISE SET 7.2B

Skill Development

In questions 1 through 12, write an equivalent equation by performing the indicated operation.

1. $w + 3 = 7$; add -3 to both sides
2. $z - 7 = 17$; add 7 to both sides
3. $-8y + 1 = 10$; add -1 to both sides
4. $6x - 12 = 12$; add 12 to both sides
5. $0.03S + 200 = 850$; add -200 to both sides
6. $0.12S + 80 = 600$; add -80 to both sides
7. $3x = 15$; divide both sides by 3
8. $-9x = 36$; divide both sides by -9
9. $6x + 18 = -12$; divide both sides by 6
10. $-2x + 6 = 0$; divide both sides by -2
11. $\frac{x}{3} = 5$; multiply both sides by 3
12. $\frac{z}{5} + 1 = 4$; multiply both sides by 5
13. The solution for the equation $y + 7 = 9$ is 2. Which of the following equations are equivalent to $y + 7 = 9$?

    a) $y + 7 = 9$ **b)** $y - 3 = -1$ **c)** $y = 2$

14. The solution for the equation $-5x = 15$ is -3. Which of the following equations are equivalent to $-5x = 15$?

    a) $x + 6 = 3$ **b)** $4x = -12$ **c)** $x + 3 = 0$

15. The solution for the equation $\frac{x}{2} = 0$ is 0. Which of the following equations are equivalent to $\frac{x}{2} = 0$?

    a) $x + 5 = -5$ **b)** $x - 3 = -3$ **c)** $x = 1$

For Some Extra Thought

16. Write an equation equivalent to $w = 4$.
17. Write an equation equivalent to $z - 5 = 6$.
18. Write an equation equivalent to $2x = 8$.

Review Questions for Maintaining Your Skills

19. Write 14,605 in words.

20. Divide 89 by 12 and find the quotient and remainder.

21. $\frac{14}{15} + \frac{9}{10} =$

22. $14\frac{2}{3} - 5\frac{1}{2} =$

23. Round 12.580674 to the nearest hundredth.

24. $13.25 \times 1{,}000 =$

25. To the nearest hundredth of a percent, 23 is what percent of 30?

26. 500% of what amount equals $4,000?

27. Use cross-multiplication to solve $\frac{2.4}{T} = \frac{10}{25}$

28. $-2.3 + 5.75 + 4 =$

29. $(-200) - (+900) =$

30. $(7.5)(-1)(-1)(-1) =$

31. $(-16.4) \div (-0.002) =$

32. $[20 - (-4)(-5)] + [(-2)(-3) - 30] =$

33. Find the number of terms in the expression $(5A + 3B) + 7C$

34. Find the numerical coefficient for $0.03H$

35. Add the like terms: $-7N + 3N - 7K - 9K$

36. Remove the parentheses and simplify: $-5(8x - 1) + 19$

37. Write the expression $7K - 3R + 9$ in words.

7.3 SOLVING EQUATIONS BY ADDITION

Preview Test

1. Solve $x - 13 = 11$
2. Solve $w + 3 = -5$
3. Solve $z + 2.3 = -6.7$
4. Solve $y + 9 = 9$
5. The Select Pro Printer II prints 60 characters per second. This exceeds the speed of the Pro Printer I by 13 characters per second. Find the speed of the Pro Printer I by writing and solving an equation.

In the previous section we noted that a solution for an equation is the number we can substitute for the variable to make the left side and right side equal. Some equations can be solved mentally. For example, the equation $x - 3 = 8$ translates to the question, "What number minus 3 equals 8?" It is not difficult to see that the number we are looking for is 11, because $11 - 3 = 8$.

Since all equations are not as easy to solve mentally as $x - 3 = 8$, it is wise at this point to learn some pencil and paper steps that can be used to solve equations. Recall that in section 7.2 we said that if we add the same number to both sides of an equation the resulting equation is equivalent (that is, it has the same solution). By adding 3 to both sides of $x - 3 = 8$ we get an equivalent equation $x = 11$ whose solution is self-evident.

Let's think about how we are able to go from $x - 3 = 8$ to $x = 11$. We pick the number we can add to -3 to get 0.

$$x - 3 = 8$$

What number added to -3 gives us 0?
Answer: $+3$
Then add $+3$ to both sides.

$$x - 3 + 3 = 8 + 3$$
$$x + 0 = 11$$
$$x = 11$$

Now x is isolated with a coefficient of 1 and the solution is self-evident. It's 11.

EXAMPLE 1: Solve $w + 7 = 9$

Solution:
Since $7 - 7 = 0$, add -7 to both sides.

$$w + 7 - 7 = 9 - 7$$
$$w + 0 = 2$$
$$w = 2$$

ANSWERS FOR PREVIEW TEST

1. 24 **2.** -8 **3.** -9 **4.** 0

5. $C + 13 = 60$, $C = 47$ characters per second

PRACTICE PROBLEM 1: Solve $z + 6 = 20$

EXAMPLE 2: Solve $y - 5 = -8$

Solution:
Since $-5 + 5 = 0$, add $+5$ to both sides.

$$y - 5 + 5 = -8 + 5$$
$$y + 0 = -3$$
$$y = -3$$

PRACTICE PROBLEM 2: Solve $x - 9 = -25$

EXAMPLE 3: Solve $v - 2.1 = 3.8$

Solution:
Since $-2.1 + 2.1 = 0$, add $+2.1$ to both sides.

$$v - 2.1 + 2.1 = 3.8 + 2.1$$
$$v + 0 = 5.9$$
$$v = 5.9$$

PRACTICE PROBLEM 3: Solve $z - 6.7 = -3.5$

The solution of equations can play a very important role in problem solving. In this section most of the verbal problems can be solved mentally without writing and solving an equation. While problem solving through mental computation is certainly admirable, we ask you now to be patient and learn the pencil and paper procedures that we present in the examples that follow. These procedures will help you in cases where a mental approach is not practical.

Let's consider the following problem.

> Mark's score on the verbal section of the SAT exam is 43 points higher than Kim's score. If Mark's score is 517, what is Kim's score?

Solving this problem through the solution of an equation requires several steps. The steps outlined below for finding Kim's SAT score can be used as a problem solving technique for the remainder of this chapter.

Step 1: Identify what you are asked to find and assign a variable to this quantity.

> Kim's SAT score is to be found. Let's assign the variable K to Kim's score.

Step 2: Identify all given numerical information.

> Mark's score is 43 points higher than Kim's.
> Mark's score is 517.

Step 3: Write an equation using the variable from step 1 and the given information from step 2:

> Since Mark's score exceeds Kim's score by 43 we can write the equation

$$K + 43 = 517$$

Step 4: Solve the equation and answer the question.

Since $43 - 43 = 0$, add -43 to both sides.

$$K + 43 - 43 = 517 - 43$$
$$K + 0 = 474$$
$$K = 474$$

So, Kim's score is 474

EXAMPLE 4: Today's closing Dow Jones Industrial Average is 1972. This represents an 18 point decrease from yesterday's closing average. Find yesterday's closing average.

Solution:

Step 1: Let Y represent yesterday's closing average.

Step 2: Today's closing average is 1972, which is 18 points lower than yesterday's closing average.

Step 3: An equation which relates Y to 1972 and 18 is $Y - 18 = 1972$

Step 4: Since $-18 + 18 = 0$, add $+18$ to both sides.

$$Y - 18 + 18 = 1972 + 18$$
$$Y + 0 = 1990$$
$$Y = 1990$$

So, yesterday's closing average was 1990.

PRACTICE PROBLEM 4: In September the rainfall measured 4.7 inches. This was 1.8 inches below last year's rainfall for September. Find the rainfall for September of last year.

ANSWERS FOR PRACTICE PROBLEMS

1. 14 **2.** -16 **3.** 3.2 **4.** 6.5 inches

EXERCISE SET 7.3A

Skill Development

In questions 1 through 15, solve each equation.

1. $x + 3 = 13$ **2.** $y - 8 = 12$ **3.** $z + 7 = -4$

4. $y - 4 = -6$ **5.** $w + 7 = 7$ **6.** $x - 9 = -20$

7. $z + 8 = 0$ **8.** $x + 2.5 = 5.5$ **9.** $w - 0.3 = 0$

10. $y - 1.2 = 3.6$ **11.** $z + 6 = -10.5$ **12.** $x - 0.06 = 0.1$

13. $x + \frac{1}{4} = \frac{1}{8}$ **14.** $y - \frac{2}{5} = \frac{3}{4}$ **15.** $z - 4\frac{1}{7} = -5$

Applications

In questions 16 through 18, answer each question by writing and solving an equation.

16. The SR60 Sports Sedan averages 2 miles per gallon more on open road driving than the SR70 Sports sedan. If the SR70 averages 26 miles per gallon find the miles per gallon for the SR60.

17. Joan Marden's earnings for the week exceeded her sales commission by $250. If her earnings were $840, what was her sales commission?

18. Mathew's math SAT score was 27 points higher than his verbal SAT score. If his math score was 568, what was his verbal score?

For Some Extra Thought

19. Solve $x - 0.13 = \frac{1}{2}$ **20.** Solve $w + 5\% = \frac{7}{100}$

Review Problems for Maintaining Your Skills

21. Round 5,604 to the nearest hundred.

22. $7\frac{1}{2} - 4\frac{3}{4} =$

23. $\frac{17}{20} \times \frac{35}{51} =$

24. Write in words 5.034

25. Divide 2.46 by 1.3 and round the quotient to the nearest tenth.

26. Find 8.7% of \$650.

27. Find the percent increase to the nearest tenth of a percent if a price increases from \$1.06 to \$1.17.

28. $-5.4 - 9.7 =$

29. $(-5)(-6)(-100) =$

30. $[17 - (6)(-7)] \times [80 + (-10)(-8)] =$

31. Add like terms: $-7.4x + 3.6y - 6.8x + 9.1y$

32. Remove parentheses and simplify: $6(9N - 2) - 9(7 - 3N)$

33. Write an equation equivalent to $-5x = 30$ by dividing both sides by -5

34. The solution for the equation $4x - 3 = 5$ is 2. Which of the following equations are equivalent to $4x - 3 = 5$?

a) $x - 5 = -3$ **b)** $5x + 6 = 11$

EXERCISE SET 7.3B

Skill Development

In questions 1 through 15, solve each equation.

1. $x + 7 = 15$ **2.** $w - 6 = 3$ **3.** $z - 4 = 4$

4. $y - 3 = 20$ **5.** $x + 15 = 8$ **6.** $w + 8 = -5$

7. $x - 9 = -20$ **8.** $w + 13 = 0$ **9.** $y + 3 = 3$

10. $x - 2.3 = 8$ **11.** $y + 3.6 = 5.8$ **12.** $z - 6.02 = -7.1$

13. $y - \frac{2}{5} = \frac{3}{5}$ **14.** $w + \frac{7}{8} = \frac{1}{2}$ **15.** $z + 4 = -3\frac{2}{3}$

Applications

In questions 16 through 18, answer each question by writing and solving an equation.

16. The average math SAT score for the 1992 freshman class at Byron college was 7 points higher than the 1982 freshman class. If the average score for the 1992 class was 496 what was the average score for the 1982 class?

17. The cost of a ten-inch black and white TV set in 1947 was $170 more than the cost of a twelve-inch set in 1991. If the 1991 set cost $89, what was the cost of the 1947 set?

18. In 1984 the New York Mets won 18 fewer regular season games than they won in 1986. If they won 90 games in 1984, how many games did they win in 1986?

For Some Extra Thought

19. Solve $x - 0.1 = \frac{3}{4}$

20. Solve $z - 200\% = -2$

Review Problems for Maintaining Your Skills

21. Round 10,807,543 to the nearest hundred thousand.

22. $5\frac{3}{7} - 2\frac{1}{2} =$

23. $\frac{9}{10} \times \frac{2}{3} \times \frac{6}{7} =$

24. Write 0.0457 in words.

25. Divide 45.8 by 1.33 and round the quotient to the nearest hundredth.

26. Find 400% of \$65.

27. Find the percent decrease to the nearest tenth of a percent if a price decreases from \$10.89 to \$9.75.

28. $10.62 - 3.4 - 1.8 =$

29. $(-16)(-7)(-2) =$

30. $-900 - [40 + (-6)(-7)] =$

31. Add like terms: $-5.6A - 3.7B + 4.8B - 10A$

32. Remove parentheses and simplify: $-5(7 - 8x) + 4(5x - 1)$

33. Write an equation equivalent to $4x - 7 = 13$ by adding 7 to both sides.

34. The solution for the equation $x + 8 = 8$ is 0. Which of the following equations are equivalent to $x + 8 = 8$?

a) $2x - 3 = 3$ **b)** $x - 19 = -19$

7.4 SOLVING EQUATIONS BY MULTIPLICATION OR DIVISION

Preview Test

1. Solve $4x = 24$
2. Solve $-2x = 33$
3. Solve $0.04S = 200$
4. Solve $\frac{x}{4} = -5$
5. Solve $\frac{2}{3}w = 6$
6. A computer printer can print 512 characters in 8 seconds. Find the printer's speed in characters per second.

When we solve the equation $5x = 20$ we are answering the question, "Five times what number equals twenty?" It is not difficult to conclude that the solution is 4, since $5(4) = 20$. A pencil and paper technique that can be used to solve $5x = 20$ involves division. In section 7.2 we said that if we divide both sides of an equation by the same number, the resulting equation is equivalent (has the same solution). By dividing both sides of the equation $5x = 20$ by 5 we get an equivalent equation, $x = 4$, whose solution is self-evident.

Let's think about how we are able to go from $5x = 20$ to $x = 4$. If we divide both sides of the equation by the coefficient of x, this will give us $(1)x$ or x on the left side.

$$5x = 20$$

Five divided by what number equals 1.
Answer: 5
Then divide both sides by 5.

$$\frac{5x}{5} = \frac{20}{5}$$
$$x = 4$$

Now x is isolated with a coefficient of 1 and the solution is self-evident. It's 4.

EXAMPLE 1: Solve $-3z = 24$

Solution:
Since $(-3) \div (-3) = 1$, divide both sides by -3.

$$\frac{-3z}{-3} = \frac{24}{-3}$$
$$z = -8$$

ANSWERS FOR PREVIEW TEST

1. 6 **2.** $-16\frac{1}{2}$ **3.** 5000 **4.** -20 **5.** 9

6. 64 characters per second

PRACTICE PROBLEM 1: Solve $-7y = -42$

EXAMPLE 2: Solve $7x = 10$

Solution:
Since $7 \div 7 = 1$, divide both sides by 7.

$$\frac{7x}{7} = \frac{10}{7}$$
$$x = 1\frac{3}{7}$$

PRACTICE PROBLEM 2: Solve $-9w = -43$

EXAMPLE 3: Solve $0.04S = 100$

Solution:
Since $0.04 \div 0.04 = 1$, divide both sides by 0.04.

$$\frac{0.04S}{0.04} = \frac{100}{0.04}$$
$$S = 2500$$

PRACTICE PROBLEM 3: Solve $0.15M = 9$

When we solve the equation $\frac{x}{5} = 8$ we are answering the question, "What number divided by 5 equals 8?" It is not difficult to see that the solution is 40, since $40 \div 5 = 8$. A pencil and paper technique that can be used to solve $\frac{x}{5} = 8$ involves multiplication. In section 7.2 we said that if we multiply both sides of an equation by the same number, the resulting equation is equivalent. By multiplying both sides of the equation $\frac{x}{5} = 8$ by 5 we get an equivalent equation, $x = 40$, whose solution is self-evident.

Since $\frac{x}{5}$ can be written as $\frac{1}{5}x$ $\left(\text{dividing by 5 is equivalent to multiplying by } \frac{1}{5}\right)$ the multiplier is dictated by the question, "What number times $\frac{1}{5}$ equals 1?" The answer to this question is 5. Therefore,

$$\frac{x}{5} = 8$$

Multiply both sides by 5

$$5\left(\frac{x}{5}\right) = 5(8)$$
$$x = 40$$

EXAMPLE 4: Solve $\frac{z}{6} = -3$

Solution:

Since $\frac{z}{6}$ can be written as $\frac{1}{6}z$, and $6\left(\frac{1}{6}\right) = 1$, multiply both sides of the equation by 6.

$$6\left(\frac{z}{6}\right) = 6(-3)$$
$$z = -18$$

PRACTICE PROBLEM 4: Solve $\frac{y}{100} = 1$

To solve the equation $\frac{3}{4}x = 12$, we ask the question, "$\frac{3}{4}$ times what number equals 1?" The answer to this question is $\frac{4}{3}$, which is called the **reciprocal** of $\frac{3}{4}$. So,

$$\frac{3}{4}x = 12$$

Multiply both sides by $\frac{4}{3}$

$$\frac{4}{3}\left(\frac{3}{4}\right)x = \frac{4}{3}(12)$$
$$x = 16$$

EXAMPLE 5: Solve $\frac{6}{11}w = -12$

Solution:

Multiply both sides by $\frac{11}{6}$.

$$\frac{11}{6}\left(\frac{6}{11}\right)w = \frac{11}{6}(-12)$$
$$w = -22$$

PRACTICE PROBLEM 5: Solve $\frac{2}{5}z = 6$

EXAMPLE 6: Solve $-\frac{3}{8}x = 5$

Solution:

Multiply both sides by $-\frac{8}{3}$.

$$-\frac{8}{3}\left(-\frac{3}{8}\right)x = -\frac{8}{3}(5)$$
$$x = -13\frac{1}{3}$$

PRACTICE PROBLEM 6: Solve $-\frac{5}{7}z = -\frac{3}{8}$

In Section 7.3 we identified a four-step procedure for solving problems through the solution of an equation. These four steps also apply to the types of equations we have been considering in this section. For your convenience we are repeating the four steps as we illustrate the solution in Example 7.

EXAMPLE 7: The cost of 7.5 gallons of gas is \$6.81. Find the price for one gallon of gas to the nearest tenth of a cent.

Solution:

Step 1: Identify what you are asked to find and assign a variable to this quantity.

The price for one gallon is to be found.

Let's assign the variable P to the price.

Step 2: Identify all given numerical information.

7.5 gallons are purchased.

The total cost is \$6.81.

Step 3: Write an equation using the variable from step 1 and the given information from step 2.

Since the number of gallons purchased times the price per gallon equals the total cost, we can write the equation

$$7.5P = 6.81$$

Step 4: Solve the equation and answer the question.

Since $7.5 \div 7.5 = 1$, divide both sides by 7.5.

$$\frac{7.5P}{7.5} = \frac{6.81}{7.5}$$

$$P = 0.908$$

So, the price per gallon is \$0.908.

PRACTICE PROBLEM 7: The 18 state scholarships won by the senior class at Rockland High School represents $\frac{2}{5}$ of the scholarships awarded to high school seniors in the county. How many scholarships were awarded to high school seniors in the county?

ANSWERS TO PRACTICE PROBLEMS

1. 6 **2.** $4\frac{7}{9}$ **3.** 60 **4.** 100 **5.** 15 **6.** $\frac{21}{40}$ **7.** 45

EXERCISE SET 7.4A

Skill Development

In questions 1 through 15, solve each equation.

1. $4w = 56$
2. $-3x = 12$
3. $6z = 6$
4. $-7y = 42$
5. $5x = -100$
6. $-3x = -40$
7. $0.15x = 60$
8. $0.6z = 12$
9. $0.05S = 3{,}000$
10. $\frac{x}{4} = 3$
11. $\frac{y}{-6} = -5$
12. $\frac{z}{9} = 1$
13. $\frac{3}{4}w = 6$
14. $-\frac{4}{7}x = 16$
15. $\frac{3}{8}z = \frac{2}{7}$

Applications

In questions 16 through 18, answer each question by writing and solving an equation.

16. The enrollment at Rollins University is 8400. This number is 1.2 times as large as the enrollment ten years ago. Find the enrollment ten years ago.

17. If the cost for one dozen video tapes is $14.04, find the cost for one tape.

18. The 35 points scored by Mary Harris for her college basketball team represent five-ninths of her team's points in the game. How many points did her team score?

For Some Extra Thought

19. Solve $2\%S = 500$

20. Solve $\frac{z}{5} = 0$

Review Problems for Maintaining Your Skills

21. Round 5,406,987 to the nearest hundred thousand.

22. Find the quotient and remainder if 127 is divided by 25.

23. $\frac{23}{40} - \frac{3}{10} =$

24. $3\frac{3}{5} \times 2\frac{1}{4} =$

25. Round 5.0378942 to the nearest hundred-thousandth.

26. To the nearest tenth of a percent, 14 is what percent of 51?

27. Use cross-multiplication to solve $\frac{B}{8} = \frac{3}{2}$

28. $(-5.68) - (-9.3) =$

29. $(-7)(-9)(-1) =$

30. Add like terms: $-9w - 5.4w + 9y$

31. Remove the parentheses and simplify: $-8(3 - 6H) + 10H$

32. Solve: $x - 17 = -25$

33. Solve: $w + 2.5 = 1$

EXERCISE SET 7.4B

Skill Development

In questions 1 through 15, solve each equation.

1. $6x = 54$ **2.** $8z = 72$ **3.** $-3y = -24$

4. $-w = 9$ **5.** $-4w = -7$ **6.** $5z = -75$

7. $0.1x = 8$ **8.** $0.04S = 800$ **9.** $0.25y = 0.1$

10. $\frac{x}{4} = 5$ **11.** $\frac{w}{7} = -3$ **12.** $\frac{y}{-6} = -4$

13. $\frac{2}{5}w = 20$ **14.** $-\frac{7}{8}z = -63$ **15.** $\frac{8}{9}w = \frac{3}{5}$

Applications

In questions 16 through 18, answer each question by writing and solving an equation.

16. Twenty-five credits times Mike's grade point average equals seventy-eight. Find Mike's grade point average.

17. The cost for 30 computer disks is \$43.80. What is the cost for one disk?

18. A current stock price of \$24 per share is two and one-half times the price per share six months ago. Find the price per share six months ago.

For Some Extra Thought

19. Solve $400\%z = 20$ **20.** Solve $x + x + x = 30$

Review Problems for Maintaining Your Skills

21. Round 6,874 to the nearest ten.

22. Find the quotient and remainder if 53 is divided by 16.

23. $\frac{14}{25} - \frac{3}{20} =$ **24.** $6\frac{4}{5} \times 10 =$

25. Round 0.04789 to the nearest hundredth.

26. To the nearest hundredth of a percent, 138 is what percent of 259?

27. Use cross-multiplication to solve $\frac{1.2}{w} = \frac{3}{5}$

28. $(-10.3) - (+6.5) =$ **29.** $(-5)(+2.6)(-4) =$

30. Add like terms: $-10A - 6B + 1.4A$

31. Remove the parentheses and simplify: $4(5x - 1) + 17$

32. Solve: $z - 13 = 21$ **33.** $y + 3.6 = -10.7$

7.5 SOLVING EQUATIONS BY MULTIPLE OPERATIONS

Preview Test

1. Solve $3x + 5 = 17$

2. Solve $4x + 3 = 7x - 2$

3. Solve $3(4x - 5) = 12$

4. Solve $-2(5x + 4) - (7 - 2x) = 9 + 3(x - 2)$

5. Solve $\frac{x}{2} + \frac{x}{5} = 6$

6. The invoice amount for four dozen copies of a book is $1200. If this amount includes $48 for shipping, find the price for each book.

In Sections 7.3 and 7.4 we solved equations by performing a single operation (addition, multiplication, or division) on both sides of the equation. In this section we will consider equations whose solution requires two or more operations. Consider $2x + 7 = 13$. We are asking, "Two times what number plus seven equals thirteen?" To solve this equation we handle the 7 and 2 the same way they would have been handled in sections 7.3 and 7.4. That is, we will add -7 to both sides of the equation and divide both sides of the equation by 2. But what should we do first, add -7 or divide by 2? Adding -7 first makes more sense because this operation leaves $2x$ intact. If we divide by 2 first then we have to divide 7 and 13 by 2, and we will have an equivalent equation with mixed numbers. So, adding -7 to each side we have

$$2x + 7 - 7 = 13 - 7$$
$$2x + 0 = 6$$
$$2x - 6$$

Now divide both sides by 2

$$\frac{2x}{2} = \frac{6}{2}$$
$$x = 3$$

We can check the solution, $x = 3$, by substituting 3 for x in the original equation.

$$2x + 7 = 13$$

Substitute 3 for x

$$2(3) + 7 = 13$$
$$6 + 7 = 13$$
$$13 - 13 \text{ is true}$$

ANSWERS FOR PREVIEW TEST

1. 4 **2.** $1\frac{2}{3}$ **3.** $2\frac{1}{4}$ **4.** $-\frac{2}{11}$ **5.** $8\frac{4}{7}$ **6.** $24

EXAMPLE 1: Solve $6x - 8 = 17$

Solution:
Add $+8$ to both sides.

$$6x - 8 + 8 = 17 + 8$$
$$6x = 25$$

Divide both sides by 6.

$$\frac{6x}{6} = \frac{25}{6}$$
$$x = 4\frac{1}{6}$$

PRACTICE PROBLEM 1: Solve $-4x + 1 = 29$

Let's consider the equation $8x + 3 = 6x - 17$. How does this equation differ from the equations in Example 1 and Practice Problem 1? The equation $8x + 3 = 6x - 17$ has a variable term on the left and right sides ($8x$ on the left side and $6x$ on the right side). Through an addition operation we can eliminate one of the variable terms from the equation. For example, if we add $-6x$ to both sides we have

$$8x - 6x + 3 = 6x - 6x - 17$$
$$2x + 3 = -17$$

Or, if we add $-8x$ to both sides we have

$$8x - 8x + 3 = 6x - 8x - 17$$
$$3 = -2x - 17$$

It does not matter whether we eliminate $6x$ or $8x$. The equations that result will have the same solution. Let's work with $2x + 3 = -17$.

Add -3 to both sides.

$$2x + 3 - 3 = -17 - 3$$
$$2x = -20$$

Divide both sides by 2.

$$\frac{2x}{2} = \frac{-20}{2}$$
$$x = -10$$

For practice show that $3 = -2x - 17$ has the solution, $x = -10$.

EXAMPLE 2: Solve $5x - 7 = -8x + 6$

Solution:
Add $+8x$ to both sides.

$$5x + 8x - 7 = -8x + 8x + 6$$
$$13x - 7 = 6$$

Add $+7$ to both sides.

$$13x - 7 + 7 = 6 + 7$$
$$13x = 13$$

Divide both sides by 13.

$$\frac{13x}{13} = \frac{13}{13}$$
$$x = 1$$

PRACTICE PROBLEM 2: Solve $-3x + 2 = 5x + 9$

In Section 7.1 you learned how to remove parentheses through multiplication. This simplification operation is sometimes required as one of the steps in the solution of an equation. This is illustrated in the next example.

EXAMPLE 3: Solve $5(6x - 3) = -17$

Solution:
Remove the parentheses.

$$5(6x) + 5(-3) = -17$$
$$30x - 15 = -17$$

Add $+15$ to both sides.

$$30x - 15 + 15 = -17 + 15$$
$$30x = -2$$

Divide both sides by 30.

$$\frac{30x}{30} = \frac{-2}{30}$$
$$x = -\frac{2}{30} \quad \text{or} \quad -\frac{1}{15}$$

PRACTICE PROBLEM 3: Solve $10 = -5(4 - 3x)$

After removing the parentheses, simplification of each side of the equation should be completed before performing any operation on both sides of the equation. That is, like terms and numbers that appear on the same side of the equation should be added. This is illustrated in Example 4.

EXAMPLE 4: Solve $3(4x - 5) - 7x = 20 - (3 - 2x)$

Solution:
Remove the parentheses.

$$(3)(4x) + (3)(-5) - 7x = 20 + (-1)(3) + (-1)(-2x)$$
$$12x - 15 - 7x = 20 - 3 + 2x$$

Simplify by adding like terms and numbers.

$$5x - 15 = 17 + 2x$$

Add $-2x$ to both sides.

$$5x - 2x - 15 = 17 + 2x - 2x$$
$$3x - 15 = 17$$

Add $+15$ to both sides.

$$3x - 15 + 15 = 17 + 15$$
$$3x = 32$$

Divide both sides by 3.

$$\frac{3x}{3} = \frac{32}{3}$$
$$x = 10\frac{2}{3}$$

PRACTICE PROBLEM 4: Solve $6 - 7(3 - 4x) = 2(x - 1) + 5(3x + 2)$

If an equation contains fractions we can clear the denominators in the first step by multiplying each term by the LCD. This is shown in Example 5.

EXAMPLE 5: Solve $\frac{4x}{5} + \frac{7x}{10} = -3$

Solution:
Multiply each term by 10, the LCD for 5 and 10.

$$(10)\left(\frac{4x}{5}\right) + (10)\left(\frac{7x}{10}\right) = (10)(-3)$$

Divide the first 10 by the denominator 5 and the second 10 by the denominator 10.

$$(2)(4x) + (1)(7x) = (10)(-3)$$

Multiply and add like terms.

$$8x + 7x = -30$$
$$15x = -30$$

Divide both sides by 15.

$$\frac{15x}{15} = \frac{-30}{15}$$
$$x = -2$$

PRACTICE PROBLEM 5: Solve $\frac{x}{2} - \frac{3x}{5} = 4$

EXAMPLE 6: Jane Laskey purchased 8 shares of stock for \$183. If the \$183 includes a \$23 brokerage commission, what did she pay for each share of stock?

Solution:
Step 1: Identify what you are to find and assign a variable to this quantity.

The price for one share of stock is to be found. Assign the variable P to the price.

Step 2: Identify all given numerical information.

8 shares are purchased.

\$183 is the total amount paid.

\$23 brokerage commission is included in the \$183.

Step 3: Write an equation using the variable from step 1 and the given information from step 2.

Since the number of shares purchased times the price per share plus \$23 equals \$183, we can write the equation

$$8P + 23 = 183$$

Step 4: Solve the equation and answer the question.
Add -23 to both sides.

$$8P + 23 - 23 = 183 - 23$$
$$8P = 160$$

Divide both sides by 8.

$$\frac{8P}{8} = \frac{160}{8}$$
$$P = 20$$

So, the price per share is \$20.

PRACTICE PROBLEM 6: John Perkin's total tuition bill for 8 credits is \$1045. If the \$1045 includes a student activities fee of \$45, what is the cost per credit?

ANSWERS FOR PRACTICE PROBLEMS

1. -7 **2.** $-\frac{7}{8}$ **3.** 2 **4.** $2\frac{1}{11}$ **5.** -40 **6.** \$125

EXERCISE SET 7.5A

Skill Development

In questions 1 through 18, solve each equation.

1. $4x - 5 = 19$

2. $-3x + 6 = 9$

3. $13 - 2x = 20$

4. $7x + 10 = 100$

5. $3x + 2 = 9x - 5$

6. $-8x - 1 = 4 + 3x$

7. $2x + 7 = -6x - 9$

8. $x - 7 = -6x + 1$

9. $5(2x - 7) = 10$

10. $-6(3 - 2x) = 4$

11. $6(2x + 1) = -20$

12. $-3(4x - 3) = 31$

13. $4 - 5(2x - 3) = 2x - 7(3 - 5x)$

14. $7(5x - 9) - (x - 2) = 3(4x - 5) + 16$

15. $\frac{x}{4} + \frac{x}{3} = 4$

16. $\frac{4x}{7} + \frac{x}{2} = -3$

17. $\frac{3x}{2} + \frac{5x}{8} = -6$

18. $\frac{9x}{8} - \frac{x}{3} = 2$

Applications

In questions 19 through 21, answer each question by writing and solving an equation.

19. The total cost for 15 tickets to a basketball game is \$162. This total includes a \$12 service charge. Find the price per ticket.

20. The Celsius temperature can be found by subtracting 32 from the Fahrenheit temperature and multiplying the difference by 5/9. What Fahrenheit temperature is associated with a Celsius temperature of 20 degrees?

21. A salesperson earns \$225 for each car she sells plus a monthly base salary of \$500. How many cars must the salesperson sell in one month in order to earn \$3650?

For Some Extra Thought

22. Solve $4(3 - (2 - 7x)) = 20$

23. Solve $x + 3 = x$

Review Problems for Maintaining Your Skills

24. Round 4,568 to the nearest thousand.

25. $4\frac{3}{4} + 1\frac{2}{5} =$

26. $10 - 3\frac{1}{3} =$

27. Divide 14.6 by 1.03 and round the quotient to the nearest hundredth.

28. Find 4.5% of \$16,000

29. 20% of what amount equals \$300?

30. $-8.5 + 4.6 + 1.2 - 17.3 =$

31. $(-24.6) - (+0.02) =$

32. Remove the parentheses and simplify: $-9(3 - 4x) + 5(2x - 7)$

33. Solve: $x + 13.2 = -14.9$

34. Solve: $z - 1 = -20$

35. Solve: $7x = -5.6$

36. Solve: $\frac{w}{8} = -6$

EXERCISE SET 7.5B

Skill Development

In questions 1 through 18, solve each equation.

1. $5x - 7 = 38$

2. $-2x + 9 = 15$

3. $5x - 8 = 0$

4. $20 - 6x = 13$

5. $9x - 2 = -3x + 4$

6. $7x - 2 = -3x + 4$

7. $x - 8 = -3x + 17$

8. $6x + 3 = 5x - 3$

9. $4(3x - 2) = -13$

10. $-2(5 - x) = 17$

11. $8(4x - 5) = 30$

12. $-4(7x + 2) = -1$

13. $8x - 7(3 - 2x) = 20 - (3x + 2)$

14. $8(3 - 5x) + (5x - 2) = -4(2x - 7) + 9x$

15. $\frac{x}{5} + \frac{x}{3} = 4$

16. $\frac{3x}{7} + \frac{5x}{2} = -2$

17. $\frac{2x}{3} - \frac{x}{6} = 5$

18. $\frac{x}{2} - \frac{3x}{10} = 1$

Applications

In questions 19 through 21, answer each question by writing and solving an equation.

19. Twenty-four cans of soup are packed in a carton. If the total weight of the full carton is 140 ounces and the empty carton weighs 20 ounces, find the weight of each can of soup.

20. Bob Richards purchases 200 shares of stock. If the total bill of $5,878 includes a brokerage commission of $178, what is the price for each share of stock?

21. The Fahrenheit temperature can be found by adding 32 to 9/5 times the Celsius temperature. What Celsius temperature is associated with a Fahrenheit temperature of 50 degrees?

For Some Extra Thought

22. Solve $5(20 - 2(3 - 4x)) = 0$

23. Solve $3(x - 2) = 3x + 6$

Review Questions for Maintaining Your Skills

24. Round 1,000,564 to the nearest hundred thousand.

25. $5\frac{2}{5} + 2\frac{1}{2} =$ **26.** $23 - 5\frac{3}{4} =$

27. Divide 3.6 by 5.1 and round the quotient to the nearest thousandth.

28. Find 12.75% of $8,000.

29. 1.5% of what amount equals $45?

30. $-20 + 3.4 - 6 =$ **31.** $(+26.5) - (-0.05) =$

32. Remove the parentheses and simplify: $-4(3 - 2z) - 7(5 + 4z)$

33. Solve: $w - 17 = -13$ **34.** Solve: $y - 6.25 = 7.25$

35. Solve: $-5x = 85$ **36.** Solve: $\frac{w}{2} = -3$

PRACTICE TEST 7A

1. Find the number of terms: $(3x - 5) + 14$
2. Remove the parentheses and simplify by combining like terms: $-5(4 - 3x) + 7(6x - 2)$
3. Write in words: $4A - 5B + 2$
4. The solution for the equation $4x - 5 = 3x + 2$ is 7. Which of the following equations are equivalent to $4x - 5 = 3x + 2$?

   a) $-2x = -14$ **b)** $3(2x + 6) = x - 17$
5. Solve: $w - 0.07 = 1.5$
6. Solve: $-5x = 45$
7. Solve: $\frac{3}{8}x = 12$
8. Solve: $5x - 7 = 3x + 19$
9. Solve: $4(x - 3) - (4 - 2x) = 7x + 3$
10. Mark Price purchased 12 tickets for a baseball game. The total cost of \$152 includes a \$2.00 service charge. Write and solve an equation to find the cost for each ticket.

PRACTICE TEST 7B

1. Find the number of terms: $\frac{3w + 7}{9} - 5$

2. Remove the parentheses and simplify by combining like terms: $4(5x - 2) - 6(5 - 4x)$

3. Write in words: $7x - 3w + 5$

4. The solution for the equation $5x - 3 = 7x + 9$ is -6. Which of the following equations is equivalent to $5x - 3 = 7x + 9$?

   a) $7x - 9 = 8x + 3$ **b)** $4(5 + x) = x + 2$

5. Solve: $A + 100 = 100$

6. Solve: $-0.02w = 92$

7. Solve: $\frac{2}{5}z = \frac{3}{8}$

8. Solve: $-7x + 2 = 3x - 8$

9. Solve: $3(1 - x) + 9x = 4 - (5x + 1)$

10. The 14 cars that Mary Scott sold last week represents two-fifths of the sales for the car dealership. Write and solve an equation to find the number of cars sold by the car dealership.

CUMULATIVE REVIEW EXERCISES FOR CHAPTERS 1 THROUGH 7

1. Write 1,203,500 in words.

2. Round 1,506,700 to the nearest thousand.

In 3 through 6, reduce each fraction to its lowest terms and express any result that is an improper fraction as a mixed number.

3. $4\frac{1}{3} \times 5\frac{1}{4} =$ 4. $\frac{10}{13} \div \frac{1}{3} =$

5. $3\frac{3}{4} + 2\frac{1}{5} =$ 6. $\frac{15}{16} - \frac{8}{15} =$

7. Round 1.20378 to the nearest ten-thousandth.

8. Round \$1.23804 to the nearest cent.

9. $3.04 \times 10{,}000 =$

10. Divide 42.6 by 1.7 and round the quotient to the nearest hundredth.

11. $5.2 + 1.6 \times 10 =$

12. Convert 5.45% to a decimal.

13. Convert 0.025 to a percent.

14. Convert 36% to a fraction reduced to its lowest terms.

15. \$2.50 is what percent of \$50?

16. Find 45% of \$1,500.

17. 4% of what amount equals \$5.20?

18. The price of a computer increases from \$640 to \$679. Find the percent increase to the nearest tenth of a percent.

19. Solve the proportion $\frac{5}{2} = \frac{R}{7}$

20. Insert the appropriate symbol, $<$ or $>$, in the blank.
   a) -7 ____ -3 b) 5.4 ____ 5.07

21. Find each sum.
   a) $-4 + 9 - 3 =$ b) $(-4.02) + (-6.73) =$

22. Find each difference.
   a) $(-19) - (-23) =$ b) $(-0.05) - (1.032) =$

23. Find each product.
   a) $(-4) \times (-7) =$ b) $(-1.2)(\ 6.4)(-10) =$

24. Find each quotient.
   a) $(-120) \div (+15) =$ b) $\left(-1\frac{1}{2}\right) \div \left(-2\frac{1}{2}\right) =$

25. $(-7) \times (-9) + (-5) \times (10) =$

26. Find the number of terms: $3A - 2B + 8$

27. Write in words: $2x - 7w$

28. The solution for the equation $2x = 12$ is 6. Which of the following equations are equivalent to $2x = 12$?

a) $x - 8 = 14$ **b)** $2x - 1 = 13$

29. Solve: $x - 9 = 17$

30. Solve: $4x = -24$

31. Solve: $-3w - 10 = -30$

32. Solve: $4(10 - 2x) = x + 13$

33. The total cost for 24 tires is \$1,400. If the \$1,400 includes a \$56 delivery charge, find the cost for each tire.

C H A P T E R 8

POWERS AND ROOTS

8.1 FINDING POWERS

8.2 PROPERTIES OF EXPONENTS

8.3 SCIENTIFIC NOTATION

8.4 FINDING SQUARE ROOTS

The multiplication problem

$$3 \times 3 \times 3 \times 3 \times 3 \times 3 \times 3$$

can be written in a form called **exponential form** as 3^7. In this notation 3 is called the **base** and 7 is called the **exponent**. The exponent tells us how many times the base appears as a factor. So, 3^7 (which is verbalized as "three to the seventh power") identifies a multiplication of seven factors of three.

In this chapter we will operate with numbers written in exponential form. We will also find a number, called a **square root**, whose second power is identified. For example, the square roots of 25 are 5 and -5 because $5^2 = 25$ and $(-5)^2 = 25$.

8.1 FINDING POWERS

Preview Test

1. Write $9 \times 9 \times 9 \times 9 \times 9 \times 9 \times 9 \times 9$ in exponential form.
2. Write $(-5)^7$ as a product of factors.

In questions 3 through 12, evaluate each term or expression.

3. $4^2 =$ **4.** $6^3 =$ **5.** $1^4 =$

6. $(-7)^2 =$ **7.** $(-10)^3 =$ **8.** $(-1)^5 =$

9. $5^2 - 3^2 =$ **10.** $2^3 \cdot 3^2 - 4^2 =$

11. $6^3 - 3^2 - 5^3 =$ **12.** $10^2 - (3^2 + 4^2) =$

In general, if n is a positive integer

$$b^n = \underbrace{b \times b \times b \times \ldots \times b}_{\text{a product of } n \text{ factors of } b}$$

EXAMPLE 1: Write each in exponential form.

a) $4 \times 4 \times 4 \times 4$

b) $(-2) \times (-2) \times (-2) \times (-2) \times (-2) \times (-2)$

c) $0.3 \times 0.3 \times 0.3 \times 0.3 \times 0.3 \times 0.3 \times 0.3 \times 0.3$

Solution:

a) 4^4 **b)** $(-2)^6$ **c)** 0.3^8

PRACTICE PROBLEM 1: Write each in exponential form.

a) $7 \times 7 \times 7$

b) $10 \times 10 \times 10 \times 10 \times 10 \times 10 \times 10 \times 10$

c) $(-1) \times (-1) \times (-1) \times (-1)$

ANSWERS FOR PREVIEW TEST

1. 9^8 **2.** $(-5) \times (-5) \times (-5) \times (-5) \times (-5) \times (-5) \times (-5)$

3. 16 **4.** 216 **5.** 1 **6.** 49 **7.** $-1{,}000$

8. -1 **9.** 16 **10.** 56 **11.** 82 **12.** 75

EXAMPLE 2: Write each as a product of factors.

a) 6^5 **b)** $(-5)^2$ **c)** 12^9

Solution:

a) $6 \times 6 \times 6 \times 6 \times 6$

b) $(-5) \times (-5)$

c) $12 \times 12 \times 12 \times 12 \times 12 \times 12 \times 12 \times 12 \times 12$

PRACTICE PROBLEM 2: Write each as a product of factors.

a) 8^3 **b)** 100^5 **c)** $(-8)^4$

To find the value of a term that is written in exponential form it is helpful to first write it as a product of factors. This is illustrated in Example 3.

EXAMPLE 3: Find the value of each.

a) 7^2 **b)** $(-4)^3$ **c)** 10^4

Solution:

a) $7^2 = 7 \times 7 = 49$

b) $(-4)^3 = (-4) \times (-4) \times (-4) = -64$

c) $10^4 = 10 \times 10 \times 10 \times 10 = 10{,}000$

PRACTICE PROBLEM 3: Find the value of each.

a) 8^3 **b)** $(-1)^9$ **c)** 5^4

The rules for order of operations can be extended to exponents. All exponential terms and factors must be evaluated before carrying out any other operations. Then the rules stated in previous chapters in this book for order of operations should be followed. Consider

$$5^3 - 6 \times 4^2.$$

Evaluating 5^3 and 4^2 first, we have $125 - 6 \times 16$. Following the order of operations, we multiply, and then subtract, and get

$$125 - 6 \times 16 = 125 - 96 = 29.$$

EXAMPLE 4: Find the value for $10^2 - (2^3 + 5^2 \cdot 3)$

Solution:

Begin by evaluating 10^2, 2^3, and 5^2.

$$100 - (8 + 25 \cdot 3)$$

Now multiply 25 times 3 and add the product to 8.

$$= 100 - (8 + 75)$$

$$= 100 - 83$$

Then subtract.

$$= 17$$

PRACTICE PROBLEM 4: Find the value for $40 + (5^3 - 3 \times 2^4)$

If we deposit P dollars in a savings account paying interest at a yearly rate of r cents per dollar, the amount A in the account after t years is given by the formula

$$A = P(1 + r)^t$$

EXAMPLE 5: A deposit of \$800 is made in an account paying annual interest at a rate of 6 cents per dollar. Find the amount in the account at the end of 3 years. Round your result to the nearest cent.

Solution:
Substitute \$800 for P, 0.06 for r, and 3 for t in the formula $A = P(1 + r)^t$.

$$A = 800(1 + 0.06)^3 = 800(1.06)^3$$

Using the rule for order of operations, we have to find the value of $(1.06)^3$ first, then multiply that result by 800.

$$(1.06)^3 = (1.06) \times (1.06) \times (1.06) = 1.191016$$

Then,

$$A = 800(1.191016) = \$952.8128$$
$$= \$952.81 \text{ rounded to the nearest cent}$$

PRACTICE PROBLEM 5: A deposit of \$2,000 is made in an account paying annual interest at a rate of 8 cents per dollar. Find the amount in the account after 2 years. Round your result to the nearest cent.

ANSWERS FOR PRACTICE PROBLEMS

1. **a)** 7^3 **b)** 10^8 **c)** $(-1)^4$

2. **a)** $8 \times 8 \times 8$ **b)** $100 \times 100 \times 100 \times 100 \times 100$

c) $(-8) \times (-8) \times (-8) \times (-8)$

3. **a)** 512 **b)** -1 **c)** 625 **4.** 117 **5.** \$2,332.80

EXERCISE SET 8.1A

Skill Development

In questions 1 through 4, write each product in exponential form.

1. $5 \times 5 \times 5 \times 5$

2. $8 \times 8 \times 8 \times 8 \times 8 \times 8 \times 8$

3. $(-4) \times (-4) \times (-4)$

4. $(1.01) \times (1.01)$

In questions 5 through 8, write each exponential expression as a product of factors.

5. 6^4

6. 2^7

7. $(-3)^5$

8. $(1.02)^3$

In questions 9 through 20, evaluate each term or expression.

9. 4^3

10. 7^2

11. 1^9

12. 6^4

13. $(-2)^5$

14. $(-1)^9$

15. $(0.2)^3$

16. $(-0.7)^3$

17. $8^3 - 5^3$

18. $10 \cdot 2^5 - 7^4$

19. $20 - (3^2 - 4^2)$

20. $13^2 - (12^2 + 5^2)$

Applications

21. A deposit of $450 is made in an account paying annual interest at a rate of 7 cents per dollar. Find the amount in the account after 3 years. Round your result to the nearest cent.

22. The grade on a college algebra exam is related to the number of hours of study by the formula $y = x^2 - 2x + 40$. If a student studies 8 hours, what is the expected grade?

23. A rocket built by the members of a model rocket club is launched upward from ground level at time $t = 0$ seconds. The distance h above the ground after t seconds is given by the formula $h = -16t^2 + 112t$. Find the distance above the ground after 3 seconds if h is measured in feet.

For Some Extra Thought

24. $(-1)^{99}$ plus what number equals zero?

25. Refer to question 23 and determine after how many seconds the rocket strikes the ground.

Review Problems for Maintaining Your Skills

26. Write 2,305,900 in words.

27. Divide 241 by 100 and find the quotient and remainder.

28. $\frac{17}{25} - \frac{3}{50} =$

29. $4\frac{3}{4} \times 2\frac{1}{5} =$

30. Write 5.0267 in words.

31. To the nearest hundredth of a percent, 13 is what percent of 19?

32. 2.5% of what amount equals $16?

33. $(-1.29) - (+3.4) =$

34. $(-34)(-2)(-2) =$

In 35 through 40, solve each equation.

35. $x + 9 = -17$

36. $-3w = 5.4$

37. $4x + 3 = 35$

38. $-2z + 9 = -27$

39. $x + 9 = -4x - 41$

40. $5(1 - 3x) = 35$

EXERCISE SET 8.1B

Skill Development

In questions 1 through 4, write each product in exponential form.

1. $3 \times 3 \times 3 \times 3 \times 3 \times 3 \times 3$
2. $10 \times 10 \times 10 \times 10$
3. $(-5) \times (-5) \times (-5) \times (-5)$
4. $(1.06) \times (1.06) \times (1.06)$

In questions 5 through 8, write each exponential expression as a product of factors.

5. 4^3 **6.** 8^2 **7.** $(-2)^5$ **8.** $(1.01)^2$

In questions 9 through 20, evaluate each term or expression.

9. 5^4 **10.** 4^3 **11.** 0^6 **12.** 10^3

13. $(-3)^5$ **14.** $(-1)^7$ **15.** $(0.3)^2$ **16.** $(-0.3)^3$

17. $6^3 - 1^5$ **18.** $8 \cdot 2^2 - 5 \cdot 3^2$

19. $40 - (4^2 + 3^2)$ **20.** $3^2 - (5^2 - 4^2)$

Applications

21. A deposit of \$700 is made in an account paying annual interest at a rate of 7.5 cents per dollar. Find the amount in the account after 2 years. Round your result to the nearest cent.

22. The relationship between the net profit P of a new product and the amount x spent on advertising is given by the formula $P = 10{,}000x - x^2$. If \$2,000 is spent in advertising, find the net profit.

23. A projectile is fired straight upward at time $t = 0$ seconds. The distance h in feet above the ground after t seconds is given by the formula $h = -16t^2 + 800t$. Find the distance above the ground after 10 seconds.

For Some Extra Thought

24. $(0)^5$ plus what number equals 2?
25. Refer to question 22 and find the amount greater than zero spent on advertising that will make the net profit zero.

Review Problems for Maintaining Your Skills

26. Write 40,005 in words.

27. Divide 371 by 5 and find the quotient and remainder.

28. $\frac{5}{8} - \frac{1}{10} =$

29. $1\frac{2}{5} \times 10\frac{1}{2} =$

30. Write 0.00504 in words.

31. To the nearest tenth of a percent, 2 is what percent of 30?

32. 400% of what amount equals \$28?

33. $(-2.57) - (-1.8) =$

34. $(-6)(-5)(+6)(-2) =$

In 35 through 40, solve each equation.

35. $x - 4 = -15$

36. $15w = -7.5$

37. $6x - 7 = -55$

38. $8y + 10 = 10$

39. $5x - 3 = 3x + 13$

40. $4(3x - 2) = 4$

8.2 PROPERTIES OF EXPONENTS

Preview Test

In questions 1 through 8, find the value of each.

1. $(-3)^2(-3)^3 =$ **2.** $(2^3)^2 =$ **3.** $\frac{7^8}{7^6} =$

4. $5^{-2} =$ **5.** $3^{-3} \cdot 3^7 =$ **6.** $8^5 \cdot 8^{-5} =$

7. $(3 \cdot 2)^3 =$ **8.** $\left(\frac{2}{3}\right)^3 =$ **9.** $4^0 =$

The expression $2^3 \cdot 2^4$ can be evaluated two different ways. We can use the rule in Section 8.1 for order of operations and first evaluate each exponential factor. Then

$$2^3 \cdot 2^4 = 8 \cdot 16 = 128$$

Another approach we can use is to recognize that 2^3 represents a product of 3 factors of 2 and 2^4 represents a product of 4 factors of 2. Therefore, $2^3 \cdot 2^4$ represents a product of $3 + 4$ or 7 factors of 2. That is, if the bases are the same we can simplify the expression by adding the exponents.

$$\underbrace{2^3 \cdot 2^4}_{\text{The bases are the same.}} = \overbrace{2^{3+4}}^{\text{Add the exponents.}} = 2^7$$

You may ask, "What have we accomplished by writing $2^3 \cdot 2^4$ as 2^7? Is 2^7 easier to evaluate than $2^3 \cdot 2^4$?" Well, not really, unless you have memorized that $2^7 = 128$, or you have a calculator that can be used to evaluate 2^7. However, our main reason for pointing out that you can add exponents if the bases are equal is to introduce you to the first of five properties of exponents. Also, as you progress through the material in this section you will find that in some cases a property of exponents makes an evaluation easier (See Example 3).

The rule for adding exponents is stated below.

Property 1 for Exponents

If we are multiplying two exponential factors with the same base, add the exponents.

$$a^m \cdot a^n = a^{m+n}$$

EXAMPLE 1: Evaluate after using Property 1 to simplify.

a) $3^4 \cdot 3^2$ **b)** $(-2)^3(-2)$

ANSWERS FOR PREVIEW TEST

1. -243 **2.** 64 **3.** 49 **4.** $\frac{1}{25}$ **5.** 81

6. 1 **7.** 216 **8.** $\frac{8}{27}$ **9.** 1

Solution:

a) $3^4 \cdot 3^2 = 3^{4+2} = 3 \cdot 3 \cdot 3 \cdot 3 \cdot 3 \cdot 3 = 729$

b) Since there is no exponent after the second factor of -2, an exponent of 1 is assumed.

$$(-2)^3(-2) = (-2)^3(-2)^1 = (-2)^{3+1}$$
$$= (-2)(-2)(-2)(-2) = 16$$

PRACTICE PROBLEM 1: Evaluate after using Property 1 to simplify.

a) $5^4 \cdot 5$ **b)** $(-4)^2(-4)^3$

> **Be Careful About This**
>
> When you are using Property 1, do not multiply the bases. That is, $3^4 \cdot 3^3$ is not equal to 9^7, but is equal to 3^7.

For the expression $(2^3)^2$ the 2 outside the parentheses tells us that 2^3 appears twice as a factor.

$$(2^3)^2 = 2^3 \cdot 2^3$$

2^3 appears twice as a factor.

We can use Property 1 to write $2^3 \cdot 2^3$ as 2^{3+3} or 2^6. However, since $3 + 3 = 2 \cdot 3$, we can reduce the procedure to one step if we multiply the exponent 3 by 2. So,

$$(2^3)^2 = 2^{3 \cdot 2} = 2^6$$

What we have done in this specific example is stated in general as Property 2.

> **Property 2 for Exponents**
>
> If we are raising an exponential term to a power, multiply the exponents.
>
> $$(a^m)^n = a^{m \cdot n}$$

EXAMPLE 2: Evaluate after using Property 2 to simplify.

a) $(10^2)^4$ **b)** $(3^3)^2$

Solution:

a) $(10^2)^4 = 10^{2 \cdot 4} = 10^8 = 10 \cdot 10 \cdot 10 \cdot 10 \cdot 10 \cdot 10 \cdot 10 \cdot 10 = 100{,}000{,}000$

b) $(3^3)^2 = 3^{3 \cdot 2} = 3^6 = 3 \cdot 3 \cdot 3 \cdot 3 \cdot 3 \cdot 3 = 729$

PRACTICE PROBLEM 2: Evaluate after using Property 2 to simplify.

a) $(4^2)^2$ **b)** $(1^6)^5$

Consider $\frac{2^7}{2^4}$. If we evaluate the numerator and denominator separately, $2^7 = 2 \cdot 2 \cdot 2 \cdot 2 \cdot 2 \cdot 2 \cdot 2 = 128$ and $2^4 = 2 \cdot 2 \cdot 2 \cdot 2 = 16$. Then, $\frac{128}{16} = 8$. However, we can save some time by dividing out (cancelling out) the common factors of 2.

$$\frac{2^7}{2^4} = \frac{\cancel{2} \cdot \cancel{2} \cdot \cancel{2} \cdot \cancel{2} \cdot 2 \cdot 2 \cdot 2}{\cancel{2} \cdot \cancel{2} \cdot \cancel{2} \cdot \cancel{2}} = 2 \cdot 2 \cdot 2 = 8$$

We can even save more time if we recognize that for $\frac{2^7}{2^4}$ the four factors of 2 in the denominator will divide out 4 of the 7 factors of 2 in the numerator, leaving us with 3 factors of 2 in the numerator. The cancelling process can then be carried out by subtracting the exponents.

$$\frac{2^7}{2^4} = 2^{7-4} = 2^3$$

This suggests the third property of exponents.

Property 3 for Exponents

If we are dividing two exponential terms with the same base, subtract the exponent in the denominator from the exponent in the numerator.

$$\frac{a^m}{a^n} = a^{m-n}$$

EXAMPLE 3: Evaluate after using Property 3 to simplify.

a) $\frac{3^5}{3^3}$ **b)** $\frac{(-2)^8}{(-2)^7}$

Solution:

a) $\frac{3^5}{3^3} = 3^{5-3} = 3^2 = 3 \cdot 3 = 9$

b) $\frac{(-2)^8}{(-2)^7} = (-2)^{8-7} = (-2)^1 = -2$

After studying Example 3, you can see the advantage of using Property 3 for the evaluation. In part (a) if we had evaluated the numerator and denominator separately, we would have carried out the following steps.

$$3^5 = 3 \cdot 3 \cdot 3 \cdot 3 \cdot 3 = 243, \quad 3^3 = 3 \cdot 3 \cdot 3 = 27$$

$$\frac{3^5}{3^3} = \frac{243}{27} = 9$$

PRACTICE PROBLEM 3: Evaluate after using Property 3 to simplify.

a) $\frac{5^{10}}{5^7}$ **b)** $\frac{(-3)^6}{(-3)^2}$

What happens if we apply Property 3 to a case where the exponents are equal? Let's consider $\frac{5^2}{5^2}$. If we subtract the exponents we get $5^{2-2} = 5^0$. What does 5^0 equal? If we return to $\frac{5^2}{5^2}$ and evaluate it by calculating the numerator and denominator separately, we get $\frac{5^2}{5^2} = \frac{5 \cdot 5}{5 \cdot 5} = \frac{25}{25} = 1$. Therefore, 5^0 must equal 1. In fact, any nonzero number raised to the power zero equals 1.

Property 4 for Exponents

If the base is not equal to zero, $a^0 = 1$

EXAMPLE 4: Use Property 4 to evaluate each.

a) 7^0 **b)** $(-3)^0$

Solution:

a) $7^0 = 1$ **b)** $(-3)^0 = 1$

PRACTICE PROBLEM 4: Use Property 4 to evaluate each.

a) $(-1)^0$ **b)** 17^0

Let's now apply Property 3 to a case where the exponent in the numerator is smaller than the exponent in the denominator. Consider $\frac{2^3}{2^7}$. If we apply Property 3 and subtract 7 from 3 we have $2^{3-7} = 2^{-4}$. What does 2^{-4} equal? If we divide out the common factors of 2, we have

$$\frac{2^3}{2^7} = \frac{\cancel{2} \cdot \cancel{2} \cdot \cancel{2}}{\cancel{2} \cdot \cancel{2} \cdot \cancel{2} \cdot 2 \cdot 2 \cdot 2 \cdot 2} = \frac{1}{2^4}.$$

Therefore, 2^{-4} must equal $\frac{1}{2^4}$. This suggests a fifth property of exponents.

Property 5 for Exponents

A base raised to the power $-n$ equals 1 divided by the base raised to the power n.

$$a^{-n} = \frac{1}{a^n}$$

Be Careful About This

There is no connection between the sign of the exponent and the sign of the number. That is, 2^{-3} is not equal to -8.

EXAMPLE 5: Evaluate after using Property 5 to simplify.

a) 4^{-3} **b)** 10^{-2}

Solution:

a) $4^{-3} = \frac{1}{4^3} = \frac{1}{4 \cdot 4 \cdot 4} = \frac{1}{64}$

b) $10^{-2} = \frac{1}{10^2} = \frac{1}{10 \cdot 10} = \frac{1}{100}$

PRACTICE PROBLEM 5: Evaluate after using Property 5 to simplify.

a) 2^{-5} **b)** $(-3)^{-3}$

We are now in a position to use Properties 1 through 5 to simplify and evaluate more complicated expressions.

EXAMPLE 6: Use Properties 1 through 5 to simplify and evaluate each.

a) $3^{-4} \cdot 3^7$ **b)** $(5^{-2})^2$ **c)** $7^{-6} \cdot 7^6$

Solution:

a) $3^{-4} \cdot 3^7 = 3^{-4+7} = 3^3 = 3 \cdot 3 \cdot 3 = 27$

b) $(5^{-2})^2 = 5^{(-2)(2)} = 5^{-4} = \frac{1}{5^4} = \frac{1}{5 \cdot 5 \cdot 5 \cdot 5} = \frac{1}{625}$

c) $7^{-6} \cdot 7^6 = 7^{-6+6} = 7^0 = 1$

PRACTICE PROBLEM 6: Use Properties 1 through 5 to simplify and evaluate each.

a) $5^4 \cdot 5^{-5}$ **b)** $(4^3)^{-2}$ **c)** $6^{-1} \cdot 6$

Two additional properties of exponents relate to the distribution of a power over multiplication or division.

Property 6 for Exponents:

$$(a \cdot b)^n = a^n \cdot b^n$$

Property 7 for Exponents:

$$\left(\frac{a}{b}\right)^n = \frac{a^n}{b^n}$$

EXAMPLE 7: Use Properties 6 or 7 to simplify and evaluate each.

a) $(3 \cdot 4)^3$ **b)** $\left(\frac{3}{4}\right)^2$

Solution:

a) $(3 \cdot 4)^3 = 3^3 \cdot 4^3 = (3 \cdot 3 \cdot 3)(4 \cdot 4 \cdot 4) = 27 \cdot 64 = 1728$

b) $\left(\frac{3}{4}\right)^2 = \frac{3^2}{4^2} = \frac{3 \cdot 3}{4 \cdot 4} = \frac{9}{16}$

PRACTICE PROBLEM 7: Use Properties 6 and 7 to simplify and evaluate each.

a) $(5 \cdot 2)^4$ **b)** $\left(\frac{1}{2}\right)^3$

ANSWERS TO PRACTICE PROBLEMS

1. a) 3,125 **b)** $-1{,}024$ **2. a)** 256 **b)** 1

3. a) 125 **b)** 81 **4. a)** 1 **b)** 1

5. a) $\frac{1}{32}$ **b)** $-\frac{1}{27}$ **6. a)** $\frac{1}{5}$ **b)** $\frac{1}{4{,}096}$ **c)** 1

7. a) 10,000 **b)** $\frac{1}{8}$

EXERCISE SET 8.2A

Skill Development

In questions 1 through 24, evaluate after using the appropriate properties to simplify.

1. $4^2 \cdot 4^4$
2. $10^5 \cdot 10$
3. $(-6)^3(-6)^2$
4. $(3^3)^2$
5. $(5^2)^2$
6. $(1^5)^{10}$
7. $\dfrac{3^9}{3^5}$
8. $\dfrac{10^5}{10^4}$
9. $\dfrac{2^8}{2^2}$
10. 6^0
11. $(-7)^0$
12. $(-1)^0$
13. 7^{-2}
14. $(-5)^{-3}$
15. 3^{-1}
16. $5^{-6} \cdot 5^8$
17. $(-2)^4(-2)^{-5}$
18. $10^{-6} \cdot 10^6$
19. $(2 \cdot 6)^2$
20. $(10 \cdot 2)^3$
21. $(3 \cdot 5)^{-2}$
22. $\left(\dfrac{2}{5}\right)^3$
23. $\left(\dfrac{1}{3}\right)^5$
24. $\left(\dfrac{3}{4}\right)^4$

For Some Extra Thought

25. 3^{-5} times what equals 1?
26. Evaluate $(((2)^2)^2)^2$

Review Problems for Maintaining Your Skills

27. Round 1,236 to the nearest ten.
28. $\dfrac{13}{15} + \dfrac{2}{25} =$
29. $4\dfrac{2}{7} - 3 =$
30. Divide 17.2 by 1.06 and round the quotient to the nearest hundredth.
31. Find 17.4% of 3,500.
32. Find the percent increase to the nearest tenth of a percent if a price increases from \$1.59 to \$1.83.
33. $-9.6 - 15.4 + 8.36 =$
34. $(-284) \div (-0.4) =$
35. Solve: $3x - 1 = -25$
36. Solve: $7(4 - 3x) = 7x$
37. Write $5 \times 5 \times 5 \times 5 \times 5 \times 5$ in exponential form.
38. Write $(-7)^4$ as a product of factors.
39. Evaluate 2^6
40. Evaluate $(-3)^5$

EXERCISE SET 8.2B

Skill Development

In questions 1 through 24, evaluate after using the appropriate properties to simplify.

1. $6^3 \cdot 6^2$ **2.** $9^3 \cdot 9$ **3.** $(-4)^2(-4)^2$

4. $(5^2)^3$ **5.** $(2^5)^2$ **6.** $(0^6)^3$

7. $\dfrac{5^{10}}{5^8}$ **8.** $\dfrac{2^{12}}{2^{11}}$ **9.** $\dfrac{10^5}{10}$

10. 8^0 **11.** 1^0 **12.** $(-3)^0$

13. 6^{-1} **14.** $(-2)^{-7}$ **15.** 9^{-2}

16. $8^5 \cdot 8^{-7}$ **17.** $10^{-12} \cdot 10^{12}$ **18.** $(-3)^9(-3)^{-5}$

19. $(3 \cdot 4)^2$ **20.** $(5 \cdot 6)^{-1}$ **21.** $(4 \cdot 2)^3$

22. $\left(\dfrac{1}{2}\right)^5$ **23.** $\left(\dfrac{3}{7}\right)^2$ **24.** $\left(\dfrac{5}{6}\right)^3$

For Some Extra Thought

25. 4 raised to what power equals $\dfrac{1}{64}$?

26. Evaluate $(((2)^{-1})^{-2})^3$

Review Problems for Maintaining Your Skills

27. Round 2,478,621 to the nearest ten thousand.

28. $\dfrac{3}{4} + \dfrac{10}{11} =$ **29.** $20 - 7\dfrac{1}{2} =$

30. Divide 23.71 by 2.8 and round the quotient to the nearest hundredth.

31. Find 125% of \$44.

32. Find the percent decrease to the nearest tenth of a percent if a price decreases from \$12.38 to \$11.03

33. $-12.5 + 3.78 - 2.4 =$ **34.** $(-36) \div (+1.2) =$

35. Solve: $-2w + 3 = -39$ **36.** Solve: $5(2x - 3) = 7x + 6$

37. Write $(-2) \times (-2) \times (-2) \times (-2)$ in exponential form.

38. Write 9^5 as a product of factors.

39. Evaluate 5^4 **40.** Evaluate $(-10)^3$

8.3 SCIENTIFIC NOTATION

Preview Test

In questions 1 through 4, write each number in scientific notation.

1. 78,200 **2.** 9,000,000,000

3. 0.00052 **4.** 0.00000842

In questions 5 through 8, write each as a whole number or decimal.

5. 1.7×10^4 **6.** 7.83×10^6

7. 4.1×10^{-3} **8.** 8.95×10^{-1}

In questions 9 and 10, multiply and express your result in scientific notation.

9. $(9.0 \times 10^3) \times (4.1 \times 10^{-5})$ **10.** $(2.1 \times 10^9) \times (6 \times 10^{-2})$

In questions 11 and 12, divide and express your result in scientific notation.

11. $\dfrac{2.4 \times 10^9}{3 \times 10^4}$ **12.** $\dfrac{1.5 \times 10^{-3}}{4.0 \times 10^{-8}}$

Consider the following multiplication exercise. Start with 3.4 and multiply by 10, 100, 1,000, and 10,000.

$$3.4 \times 10 = 34$$
$$3.4 \times 100 = 340$$
$$3.4 \times 1{,}000 = 3{,}400$$
$$3.4 \times 10{,}000 = 34{,}000$$

Each multiplier, 10, 100, 1,000 and 10,000, can be written as a power of 10.

$$10 = 10^1$$
$$100 = 10 \cdot 10 = 10^2$$
$$1{,}000 = 10 \cdot 10 \cdot 10 = 10^3$$
$$10{,}000 = 10 \cdot 10 \cdot 10 \cdot 10 = 10^4$$

Let's rewrite the above multiplication of 3.4 using the powers of 10.

$$3.4 \times 10^1 = 34$$
$$3.4 \times 10^2 = 340$$
$$3.4 \times 10^3 = 3{,}400$$
$$3.4 \times 10^4 = 34{,}000$$

ANSWERS FOR PREVIEW TEST

1. 7.82×10^4 **2.** 9.0×10^9 **3.** 5.2×10^{-4} **4.** 8.42×10^{-6}

5. 17,000 **6.** 7,830,000 **7.** 0.0041 **8.** 0.895

9. 3.69×10^{-1} **10.** 1.26×10^8 **11.** 8.0×10^4 **12.** 3.75×10^4

Notice that if we start with 3.4, the positive integer exponent of 10 tells us how many places to move the decimal point to the right. For 3.4×10^4 we have

$$3.4 \times 10^4 = 3.4000. = 34{,}000$$

Move the decimal point 4 places to the right and fill in the empty spaces with zeros.

We can drop the decimal point since it is on the right of the whole number.

Now consider the following multiplication exercise. Start with 3.4 and multiply by $\frac{1}{10}$, $\frac{1}{100}$, $\frac{1}{1{,}000}$, and $\frac{1}{10{,}000}$.

$$3.4 \times \frac{1}{10} = 0.34$$

$$3.4 \times \frac{1}{100} = 0.034$$

$$3.4 \times \frac{1}{1{,}000} = 0.0034$$

$$3.4 \times \frac{1}{10{,}000} = 0.00034$$

Each multiplier, $\frac{1}{10}$, $\frac{1}{100}$, $\frac{1}{1{,}000}$, $\frac{1}{10{,}000}$, can be written as a negative power of 10.

$$\frac{1}{10} = 10^{-1}$$

$$\frac{1}{100} = 10^{-2}$$

$$\frac{1}{1{,}000} = 10^{-3}$$

$$\frac{1}{10{,}000} = 10^{-4}$$

Now let's rewrite the multiplication of 3.4 using the negative powers of 10.

$$3.4 \times 10^{-1} = 0.34$$

$$3.4 \times 10^{-2} = 0.034$$

$$3.4 \times 10^{-3} = 0.0034$$

$$3.4 \times 10^{-4} = 0.00034$$

Notice that if we begin with 3.4, the negative exponent of 10 tells us how many places to move the decimal point to the left. For 3.4×10^{-4} we have

$$3.4 \times 10^{-4} = 0.0003.4 = 0.00034$$

Move the decimal point four places to the left and fill in the empty spaces with zeros.

When we write a whole number or decimal as the product of a number between 1 and 10 with some power of ten we are expressing the whole number or decimal in **scientific notation**.

SCIENTIFIC NOTATION: (Number between 1 and 10) $\times$ 10^{power}

Scientific notation allows us to write very large whole numbers or very small decimal numbers in condensed form. For example,

$$860{,}000{,}000{,}000{,}000 = 8.6 \times 10^{14}$$
$$0.0000000086 = 8.6 \times 10^{-9}$$

We suggest you check the above using the rules we have stated for moving the decimal point to the right for positive integer exponents and to the left for negative integer exponents.

Now let's have some practice in converting from scientific notation to whole numbers or decimal numbers and converting whole numbers or decimal numbers to scientific notation.

EXAMPLE 1: Convert each to a whole number or decimal number.

a) 5.3×10^5 **b)** 8.7×10^{-3}

Solution:

a) We have to move the decimal point 5 places to the right.

$$5.3 \times 10^5 = 5.30000. = 530{,}000$$

5 places

b) We have to move the decimal point 3 places to the left.

$$8.7 \times 10^{-3} = 0.008.7 = 0.0087$$

3 places

PRACTICE PROBLEM 1: Convert each to a whole number or decimal number.

a) 3.6×10^{-2} **b)** 7.4×10^6

EXAMPLE 2: Convert each whole number or decimal number to scientific notation.

a) 63,000,000 **b)** 0.00000000754

Solution:

a) Since the multiplier of the power of 10 must be between 1 and 10, it must be 6.3. To convert 6.3 to 63,000,000 we have to move the decimal point 7 places to the right. Therefore,

$$63{,}000{,}000 = 6.3 \times 10^7$$

b) The multiplier of the power of 10 has to be 7.54. To convert 7.54 to 0.00000000754 we have to move the decimal point 9 places to the left. Therefore,

$$0.00000000754 = 7.54 \times 10^{-9}$$

PRACTICE PROBLEM 2: Convert each whole number or decimal to scientific notation.

a) 230,000,000,000 **b)** 0.0092

We can perform operations with numbers written in scientific notation. In Examples 3 and 4 multiplication and division operations will be carried out. Notice the role played by the properties of exponents (adding exponents and subtracting exponents).

EXAMPLE 3: Multiply $(4.6 \times 10^3) \times (6.8 \times 10^{-5})$ and express your result in scientific notation.

Solution:
$(4.6 \times 10^3)(6.8 \times 10^{-5})$ can be written as $(4.6 \times 6.8) \times (10^3 \times 10^{-5})$. In the first parentheses we have to multiply 4.6 times 6.8.

$$4.6 \times 6.8 = 31.28$$

In the second parentheses we have to add the exponents.

$$10^3 \times 10^{-5} = 10^{3-5} = 10^{-2}$$

So, $(4.6 \times 6.8) \times (10^3 \times 10^{-5}) = 31.28 \times 10^{-2}$. Our only problem now is that 31.28 is not between 1 and 10. Since $31.28 = 3.128 \times 10^1$, we can write 31.28×10^{-2} as $3.128 \times 10^1 \times 10^{-2} = 3.128 \times 10^{-1}$. Therefore,

$$(4.6 \times 10^3) \times (6.8 \times 10^{-5}) = 3.128 \times 10^{-1}.$$

PRACTICE PROBLEM 3: Multiply $(3 \times 10^{-6}) \times (6.7 \times 10^{11})$ and express your result in scientific notation.

EXAMPLE 4: Divide and express your result in scientific notation.

$$\frac{3.6 \times 10^7}{4.5 \times 10^{-3}}$$

Solution

$\dfrac{3.6 \times 10^7}{4.5 \times 10^{-3}}$ can be written as $\dfrac{3.6}{4.5} \times \dfrac{10^7}{10^{-3}}$

$\dfrac{3.6}{4.5} = 0.8$ and $\dfrac{10^7}{10^{-3}} = 10^{7-(-3)} = 10^{10}$.

So, $\dfrac{3.6 \times 10^7}{4.5 \times 10^{-3}} = 0.8 \times 10^{10}$.

But 0.8×10^{10} is not in scientific notation.
Since $0.8 = 8.0 \times 10^{-1}$ we can write 0.8×10^{10} as $8.0 \times 10^{-1} \times 10^{10}$, which equals 8.0×10^9.

Therefore, $\dfrac{3.6 \times 10^7}{4.5 \times 10^{-3}} = 8.0 \times 10^9$

PRACTICE PROBLEM 4: Divide and express your result in scientific notation.

$$\frac{2.4 \times 10^{-8}}{1.5 \times 10^{-5}}$$

ANSWERS FOR PRACTICE PROBLEMS

1. **a)** 0.036 **b)** 7,400,000 **2.** **a)** 2.3×10^{11} **b)** 9.2×10^{-3}
3. 2.01×10^6 **4.** 1.6×10^{-3}

EXERCISE SET 8.3A

Skill Development

In questions 1 through 8, write each number in scientific notation.

1. 8,640,000
2. 6,000
3. 240,000,000,000
4. 10,000
5. 0.000321
6. 0.0000089
7. 0.00503
8. 0.001

In questions 9 through 16, write each as a whole number or decimal.

9. 6.2×10^2
10. 3.8×10^5
11. 3.0×10^4
12. 9.6×10
13. 2.1×10^{-3}
14. 7.8×10^{-7}
15. 5.0×10^{-1}
16. 8.1×10^{-5}

In questions 17 through 20, multiply and express your result in scientific notation.

17. $(3.5 \times 10^4) \times (4.2 \times 10^{-1})$
18. $(4.0 \times 10^5) \times (6.0 \times 10^5)$
19. $(9.8 \times 10^{-2}) \times (1.3 \times 10^{-4})$
20. $(4.0 \times 10^{-8}) \times (3.4 \times 10^6)$

In questions 21 through 24, divide and express your result in scientific notation.

21. $\dfrac{6.0 \times 10^{12}}{2.0 \times 10^3}$
22. $\dfrac{7.2 \times 10^6}{8.0 \times 10^{-2}}$
23. $\dfrac{1.9 \times 10^{-4}}{9.5 \times 10^6}$
24. $\dfrac{1.0 \times 10^{-3}}{4.0 \times 10^{-8}}$

Applications

25. The radius of the earth at the equator is approximately 6,378,000 meters. Write this measurement in scientific notation.

26. Ultraviolet light emitted from a mercury arc has a wavelength of approximately 0.000254 centimeters. Write this measurement in scientific notation.

27. Use the formula DISTANCE = RATE × TIME to find the distance if the rate is 3.4×10^7 feet per second and the time is 40 seconds. Write the answer in scientific notation.

For Some Extra Thought

28. Write 2 in scientific notation.
29. Convert 0.0058×10^{-6} to scientific notation.

Review Questions for Maintaining Your Skills

30. Round 12,387,642 to the nearest hundred thousand.

31. Reduce $\frac{72}{84}$ to its lowest terms.

32. Convert $3\frac{7}{9}$ to an improper fraction.

33. To the nearest tenth of a percent, 15 is what percent of 67?

34. Use cross-multiplication to solve $\frac{N}{5} = \frac{5}{2}$

35. $(-2.45) - (-6.8) =$

36. $(-8)(-7)(+4)(-2) =$

37. Solve: $-8x + 7 = 39$

38. Solve: $5x - 3 = 7x - 17$

39. Evaluate 6^3

40. Evaluate $(-1)^{13}$

In 41 and 42, evaluate after using the appropriate properties to simplify.

41. $(-3)^6(-3)^{-4}$

42. $\frac{3^{12}}{3^8}$

EXERCISE SET 8.3B

Skill Development

In questions 1 through 8, write each number in scientific notation.

1. 67,100 **2.** 891,000,000,000

3. 2,000,000 **4.** 360

5. 0.00041 **6.** 0.00703

7. 0.5 **8.** 0.0000001

In questions 9 through 16, write each as a whole number or decimal.

9. 4.8×10^3 **10.** 6.1×10^6

11. 7.0×10 **12.** 2.0×10

13. 5.4×10^{-2} **14.** 8.3×10^{-5}

15. 3.6×10^{-7} **16.** 4.2×10^{-3}

In questions 17 through 20, multiply and express your result in scientific notation.

17. $(2.3 \times 10^2) \times (5.1 \times 10^3)$

18. $(6.0 \times 10^{-6}) \times (3.0 \times 10^{-4})$

19. $(5.2 \times 10^2) \times (8.0 \times 10^{-6})$

20. $(5.2 \times 10) \times (7.3 \times 10)$

In questions 21 through 24, divide and express your result in scientific notation.

21. $\dfrac{8.0 \times 10^9}{2.0 \times 10^5}$ **22.** $\dfrac{1.5 \times 10^{-2}}{2.5 \times 10^5}$

23. $\dfrac{5.6 \times 10^{-1}}{6.4 \times 10^7}$ **24.** $\dfrac{2.0 \times 10^{-6}}{5.0 \times 10^{11}}$

Applications

25. In a computer circuit electricity travels at approximately 0.0000186 miles per second. Write this measurement in scientific notation.

26. For the six month period ending September 30, 1987 the circulation of the Dallas Times Herald was 246,370. Write the circulation in scientific notation.

27. The diameter of a cell is 5.2×10^{-5} centimeters. If 3,000 of these cells are placed in a row, find the length in centimeters. Write the answer in scientific notation.

For Some Extra Thought

28. Write 9.54×10^0 as a decimal number.

29. Is 3.415678204521×10 written in scientific notation?

Review Questions for Maintaining Your Skills

30. Round 5,684 to the nearest ten.

31. Reduce $\frac{360}{450}$ to its lowest terms.

32. Convert $4\frac{1}{8}$ to an improper fraction.

33. To the nearest tenth of a percent, 23 is what percent of 30?

34. Use cross-multiplication to solve $\frac{16}{10} = \frac{w}{5}$

35. $(+4.5) - (-6.872) =$

36. $(-4)(+10)(-2.5) =$

37. Solve: $2x - 5 = -45$

38. Solve: $8x - 5 = -2x + 25$

39. Evaluate 10^4

40. Evaluate $(-5)^3$

In 41 and 42, evaluate after using the appropriate properties to simplify.

41. $(-2)^{-7}(-2)^8$

42. $\frac{5^3}{5^3}$

8.4 FINDING SQUARE ROOTS

Preview Test

In questions 1 through 4, find the square roots of each number.

1. 25 **2.** 9 **3.** 64 **4.** 100

In questions 5 through 8, find the value of each square root.

5. $\sqrt{49}$ **6.** $-\sqrt{4}$ **7.** $-\sqrt{900}$ **8.** $\sqrt{36}$

In questions 9 through 12, find two consecutive integers that bound each square root.

9. $\sqrt{11}$ **10.** $\sqrt{30}$ **11.** $\sqrt{55}$ **12.** $\sqrt{115}$

In questions 13 through 18, use a calculator to approximate each square root to the nearest tenth.

13. $\sqrt{17}$ **14.** $\sqrt{40}$ **15.** $\sqrt{5}$ **16.** $\sqrt{200}$

17. $\sqrt{4+9}$ **18.** $\sqrt{3^2+7^2}$

Whether we square 3 or -3, the result is 9.

$$3^2 = 3 \cdot 3 = 9 \quad \text{and} \quad (-3)^2 = (-3)(-3) = 9$$

Suppose we ask, "What numbers have a square equal to 9?" The answer of course is 3 and -3. Another way of asking this question is,

"FIND THE SQUARE ROOTS OF 9"

The square roots of 9 are 3 and -3.

To find the square roots of the number N, we are looking for those numbers whose square (number times itself) equals N. For example, the two square roots of 25 are 5 and -5, because

$$5^2 = 25 \quad \text{and} \quad (-5)^2 = 25$$

ANSWERS FOR PREVIEW TEST

1. 5 and -5 **2.** 3 and -3 **3.** 8 and -8 **4.** 10 and -10
5. 7 **6.** -2 **7.** -30 **8.** 6 **9.** 3 and 4 **10.** 5 and 6
11. 7 and 8 **12.** 10 and 11 **13.** 4.1 **14.** 6.3 **15.** 2.2
16. 14.1 **17.** 3.6 **18.** 7.6

As you can see, every positive number has two square roots, one positive and the other negative.*

To find square roots it is helpful to memorize a table of perfect squares. Of course there is a limit to the number of squares anyone can be expected to know. You will find the following list reasonable in length and very useful.

Table of Perfect Squares

$1^2 = 1$	$6^2 = 36$	$11^2 = 121$	$16^2 = 256$
$2^2 = 4$	$7^2 = 49$	$12^2 = 144$	$20^2 = 400$
$3^2 = 9$	$8^2 = 64$	$13^2 = 169$	$25^2 = 625$
$4^2 = 16$	$9^2 = 81$	$14^2 = 196$	$30^2 = 900$
$5^2 = 25$	$10^2 = 100$	$15^2 = 225$	$40^2 = 1600$

EXAMPLE 1: Find the square roots of 49.

Solution:
Since $7^2 = 49$ and $(-7)^2 = 49$, the square roots of 49 are 7 and -7.

PRACTICE PROBLEM 1: Find the square roots of 36.

There is a special notation for square roots.

$\sqrt{\text{NUMBER}}$ represents the positive square root

and

$-\sqrt{\text{NUMBER}}$ represents the negative square root.

So, $\sqrt{9} = 3$ and $-\sqrt{9} = -3$.

EXAMPLE 2: Find the value of each.

a) $\sqrt{100}$ **b)** $-\sqrt{81}$

Solution:
a) $\sqrt{100} = 10$ **b)** $-\sqrt{81} = -9$

PRACTICE PROBLEM 2: Find the value of each.

a) $\sqrt{400}$ **b)** $-\sqrt{49}$

So far we have only found the positive and negative square roots of numbers that are perfect squares. Suppose we want to find $\sqrt{40}$. We can get a rough idea of its value by noting that 40 lies between the two perfect squares 36 and 49. Then $\sqrt{40}$ must lie between $\sqrt{36}$ and $\sqrt{49}$. So, $\sqrt{40}$ is somewhere between 6 and 7. That is, we know that $\sqrt{40} = 6 +$ a decimal. There are various ways of finding the value of the decimal to one or more places. The best approach is to use a calculator.

* Negative numbers also have square roots. However, the square roots of a negative number, such as -5, are not the type of numbers we have been dealing with in this book. A number whose square equals -5 is an imaginary number. In more advanced math courses you will study imaginary numbers. For now, we will only consider square roots of positive numbers.

Before using a calculator, let's go through an example where we are required to find two consecutive integers that bound the square root of a number. This skill will help us determine whether or not our calculator answers appear reasonable. For example, if our calculator result for $\sqrt{60}$ is 7.75 to the nearest hundredth, we know this answer is reasonable because $\sqrt{60}$ lies between $\sqrt{49}$ and $\sqrt{64}$ (that is, between 7 and 8).

EXAMPLE 3: Find two consecutive integers that bound each square root.

a) $\sqrt{20}$ **b)** $\sqrt{73}$

Solution:

a) Two perfect squares that bound 20 are 16 and 25. Since $\sqrt{16} = 4$ and $\sqrt{25} = 5$, the two integers that bound $\sqrt{20}$ are 4 and 5.

b) Two perfect squares that bound 73 are 64 and 81. Since $\sqrt{64} = 8$ and $\sqrt{81} = 9$, the two integers that bound $\sqrt{73}$ are 8 and 9.

PRACTICE PROBLEM 3: Find two consecutive integers that bound each square root.

a) $\sqrt{32}$ **b)** $\sqrt{2}$

To use a calculator to approximate the square root of a number, first enter the number. Then press the key labeled $\sqrt{\ }$. Try the procedure to find $\sqrt{16}$ and see if the calculator gives you 4 as a result. Now use your calculator to find $\sqrt{10}$. If your calculator is set up to give you a maximum of 8 digits, then the calculator will display 3.1622776. If the calculator rounds automatically to the seventh decimal place the display will be 3.1622777. In either case, if we are asked to round our result to the nearest hundredth, our approximation for $\sqrt{10}$ is 3.16.

EXAMPLE 4: Use a calculator to approximate each to the nearest thousandth.

a) $\sqrt{55}$ **b)** $\sqrt{140}$ **c)** $\sqrt{13}$

Solution

a) 7.416 **b)** 11.832 **c)** 3.606

PRACTICE PROBLEM 4: Use a calculator to approximate each to the nearest hundredth.

a) $\sqrt{92}$ **b)** $\sqrt{19}$ **c)** $\sqrt{200}$

If we are finding the square root of an expression, all other operations should be completed before finding the root. Let's consider $\sqrt{9 + 16}$. First add, then find the root. So,

$$\sqrt{9 + 16} = \sqrt{25} = 5$$

Be Careful About This

The square root of a sum *is not equal to* the sum of the square roots of each term. For example, it is wrong to find $\sqrt{9 + 16}$ by saying

$$\sqrt{9 + 16} = \sqrt{9} + \sqrt{16} = 3 + 4 = 7$$

We know 7 is not correct because $\sqrt{25}$ is not equal to 7.

EXAMPLE 5: Use a calculator to approximate each to the nearest tenth.

a) $\sqrt{5+9}$ **b)** $\sqrt{7^2 - 3^2}$

Solution:

a) $\sqrt{5+9} = \sqrt{14} = 3.7$

b) $\sqrt{7^2 - 3^2} = \sqrt{49 - 9} = \sqrt{40} = 6.3$

PRACTICE PROBLEM 5: Use a calculator to approximate each to the nearest hundredth.

a) $\sqrt{125 - 80}$ **b)** $\sqrt{5^2 + 1^2}$

ANSWERS FOR PRACTICE PROBLEMS

1. 6 and -6 **2. a)** 20 **b)** -7 **3. a)** 5 and 6 **b)** 1 and 2
4. a) 9.59 **b)** 4.36 **c)** 14.14 **5. a)** 6.71 **b)** 5.10

EXERCISE SET 8.4A

Skill Development

In questions 1 through 4, find the square roots of each number.

1. 36 **2.** 100 **3.** 49 **4.** 1

In questions 5 through 12, find the value of each square root.

5. $\sqrt{25}$ **6.** $-\sqrt{4}$ **7.** $-\sqrt{900}$ **8.** $\sqrt{64}$

9. $-\sqrt{1}$ **10.** $\sqrt{196}$ **11.** $\sqrt{81}$ **12.** $-\sqrt{9}$

In questions 13 through 16, find two consecutive integers that bound each square root.

13. $\sqrt{18}$ **14.** $\sqrt{97}$ **15.** $\sqrt{33}$ **16.** $\sqrt{140}$

In questions 17 through 30, use a calculator to approximate each square root to the nearest tenth.

17. $\sqrt{21}$ **18.** $\sqrt{68}$ **19.** $\sqrt{153}$ **20.** $\sqrt{41}$

21. $\sqrt{206}$ **22.** $\sqrt{80}$ **23.** $\sqrt{135}$ **24.** $\sqrt{7}$

25. $\sqrt{100 - 73}$ **26.** $\sqrt{25 + 49}$

27. $\sqrt{3^2 + 2^2}$ **28.** $\sqrt{10^2 - 7^2}$

29. $\sqrt{19 - 17}$ **30.** $\sqrt{67 + 3}$

For Some Extra Thought

31. Find two consecutive integers that bound $-\sqrt{130}$.

32. Find the value of $\sqrt{0}$.

Review Questions for Maintaining Your Skills

33. Write 2,187,000,023 in words.

34. $\frac{5}{6} - \frac{2}{3} =$ **35.** $10\frac{1}{2} \times 3\frac{1}{2} =$

36. $2.57 \times 5.69 =$

37. Find 6.72% of \$9,200.

38. If a price increases from \$12.50 to \$13.89, find the percent increase to the nearest tenth of a percent.

39. $(+6.7) - (-8.2) =$ **40.** $(-50) - (+2.5) =$

41. Solve: $4x - 15 = -27$ **42.** Solve: $-2(3 - .4x) = x - 7$

43. Evaluate: 8^3 **44.** Evaluate: $(-2)^5$

In 45 and 46, evaluate after using the appropriate properties to simplify.

45. $(-3)^8(-3)^{-6} =$

46. $\frac{6^8}{6^5} =$

47. Write each number in scientific notation.

a) 6,870 **b)** 0.0000072

48. Write each as a whole number or decimal.

a) 6.1×10^5 **b)** 3.8×10^{-7}

49. Multiply and express your result in scientific notation.

$(5.1 \times 10^{12}) \times (3.5 \times 10^{-9})$

50. Divide and express your result in scientific notation.

$$\frac{4.0 \times 10^9}{2.0 \times 10^4}$$

EXERCISE SET 8.4B

Skill Development

In questions 1 through 4, find the square roots of each number.

1. 81 **2.** 400 **3.** 4 **4.** 25

In questions 5 through 12, find the value of each square root.

5. $\sqrt{36}$ **6.** $-\sqrt{1}$ **7.** $-\sqrt{64}$ **8.** $\sqrt{121}$

9. $-\sqrt{49}$ **10.** $\sqrt{169}$ **11.** $\sqrt{9}$ **12.** $-\sqrt{256}$

In questions 13 through 16, find two consecutive integers that bound each square root.

13. $\sqrt{23}$ **14.** $\sqrt{180}$ **15.** $\sqrt{59}$ **16.** $\sqrt{42}$

In questions 17 through 30, use a calculator to approximate each square root to the nearest thousandth.

17. $\sqrt{44}$ **18.** $\sqrt{13}$ **19.** $\sqrt{106}$ **20.** $\sqrt{39}$

21. $\sqrt{178}$ **22.** $\sqrt{71}$ **23.** $\sqrt{8}$ **24.** $\sqrt{220}$

25. $\sqrt{400 - 100}$ **26.** $\sqrt{16 + 16}$

27. $\sqrt{6^2 - 4^2}$ **28.** $\sqrt{10^2 + 9^2}$

29. $\sqrt{5 - 3}$ **30.** $\sqrt{20 + 15}$

For Some Extra Thought

31. Find two consecutive integers that bound $-\sqrt{200}$

32. Is there a positive or negative number whose square is -1?

Review Questions for Maintaining Your Skills

33. Write 2,005,000 in words.

34. $\frac{1}{2} - \frac{1}{5} =$ **35.** $6\frac{2}{5} \times 2\frac{1}{3} =$

36. $6.786 \times 100 =$

37. Find 125% of \$24.

38. If a price decreases from \$2.56 to \$2.24, find the percent decrease.

39. $-7.12 + 9.2 - 7.4 =$ **40.** $(-24.6) - (-0.004) =$

41. Solve: $0.02x + 7 = 8$ **42.** Solve: $3 - (4 - x) + 9x = 13$

43. Evaluate: 10^5 **44.** Evaluate: $(-4)^3$

45. $(-2)^9(-2)^{-5} =$ **46.** $\frac{8^5}{8^2} =$

47. Write each number in scientific notation.

a) 78,000,000 **b)** 0.0000000044

48. Write each as a whole number or decimal.

a) 7.6×10^{-4} **b)** 6.3×10^{6}

49. Multiply and express your result in scientific notation.

$(5.6 \times 10^{14}) \times (4.1 \times 10^{-2})$

50. Divide and express your result in scientific notation.

$$\frac{8.4 \times 10^{7}}{2.1 \times 10^{-4}}$$

PRACTICE TEST 8A

1. Write $7 \times 7 \times 7 \times 7 \times 7$ in exponential form.
2. Write $(-6)^7$ as a product of factors.
3. Evaluate each term or expression.
   a) 5^4 **b)** $(-2)^6$ **c)** $7^2 - 5^2 + 1^2$
4. Find the value of each.
   a) $2^4 \cdot 2^3$ **b)** $3^{-4} \cdot 3^7$ **c)** $\left(\frac{1}{2}\right)^5$
5. Write each number in scientific notation.
   a) 12,300,000 **b)** 0.0000087
6. Write as a whole number or decimal.
   a) 5.23×10^5 **b)** 2.1×10^{-3}
7. Express each result in scientific notation.
   a) $(6.4 \times 10^{-5}) \times (1.2 \times 10^{12})$ **b)** $\dfrac{3.6 \times 10^6}{1.2 \times 10^{-2}}$
8. Find the square roots of 121.
9. Find the value of **a)** $-\sqrt{64}$ **b)** $\sqrt{100}$
10. Find two consecutive integers that bound $\sqrt{29}$
11. Use a calculator to approximate each square root to the nearest tenth.
    a) $\sqrt{47}$ **b)** $\sqrt{145}$ **c)** $\sqrt{200}$

PRACTICE TEST 8B

1. Write $(-3) \times (-3) \times (-3) \times (-3)$ in exponential form.
2. Write 5^8 as a product of factors.
3. Evaluate each term or expression.

   a) 9^2 **b)** $(-5)^4$ **c)** $5^3 - 4^3$
4. Find the value of each

   a) $3^5 \cdot 3$ **b)** $(-2)^{-6}(-2)^8$ **c)** $\left(\frac{2}{5}\right)^3$
5. Write each number in scientific notation.

   a) 11,600,000,000 **b)** 0.0000000213
6. Write as a whole number or decimal.

   a) 3.78×10^8 **b)** 6.7×10^{-5}
7. Express each result in scientific notation.

   a) $(7.5 \times 10^{17}) \times (3.0 \times 10^{-2})$ **b)** $\dfrac{7.5 \times 10^{-4}}{1.5 \times 10^{-5}}$
8. Find the square roots of 81.
9. Find the value of **a)** $\sqrt{36}$ **b)** $-\sqrt{49}$
10. Find two consecutive integers that bound $\sqrt{73}$
11. Use a calculator to approximate each square root to the nearest hundredth.

    a) $\sqrt{61}$ **b)** $\sqrt{153}$ **c)** $\sqrt{17}$

CUMULATIVE REVIEW EXERCISES FOR CHAPTERS 1 THROUGH 8

1. Write 20,304,000 in words.

2. Round 20,574 to the nearest thousand.

In 3 through 6, reduce each fraction to its lowest terms and express any result that is an improper fraction as a mixed number.

3. $\frac{10}{15} \times \frac{3}{20} =$

4. $4\frac{1}{3} \div 1\frac{2}{3} =$

5. $\frac{7}{8} + \frac{5}{6} =$

6. $8 - 2\frac{1}{7} =$

7. Write 1.0005 in words.

8. Round 21.5807916 to the nearest millionth.

9. Round \$41.20643 to the nearest cent.

10. $2.3 \times 1.066 =$

11. Divide 23.6 by 1.72 and round the quotient to the nearest hundredth.

12. $1.3 \times 6.4 + 7.52 =$

13. Convert 12.75% to a decimal.

14. Convert 0.048 to a percent.

15. Convert 4% to a fraction reduced to its lowest terms.

16. Find 26% of \$14,000.

17. 12 is what percent of 400?

18. 2% of what amount equals \$42?

19. If a price increases from \$50 to \$59, find the percent increase.

20. Solve the proportion $\frac{200}{M} = \frac{4}{5}$

21. $-14 + 19 - 33 - 10 =$

22. $(-47) - (-16) =$

23. $(-2.3) \times (-6.7) =$

24. $(-400) \div (+2.5) =$

25. Find the number of terms: $(8x + 7) - 3y$

26. Write in words: $4x + 2w - 7$

27. Solve: $3x - 17 = 19$

28. Solve: $4x - 3 = 6x + 27$

29. Solve: $4(x + 3) = 3x - 9$

30. Write $(-8) \times (-8) \times (-8)$ in exponential form.

31. Write 7^5 as a product of factors.

32. Evaluate each term or expression.

a) 5^4 **b)** $(-3)^5$ **c)** $6^3 - 2^4$

33. Find the value of each.

a) $5^3 \cdot 6^2$ **b)** $9^2 \cdot 2^3$ **c)** $\left(\frac{2}{5}\right)^3$

34. Write each number in scientific notation.

a) 230,000 **b)** 0.00041

35. Write as a whole number or decimal.

a) 6.41×10^4 **b)** 7.3×10^{-2}

36. Express the result in scientific notation.

$(7.5 \times 10^3) \times (4.2 \times 10^{-1})$

37. Find the square roots of 225.

38. Find the value of **a)** $\sqrt{400}$ **b)** $-\sqrt{64}$

39. Find two consecutive integers that bound $\sqrt{78}$.

40. Use a calculator to approximate each square root to the nearest hundredth.

a) $\sqrt{22}$ **b)** $\sqrt{134}$ **c)** $\sqrt{93}$

United States Customary System of Length Measurements

The most commonly used measurements of length in the United States Customary system are listed below. The abbreviations are written in parentheses. If you are not already familiar with these conversions, you should learn them.

Length Measurements in the United States Customary System

12 inches (in) = 1 foot (ft)

3 feet = 1 yard (yd)

5,280 feet = 1 mile (mi)

To convert 48 inches to feet you can multiply 48 inches by the ratio $\frac{1 \text{ foot}}{12 \text{ inches}}$. Since 1 foot equals 12 inches, the ratio $\frac{1 \text{ foot}}{12 \text{ inches}}$ is equivalent to 1. The product of 48 inches with a ratio equivalent to 1 will not change the value of the measurement.

To carry out the conversion you can cancel like units that appear in the numerator and denominator (inches in this case) and divide 48 by 12.

$$\overset{4}{\cancel{48}} \cancel{\text{inches}} \times \frac{1 \text{ foot}}{\cancel{12} \cancel{\text{inches}}} = 4 \text{ feet}$$

EXAMPLE 1: The distance from a dormitory room to the Science Building is $\frac{1}{2}$ mile. How many feet is this?

Solution:

Since 5,280 feet equals 1 mile, the ratio $\frac{5{,}280 \text{ feet}}{1 \text{ mile}}$ is equivalent to 1. The conversion is carried out by multiplying $\frac{5{,}280 \text{ feet}}{1 \text{ mile}}$ by $\frac{1}{2}$ mile.

$$\frac{1}{2} \cancel{\text{mi}} \left(\frac{5{,}280 \text{ ft}}{1 \cancel{\text{mi}}}\right) = 2{,}640 \text{ ft}$$

Notice that in the multiplication process the "mile" measurements cancel out, leaving the "foot" measurement. Thus, $\frac{1}{2}$ mile is the same as 2,640 feet.

PRACTICE PROBLEM 1: Main Street U.S.A. is approximately 1.5 miles long. How long is Main Street in feet?

The key to converting measurements is to set up the ratios so that appropriate units cancel. Recall the logic used to set up the ratio in Example 1. We used $\frac{5{,}280 \text{ ft}}{1 \text{ mi}}$ and not $\frac{1 \text{ mi}}{5{,}280 \text{ ft}}$ because we wanted the "mile" measurement to cancel. If you're not clear about the cancellation situation, try multiplying $\frac{1}{2}$ mile by the second ratio.

EXAMPLE 2: The width of a sheet of aluminum measures 50.4 in. Determine the width of the sheet in feet.

Solution:
To convert from inches to feet we will write a ratio using 12 inches for 1 foot. Since we want the "in" measurement to cancel, we must use the ratio $\frac{1 \text{ ft}}{12 \text{ in}}$. Then cancel and multiply.

$$50.4 \not{\text{in}} \left(\frac{1 \text{ ft}}{12 \not{\text{in}}}\right) = 4.2 \text{ ft}$$

As needed, the "in" measurements cancel leaving the "ft" measurement. The width of the sheet is 4.2 ft.

Be Careful About This.

Notice that in Example 1, the conversion factor for 1 mile (5,280 ft) is located in the numerator of the ratio. In Example 2, the conversion factor for 1 foot (12 in) is located in the denominator of the fraction. Don't assume that the conversion factor for one unit (1 mile, 1 foot, etc.) will always be either a numerator or denominator term. How do you figure out where the conversion factor goes in the ratio? Let your problem direct you—set up the ratio so the appropriate measure cancels out. You'll never go wrong!

PRACTICE PROBLEM 2: A VCR measures $1\frac{3}{4}$ feet. How many yards is this?

Some calculations are "chain" conversions and require more than 1 ratio. Example 3 is a chain conversion.

EXAMPLE 3: The length of a bridge is $\frac{1}{3}$ mile. How many yards is this?

Solution:
Since we do not have a conversion from miles to yards, we will have to convert miles to feet (5,280 ft = 1 mi) and then feet to yards (3 ft = 1 yd). This can be done with the "chain" of ratios shown below. Take note of the cancellation that occurs due to the set-up of the ratios.

$$\frac{1 \not{\text{mi}}}{3}\left(\frac{5{,}280 \not{\text{ft}}}{1 \not{\text{mi}}}\right)\left(\frac{1 \text{ yd}}{3 \not{\text{ft}}}\right) = \frac{1{,}760 \text{ yd}}{3} = 586\frac{2}{3} \text{ yd}$$

Convert mi to ft   Convert ft to yd

Therefore, the length of the bridge is $586\frac{2}{3}$ yards.

PRACTICE PROBLEM 3: The length of a football field is 140 yards. What is its length to the nearest hundredth of a mile?

Metric System of Length Measurements

As you might guess from the conversions given so far, there is no pattern in the U.S. Customary system of measurements. This is the major disadvantage of the U.S. Customary system. On the other hand, the major advantage of the metric system is its patterns. The basic unit of length in the metric system is the meter. Just as we can measure distances with a yardstick in the U.S. Customary system, we can measure distances with a meterstick in the metric system. A meterstick is about 3 inches longer than a yardstick (see Figure 9.1.1).

1 yard or 3 feet or 36 inches

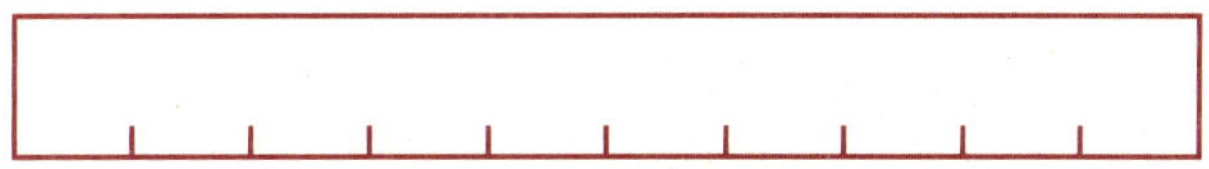

1 meter or 10 decimeters or 100 centimeters

Figure 9.1.1

The yardstick is divided into feet and inches. The meterstick is divided into **decimeters** and **centimeters**. The conversions for these and other metric length measurements and their abbreviations follow.

Length Measurements in the Metric System

$$1 \text{ kilometer (km)} = 1{,}000 \text{ meters (m)}$$
$$1 \text{ hectometer (hm)} = 100 \text{ meters (m)}$$
$$1 \text{ dekameter (dam)} = 10 \text{ meters}$$
$$1 \text{ decimeter (dm)} = \frac{1}{10} \text{ or } 0.1 \text{ meter}$$
$$1 \text{ centimeter (cm)} = \frac{1}{100} \text{ or } 0.01 \text{ meter}$$
$$1 \text{ millimeter (mm)} = \frac{1}{1000} \text{ or } 0.001 \text{ meter}$$

As you can see, the metric system is based on a pattern of tens. Not only does this give us an easy way to remember the relationships from one unit to another, it will also give us an easy way to convert within the metric system. Let's first do a problem using ratios.

EXAMPLE 4: A package is being readied for mailing. Its dimensions are 0.8 m (length) × 75.2 cm (width) × 5 mm (height). Convert the length to decimeters.

Solution:

There are 10 decimeters to 1 meter, and we want the meters to cancel out. The ratio we use is $\frac{10 \text{ dm}}{1 \text{ m}}$.

$$0.8 \cancel{\text{m}}\left(\frac{10 \text{ dm}}{1 \cancel{\text{m}}}\right) = 8 \text{ dm}$$

The length of the package is 8 dm.

PRACTICE PROBLEM 4: A package is 3.25 cm wide. Convert the width to decimeters.

Notice that in Example 4, we converted 0.8 m to 8 dm by multiplying by 10. The net effect of this is to move the decimal point one place to the right.

0.8 m = 0.8. dm = 8 dm

1 place to the right

We can do all our conversions in the metric system by moving the decimal point according to the conversion being done. To figure out how many places to move the decimal point, we can use the following chart of metric prefixes.

Metric Prefixes

kilo- hecto- deka- meter deci- centi- milli-

Notice that the basic length measurement, the meter, is in the middle of the chart. Similar charts will be made in later sections for the weight measurement (gram) and the liquid volume measurement (liter). Thus, memorizing the chart now will simplify conversions for the next sections. Let's now see how to use this chart to convert 0.8 meters to decimeters.

Count the number of blocks from meters to decimeters (deci-) on the chart. Be sure to note which direction is used (left or right).

kilo- hecto- deka- meter deci- centi- milli-

1 block to the right

Now move the decimal point in 0.8 one place to the right.

0.8 m = 0.8. dm = 8 dm

1 decimal place to the right

Notice that our answer of 8 dm matches the answer we got in Example 4.

EXAMPLE 5: Convert 0.062 hm to cm.

Solution:

Count the number of blocks from hm (hecto-) to cm (centi-) on the chart. Note the direction used.

kilo- hecto- deka- meter deci- centi- milli-

4 blocks to the right

The block for cm is 4 to the right of the block for hm. Therefore, to make the conversion we have to move the decimal point 4 places to the right

$$0.062 \text{ hm} = 0.0620. \text{ cm} = 620 \text{ cm}$$

4 decimal places to the right

Notice that a 0 had to be added to the right of 2 after the decimal point was moved. So, 0.062 hm = 620 cm.

PRACTICE PROBLEM 5: For the package in Example 4, convert the width of 75.2 cm to dm.

EXAMPLE 6: For the package in Example 4, convert the height of 5 mm to dm.

Solution:

kilo- hecto- deka- meter deci- centi- milli-

2 blocks to the left

The block for dm is 2 to the left of the block for mm. Now move the decimal point of 5 two decimal places to the left. Remember that since 5 is a whole number, it has an understood decimal point at the right end.

$$5 \text{ mm} = 5. \text{ mm} = 0.05. \text{ dm} = 0.05 \text{ dm}$$

2 places to the left

Notice that a 0 had to be added to the left of 5 after the decimal point was moved.

PRACTICE PROBLEM 6: A package has a height of 0.034 m. What is the height in mm?

You should notice from the metric conversions that the decimal point is moved varying numbers of places and in different directions. If you use the chart correctly it should indicate how many places you need to move the point and in what direction.

Be Careful About This.

When using the chart for your metric conversions, don't count the beginning block. For example, when converting from meters to kilometers, count the dekameter block, not the meter block, as your first block.

kilo- hecto- deka- meter deci- centi- milli-

3 blocks to the left

The decimal point for the meter measurement is then moved 3 places to the left to get the kilometer measurement.

ANSWER FOR PRACTICE PROBLEMS

1. 7920 ft **2.** $\frac{7}{12}$ yd **3.** 0.08 mi **4.** 0.325 dm

5. 7.52 dm **6.** 34 mm

EXERCISE SET 9.1A

Skill Development

Determine the following conversions.

1. 21 ft = ________ yd
2. 2 mi = ________ ft
3. $2\frac{5}{6}$ yd = ________ ft
4. 14.4 in = ________ ft
5. 12.07 m = ________ cm
6. 670 mm = ________ cm
7. 3 km = ________ dm
8. 12 dam = ________ hm
9. 0.08 m = ________ dam
10. 3 mi = ________ yd
11. 480 in = ________ yd

Applications

12. The ductwork from a heater to the first floor of a building is 25 feet long. How many yards is this?

13. George rides his bike 1 mile every day as part of his fitness regime. Determine the distance in feet he rides each day.

14. A moving company estimates 48 inches of tape are required to bind one box of hardbound books. How many yards are needed for 5 boxes of books?

15. The front of a farm lot measures 0.76 km. What is the measurement in meters?

16. A computer record requires 8.3 cm on magnetic tape. Determine the length in millimeters.

For Some Extra Thought

17. Find the cost of carpet stripping for a room with a perimeter of 24 yards if the stripping costs $0.50 for 8 feet.

18. Find the cost of 2,000 meters of siding at $48 for 100 decimeters.

19. The perimeter of a lot is the sum of the sides. Find the perimeter of a three-sided lot in meters with lengths of 0.7 km, 120 m, and 325 m.

Review Problems for Maintaining Your Skills

20. Round 22,854,750 to the nearest ten thousand.

21. $\frac{5}{8} + \frac{1}{7} =$

22. $2\frac{3}{4} - 1\frac{1}{2} =$

23. Divide 26.8 by 1.45 and round the quotient to the nearest hundredth.

24. Find 17.5% of \$9,400.

25. 4% of what amount equals \$7.28?

26. $5.4 - 6.8 - 7.9 =$

27. $(-3)(-7)(-2) =$

28. Solve: $4x - 3 = -95$

29. Solve: $3(2 - 7x) = 4x - 19$

30. $5^4 =$

31. $(-3)^5 =$

32. $(-7)^{-5}(-7)^7 =$

33. $\frac{9^4}{9^3} =$

34. Write 45,300,000,000 in scientific notation.

35. Write as a whole number or decimal.

    a) 3.75×10^{-2} **b)** 4.9×10^6

36. Multiply and express your result in scientific notation.
    $(3.4 \times 10^{-4}) \times (1.2 \times 10^{13}) =$

37. Divide and express your result in scientific notation. $\frac{5.4 \times 10^7}{1.8 \times 10^{-2}} =$

38. Find the square roots of 49.

39. Find the value of $-\sqrt{400}$.

40. Find two consecutive integers that bound $\sqrt{44}$.

41. Use a calculator to approximate $\sqrt{19}$ to the nearest tenth.

EXERCISE SET 9.1B

Skill Development

Determine the following conversions.

1. 15,840 ft = ________ mi

2. 12 yd = ________ ft

3. $2\frac{1}{2}$ ft = ________ in

4. 27.09 ft = ________ yd

5. 5.72 dam = ________ hm

6. 520 m = ________ cm

7. 9 dam = ________ km

8. 26 dam = ________ mm

9. 0.02 mm = ________ dm

10. 3,520 yd = ________ mi

11. 7 yd = ________ in

Applications

12. The height of a room with a cathedral ceiling is 20 feet. How many yards is this?

13. A quarterback throws a 35-yard pass. Determine the number of feet that the football was thrown.

14. The average city block is approximately 1/15 mile. How long is this in feet?

15. The inter-record gap between computer records on a magnetic tape is 1.9 centimeters. Find the length in meters.

16. A bus route is approximately 27.7 kilometers long. What is the distance in hectometers?

For Some Extra Thought

17. It costs approximately $0.23 to produce 6 inches of electrical tape. What is the cost of producing a standard roll of 17 yards?

18. An infant was 56 cm long at birth. When she was 15 years old, she was 1.9 m tall. How much has she grown?

19. The perimeter of a field is the sum of the lengths of the sides. Find the perimeter of a field in meters with sides of 0.3 km, 0.9 hm, 185 m and 42 dam.

Review Problems for Maintaining Your Skills

20. Round 687,471,200 to the nearest million.

21. $\frac{13}{16} + \frac{1}{24} =$

22. $6 - 2\frac{2}{3} =$

23. Divide 12 by 6.7 and round the quotient to the nearest tenth.

24. Find 8.67% of \$20,000.

25. 2.5% of what amount equals \$45?

26. $-3.75 - 6.9 + 4.88 =$

27. $(+2)(-7)(-5)(-1) =$

28. Solve: $0.2x - 7 = 19$

29. Solve: $5 - 9x = 3 - 8x$

30. $6^3 =$

31. $(-10)^4 =$

32. $(-3)^6(-3)^{-4} =$

33. $\frac{2^{10}}{2^6} =$

34. Write 0.000000031 in scientific notation.

35. Write as a whole number or decimal.

a) 4.6×10^6 **b)** 8.7×10^{-2}

36. Multiply and express your result in scientific notation.
$(4.6 \times 10^{15}) \times (6.4 \times 10^{-8}) =$

37. Divide and express your result in scientific notation.
$\frac{8.0 \times 10^2}{2.0 \times 10^{-5}} =$

38. Find the square roots of 400.

39. Find the value of $\sqrt{81}$.

40. Find two consecutive integers that bound $\sqrt{117}$.

41. Use a calculator to approximate $\sqrt{61}$ to the nearest hundredth.

9.2 CONVERTING LENGTHS INTO DIFFERENT SYSTEMS

Preview Test

1. Select the metric unit(s) appropriate to measure the height of a kerosene heater.
   a) km **b)** m **c)** cm
2. Select the metric unit(s) appropriate to measure the distance you walk in one day.
   a) km **b)** m **c)** mm
3. Choose the value that best estimates the average height of an adult female.
   a) 10 m **b)** 160 cm **c)** 5 km **d)** 26 cm
4. Choose the value that best estimates the cruising altitude of an airplane.
   a) 7.5 km **b)** 18 m **c)** 7500 m **d)** 120 dm
5. George has estimated his automobile use for business purposes to be 37,000 kilometers in one year. Determine his mileage to the nearest thousand miles. (1 mi = 1.609 km)
6. An IBM card is $3\frac{1}{4}$ in by $7\frac{3}{8}$ in. What are its dimensions to the nearest tenth in centimeters? (1 in = 2.54 cm)
7. The total length of a football field is approximately 140 yards. What is its length to the nearest meter? (1 ft = 0.3048 m)

If this is your first contact with the metric system, you may be uncomfortable working with meters, kilometers, decimeters, etc. This is a natural reaction when you're dealing with something unfamiliar to you. One way to become more comfortable with the metric system is to relate it to a system you already know, the U.S. Customary system. In this section, we will explore some approximations between the metric and U.S. Customary systems and use ratios to determine equivalent values in the two systems.

Let's first do some estimations using the following chart.

U.S. Customary/Metric Length Measurements

U.S. Customary Unit	*Metric Estimate*
1 inch	2.5 cm
1 foot	$\frac{1}{3}$ m
1 yard	0.9 m
1 mile	1.6 km

ANSWERS FOR PREVIEW TEST

1. (b) **2.** (a) **3.** (b) **4.** (a) or (c) **5.** 23,000 mi
6. 8.3 cm by 18.7 cm **7.** 128 m

Notice that the centimeter is used to estimate the inch, the meter is used to estimate the foot or the yard, and the kilometer is used to estimate the mile. Keep in mind that these are only estimates. Later in this section, we will work with more exact values. For now, though, all we want to do is to "guestimate" metric lengths. Let's start by identifying metric units appropriate for different situations.

EXAMPLE 1: Select the metric unit(s) appropriate to measure the distance from Philadelphia to Denver.

a) kilometer **b)** meter **c)** centimeter

Solution:
The distance from Philadelphia to Denver is a large one, so the unit we use must be large. Of the three possible answers the largest is the kilometer. The kilometer is the best unit for the distance.

EXAMPLE 2: Select the metric unit(s) appropriate to measure the height of a chain link fence.

a) kilometer **b)** meter **c)** centimeter

Solution:
The kilometer is an estimate for the mile, so it is too large to use for the height of a fence. The centimeter is an estimate for the inch, so it is too small to use for the fence. However, the meter is an estimate either for the foot or the yard measure. The meter is the best unit to measure the fence.

EXAMPLE 3: Select the metric unit(s) appropriate to measure the length of an earring.

a) kilometer **b)** meter **c)** centimeter

Solution:
Both the kilometer and the meter are too large for the earring. The centimeter, similar to the inch, is the most appropriate measure.

PRACTICE PROBLEM 1: Select the metric unit(s) appropriate to measure the distance from New York to Tokyo.

a) kilometer **b)** meter **c)** centimeter

PRACTICE PROBLEM 2: Select the metric unit(s) appropriate to measure the length of a paper clip.

a) kilometer **b)** meter **c)** centimeter

PRACTICE PROBLEM 3: Select the metric unit(s) appropriate to measure the length of a hallway.

a) kilometer **b)** meter **c)** centimeter

Now, let's use the estimations given in the previous chart to approximate some metric lengths.

EXAMPLE 4: Choose the value that best estimates the height of a door.

a) 1.2 m **b)** 0.5 km **c)** 2.3 m **d)** 20 cm

Solution:
The meter is the most appropriate measure to use, so the answers (b) and (d) can be

eliminated. Since a meter is slightly longer than a yard, answer (a) 1.2 meters is not long enough. Answer (c) 2.3 meters is the best estimate.

EXAMPLE 5: Choose the value that best estimates the length of an intravenous (IV) bottle.

a) 25 cm **b)** 0.3 m **c)** 8 km **d)** 4 dam

Solution:
The kilometer and dekameter are too large, so answers, (c) and (d) can be eliminated. An IV bottle is about a foot long, so answer (b) estimates the length. However, since an inch is about 2.5 cm, choice (a) is also a possible estimate.

EXAMPLE 6: Choose the value that best estimates the distance from New York to Boston (350 miles).

a) 3.5 mm **b)** 240 km **c)** 38 m **d)** 600 km

Solution:
The kilometer is the best unit for the long distance, so answers (a) and (c) can be eliminated. A mile is approximately 1.6 kilometers, so the answer has to be more than 350. That eliminates choice (b), and leaves (d) as the best estimate.

Use the logic of the above examples to do these practice problems. (Be careful. As in Example 5, there may be more than one answer to the problem!)

PRACTICE PROBLEM 4: Choose the value that best estimates the width of a computer screen.

a) 23 cm **b)** 0.1 km **c)** 23 m **d)** 5 cm

PRACTICE PROBLEM 5: Choose the value that best estimates the average height of an adult male.

a) 32 mm **b)** 1.8 m **c)** 180 cm **d)** 17 km

PRACTICE PROBLEM 6: Choose the value that best estimates the cruising altitude of an airplane (33,000 ft).

a) 10 km **b)** 15 m **c)** 100 m **d)** 33,000 m

Now that we've estimated metric lengths, we can determine more exact values. How do we do it? We multiply by ratios set up so that appropriate units cancel out. (Sound familiar?) The following chart lists some common conversions.

U.S. Customary/Metric Length Conversions

1 in = 2.540 cm
1 ft = 0.3048 m
1 mi = 1.609 km

In the examples and practice problems that follow, use the preceding chart to determine each conversion.

EXAMPLE 7: A computer record of 2,000 bytes requires a $3\frac{1}{4}$ in length on a magnetic tape. Determine this length in cm to the nearest hundredth.

Solution:

Using the conversion from the chart, we set up the ratio $\frac{2.540 \text{ cm}}{1 \text{ in}}$ so that the "in" will cancel. Multiplying and cancelling we have

$$3\frac{1}{4} \not{\text{in}} \left(\frac{2.540 \text{ cm}}{1 \not{\text{in}}}\right) = 8.255 \text{ cm} \cong 8.26 \text{ cm}$$

So, the $3\frac{1}{4}$ in length is approximately 8.26 cm.

PRACTICE PROBLEM 7: How many cm wide is a $\frac{1}{4}$ in drill bit?

EXAMPLE 8: The perimeter of a lot is 3.6 km. What is the perimeter to the nearest hundredth of a mile.

Solution:

Since the conversion in the chart is 1 mi = 1.609 km, the ratio should be set up as $\frac{1 \text{ mi}}{1.609 \text{ km}}$ for the "km" to cancel. Multiplying and cancelling gives us

$$3.6 \not{\text{km}} \left(\frac{1 \text{ mi}}{1.609 \not{\text{km}}}\right) = 2.237 \text{ mi} \cong 2.24 \text{ mi}$$

Therefore the perimeter of the lot is approximately 2.24 mi.

PRACTICE PROBLEM 8: A car has the gas capacity to travel 620 km on one tank of gas. To the nearest mile, how many miles can it travel on one tank of gas?

EXAMPLE 9: The height of a room is 3.2 meters. Find the height of the room to the nearest tenth of a yard.

Solution:

We do not have a direct conversion from meters to yards. We'll first convert the meters to feet (1 ft = 0.3048 m) and then convert the feet to yards. Note how the ratios are set up in the problem so that the appropriate units will cancel.

$$3.2 \not{\text{m}} \left(\frac{1 \not{\text{ft}}}{0.3048 \not{\text{m}}}\right)\left(\frac{1 \text{ yd}}{3 \not{\text{ft}}}\right) = 3.49 \text{ yd} \cong 3.5 \text{ yd}$$

The room is approximately 3.5 yd high.

PRACTICE PROBLEM 9: A student athlete can broadjump 5 yd. Determine this distance to the nearest tenth of a meter.

ANSWERS FOR PRACTICE PROBLEMS

1. (a) **2.** (c) **3.** (b) **4.** (a) **5.** (b) or (c)

6. (a) **7.** 0.635 cm **8.** 385 mi **9.** 4.6 m

EXERCISE SET 9.2A

Skill Development

Determine the following conversions. Round all results to the nearest hundredth.

1. 4 in = ________ cm
2. 7 ft = ________ m
3. 20 km = ______ mi
4. 3.6 ft = ________ m
5. 18.5 cm = ________ in
6. 126.85 mi = ________ km
7. $2\frac{1}{4}$ in = ________ cm
8. $6\frac{3}{8}$ mi = ________ km
9. 4.5 cm = ________ ft
10. 6.3 m = ________ yd

Applications

11. Select the metric unit(s) appropriate to measure the length of a nail.

    a) km **b)** m **c)** cm
12. Select the metric unit(s) appropriate to measure the height of a 20-story building.

    a) km **b)** m **c)** mm
13. Choose the value that best estimates the height of an adult male.

    a) 180 cm **b)** 18 km **c)** 14 mm **d)** 3 dm
14. Choose the value that best estimates the length of a large paper clip.

    a) 3 m **b)** 3 cm **c)** 30 mm **d)** 30 dam
15. Bobby has calculated that he drives his car about 42,000 km in a year. How many miles is that to the nearest mile?
16. The perimeter of a building is 1,200 feet. Find the perimeter in meters to the nearest meter.
17. A coffee table has a height of 20 inches. What is its height in centimeters to the nearest centimeter?

For Some Extra Thought

18. A publishing company estimates that 500 pages of text have a width of 2.5 inches. Find the width in centimeters of 700 pages of text.
19. Determine the better buy—2 feet of telephone cord for $3.25 or 60 cm of cord for $3.10.

Review Problems for Maintaining Your Skills

20. Write 21,608,003 in words.

21. $\frac{11}{12} - \frac{8}{15} =$ **22.** $5\frac{2}{7} - 1\frac{1}{2} =$

23. $8.67 \times 1.02 =$

24. To the nearest tenth of a percent, 5 is what percent of 17?

25. Find 2.25% of 12,500.

26. Use cross-multiplication to solve: $\frac{T}{5} = \frac{20}{25}$

27. $(-17) - (-23) =$ **28.** Solve: $3x - 2 = 5x + 18$

29. $4^5 =$ **30.** $(-5)^{-7}(-5)^8 =$ **31.** $\frac{8^7}{8^5} =$

32. Write 56,200,000 in scientific notation.

33. Write as a whole number or decimal.

a) 5.7×10^4 **b)** 8.1×10^{-5}

34. Multiply and express your result in scientific notation.
$(3.4 \times 10^5) \times (6.8 \times 10^{-2}) =$

35. Divide and express your result in scientific notation.
$\frac{4.5 \times 10^8}{1.5 \times 10^{-4}}$

36. Find the square roots of 64.

37. Find the value of $\sqrt{144}$.

38. Find two consecutive integers that bound $\sqrt{72}$.

39. Use a calculator to approximate $\sqrt{133}$ to the nearest tenth.

40. Determine the following conversions.

a) 18 yd = __________ ft **b)** 6 dam = __________ cm

EXERCISE SET 9.2B

Skill Development

Determine the following conversions. Round all results to the nearest hundredth.

1. 14 mi = __________ km
2. 8 m = __________ ft
3. 16 in = __________ cm
4. 3.75 ft = __________ m
5. 2.6 km = __________ mi
6. 4.1 cm = __________ in
7. $12\frac{3}{4}$ in = __________ cm
8. $10\frac{1}{8}$ mi = __________ km
9. 6.3 cm = __________ ft
10. 8.2 m = __________ yd

Applications

11. Select the metric unit(s) appropriate to measure the length of a landing strip on an airfield.

    a) km **b)** hm **c)** mm

12. Select the metric unit(s) appropriate to measure the length of a straight pin.

    a) km **b)** m **c)** mm

13. Choose the value that best estimates the length of a baseball bat.

    a) 1 m **b)** 1 km **c)** 1 mm **d)** 1 cm

14. Choose the value that best estimates the diameter of the earth.

    a) 13,000 dm **b)** 13 m **c)** 13,000 km **d)** 13 dam

15. An express mail service uses a conveyer belt with a length of 2,000 feet. What is the length to the nearest meter?

16. The length of a manufacturing plant is 1.2 miles. Find its length to the nearest tenth of a kilometer.

17. A pencil is 15.2 cm long. Find its length to the nearest inch.

For Some Extra Thought

18. An auto company can buy 8 yards of car molding for \$2.50 or 9 meters for \$2.80. Which length should the company buy?

19. If a character requires 0.4 cm width on a video display, how many characters, to the nearest unit, can be displayed on an 11-inch-wide video display?

Review Problems for Maintaining Your Skills

20. Write 1,350,000 in words.

21. $\frac{2}{5} - \frac{2}{7} =$ **22.** $4\frac{2}{5} - 3\frac{1}{3} =$

23. $1.203 \times 6.8 =$

24. To the nearest hundredth of a percent, 47 is what percent of 61?

25. Find 17.6% of $45,000.

26. Use cross-multiplication to solve: $\frac{12}{R} = \frac{1.2}{113}$

27. $(-2.6) - (+4.9) =$ **28.** Solve: $15x - 13 = 17$

29. $6^3 =$ **30.** $(-3)^{-1}(-3)^5 =$ **31.** $\frac{2^{10}}{2^5} =$

32. Write 0.000000000157 in scientific notation.

33. Write as a whole number or decimal.

a) 4.3×10^{-3} **b)** 1.9×10^5

34. Multiply and express your result in scientific notation.
$(7.2 \times 10^{12}) \times (1.3 \times 10^{-9}) =$

35. Divide and express your result in scientific notation.
$\frac{7.2 \times 10^{-3}}{1.8 \times 10^{-7}} =$

36. Find the square roots of 1.

37. Find the value of $\sqrt{49}$.

38. Find two consecutive integers that bound $\sqrt{31}$.

39. Use a calculator to approximate $\sqrt{513}$ to the nearest tenth.

40. Determine the following conversions.

a) 3 mi = ________ ft **b)** 4 km = ________ m

9.3 LIQUID VOLUME MEASUREMENTS

Preview Test

1. A saline solution requires 5 quarts of distilled water. How many gallons is this?
2. A $1\frac{1}{2}$ pt carton of orange juice contains how many quarts of orange juice?
3. A drug company packages its antacid product in 22 fl oz bottles. How many full bottles can be obtained from 10 pints of the antacid?
4. An office building is equipped with a 0.76 hL water heater. Determine the capacity in kL.
5. A family consumes 8 L of milk weekly. How many cL is this?
6. Select the metric unit appropriate to measure the volume of a full pitcher of beer.

   a) kL **b)** hL **c)** L **d)** mL
7. A hospital patient is to take 2 teaspoons of tetracycline at mealtimes. Find the equivalent amount in mL. (1 tsp = 4.9 mL).
8. A kerosene can holds 18 L of kerosene. How many gallons does it hold? (1 qt = 0.9 L).

In this section we will deal with measures such as a gallon of water and a liter of gasoline. The measures "gallon" and "liter" are for liquid volume. The procedures we will use will be very similar to those we developed for the length conversions in the first section. Following are the most common liquid volume measurements in the U.S. Customary system of measurement.

Liquid Volume Measurement (U.S. Customary)

3 teaspoons (tsp) = 1 tablespoon (tbsp)
2 tablespoons = 1 fluid ounce (fl oz)
8 fluid ounces = 1 cup (c)
2 cups = 1 pint (pt)
2 pints = 1 quart (qt)
4 quarts = 1 gallon (gal)

To convert liquid volume measurements, multiply by a ratio set up to cancel out appropriate units of measure.

ANSWERS FOR PREVIEW TEST

1. 1.25 gal **2.** $\frac{3}{4}$ qt or 0.75 qt **3.** 7 **4.** 0.076 kL **5.** 800 cL

6. (c) **7.** 9.8 mL **8.** 5 gal

EXAMPLE 1: A technician wants to pour distilled water from a 5-gallon container into 1-quart containers. How many 1-quart containers will be needed?

Solution

Reading from the preceding chart, there are 4 quarts to a gallon. Since we are converting gallons to quarts, we set up the ratio as $\frac{4 \text{ qt}}{1 \text{ gal}}$ so that "gal" will cancel out. Then by multiplying and cancelling we get

$$5 \cancel{\text{gal}} \left(\frac{4 \text{ qt}}{1 \cancel{\text{gal}}} \right) = 20 \text{ qt}$$

The technician will need 20 1-quart containers.

PRACTICE PROBLEM 1: How many 1-quart containers will the technician need for a 6-gallon container?

Chain conversions can occur with liquid volume measurements as they did with length conversions.

EXAMPLE 2: A physician has recommended that a parent give his ailing child 1 teaspoon of cough syrup four times a day for two weeks. What size bottle of cough syrup (in pints) should the parent purchase?

Solution:

Since the child needs the syrup four times a day for 14 days, he will need 4 times 14, or 56 teaspoons of syrup total. To convert teaspoons to pints, we will set up the ratios for the following sequences of conversions. (It may help you to look at the preceding chart as you follow this sequence.).

convert teaspoons to tablespoons (3 tsp to 1 tbsp)
convert tablespoons to fluid ounces (2 tbsp to 1 fl oz)
convert fluid ounces to cups (8 fl oz to 1 c)
convert cups to pints (2 c to 1 pt)

Carefully check the ratios in this calculation. Each ratio has been set up for the appropriate cancellation.

$$56 \cancel{\text{tsp}} \left(\frac{1 \cancel{\text{tbsp}}}{3 \cancel{\text{tsp}}} \right) \left(\frac{1 \cancel{\text{fl oz}}}{2 \cancel{\text{tbsp}}} \right) \left(\frac{1 \cancel{\text{c}}}{8 \cancel{\text{fl oz}}} \right) \left(\frac{1 \text{ pt}}{2 \cancel{\text{c}}} \right) = \frac{7}{12} \text{ pt}$$

The parent will need to purchase $\frac{7}{12}$ pints of cough syrup.

PRACTICE PROBLEM 2: If the parent needs to give the child cough syrup for an additional week, how much total syrup is needed in pints?

So far we've only worked with liquid volume measurements in the U.S. Customary system. But just as we can buy milk by the quart, we can also buy it by the liter, which is the basic liquid volume measurement for the metric system and is abbreviated "L". A liter is slightly larger than a quart. The following chart lists the metric liquid volume measurements. Note the similarities between this chart and the one for length measurements in the first section of this chapter.

Liquid Volume Measurements (Metric)

$$1 \text{ kiloliter (kL)} = 1{,}000 \text{ liters (L)}$$
$$1 \text{ hectoliter (hL)} = 100 \text{ liters}$$
$$1 \text{ dekaliter (daL)} = 10 \text{ liters}$$
$$1 \text{ deciliter (dL)} = \frac{1}{10} \text{ or } 0.1 \text{ liter}$$
$$1 \text{ centiliter (cL)} = \frac{1}{100} \text{ or } 0.01 \text{ liter}$$
$$1 \text{ milliliter (mL)} = \frac{1}{1000} \text{ or } 0.001 \text{ liter}$$

We can convert measurements in the metric system using ratios as we did in the U.S. Customary system. However, it's simpler to make use of the pattern of tens in the metric measurements and move the decimal point to do the conversions. Following is an example of liquid volume metric conversion. Compare this process with our conversions with metric length measurements.

EXAMPLE 3: The gas tank of a compact car has a capacity of about 460 deciliters. How many liters is this?

Solution:
Set up a chart for liter measurement similar to the chart for the meter measurements. Then count the number of blocks from deciliters to liters, noting the direction.

kilo- hecto- deka- liter deci- centi- milli-

1 block to the left

Now move the decimal point in 460 one place to the left. (Recall that the whole number 460 has an understood decimal point on the right).

46.0.

1 decimal place to the left

Thus, the capacity of the gas tank is 46 liters.

PRACTICE PROBLEM 3: The capacity for a gas tank for a medium-sized car is 8.5 daL. What is the capacity in liters?

EXAMPLE 4: A building is heated with an oil system using a 2.12-kL oil tank. How many liters of oil are needed to fill the tank?

Solution:
Set up the chart and count the number of blocks from kiloliters to liters.

kilo- hecto- deka- liter deci- centi- milli-

3 blocks to the right

Now move the decimal point in 2.12 three places to the right.

2.120.

3 places to the right

Notice that a zero is added as a placeholder. Therefore, the tank will be filled with 2,120 liters of oil.

PRACTICE PROBLEM 4: A solution requires 12 mL of water. How many centiliters is this?

Now that we've practiced converting liquid volume measurements within systems, we can convert measurements between systems. First let's become familiar with some metric estimations of U.S. Customary Unit measurements.

U.S. Customary/Metric Liquid Volume Measurements

U.S. Customary Unit	*Metric Estimate*
1 qt	1 L
1 fl oz	30 mL
1 tsp	5 mL

Notice that the liter is used to estimate quarts, and the millileter is used to estimate fluid ounces and teaspoons. We will use these conversions as estimates only. More accurate conversions will be used later.

EXAMPLE 5: Select the metric unit(s) appropriate to measure a dosage of medicine.

a) mL **b)** L **c)** cL

Solution:
A dosage of medicine is a small volume, so a liter is too large. Answer (b) can be eliminated. Answer (a) milliliter is a small unit, so it is appropriate. Answer (c) centiliter is a small unit, so it is also appropriate.

EXAMPLE 6: Select the metric unit(s) appropriate to measure the volume of gas needed to fill an automobile gas tank.

a) mL **b)** L **c)** kL

Solution:
Since a gas tank usually holds at least 10 gallons of gas, answer (a) milliliter is too small. Answer (b) liter is similar to a quart, so it is an appropriate measure. Answer (c) kiloliter is 1,000 times greater than a liter, so it is too large.

EXAMPLE 7: Select the metric unit(s) appropriate to measure the volume of air in a multi-story office building.

a) L **b)** mL **c)** kL

Solution:
The volume of air is very large, so (c) kiloliters is appropriate. Answer (a) liters could be used, but it would be a large number. Answer (b) milliliters is too small.

Now try these practice problems.

PRACTICE PROBLEM 5: Select the metric unit(s) appropriate to measure the vanilla extract used to make a dozen brownies.

a) kL **b)** mL **c)** L

PRACTICE PROBLEM 6: Select the metric unit(s) appropriate to measure the volume of a commercial size freezer.

a) L **b)** kL **c)** cL

PRACTICE PROBLEM 7: Select the metric unit(s) appropriate to measure the quantity of milk consumed by a family in one day.

a) mL **b)** L **c)** kL

Now let's use the estimations given in the preceding chart to approximate some metric volumes.

EXAMPLE 8: Choose the value that best estimates the volume of a bottle of bleach.

a) 4 kL **b)** 50 L **c)** 2 L

Solution:
A bottle of bleach usually contains about 2 quarts of liquid. Since a liter is approximately one quart, answer (c) is the best estimate.

EXAMPLE 9: Choose the value that best estimates the volume of a cup of coffee.

a) 8 L **b)** 8 mL **c)** 240 mL

Solution
Since a cup of coffee is about 8 ounces and each ounce is about 30 mL, the best estimate is 240 mL, answer (c).

PRACTICE PROBLEM 8: Choose the value that best estimates the volume of a jug of cider.

a) 2 L **b)** 5 mL **c)** 0.01 cL

PRACTICE PROBLEM 9: Choose the value that best estimates the adult dose of cough syrup.

a) 30 mL **b)** 3 cL **c)** 3 L

The U.S. Customary/metric conversions we've been using thus far are approximations. The following chart shows more accurate conversions.

U.S. Customary/Metric Liquid Volume Conversions

1 qt = 0.9 L

1 fl oz = 29.6 mL

1 tsp = 4.9 mL

With the information from this chart we can use ratios to find more accurate conversions between U.S. Customary and metric measurements.

EXAMPLE 10: A metal working plant receives machine lubricant in 120 mL containers. How many fluid ounces is this to the nearest tenth?

Solution:

Since 1 fl oz = 29.6 mL, use the ratio $\frac{1 \text{ fl oz}}{29.6 \text{ mL}}$ so that the mL will cancel. Multiplying and cancelling, we get

$$120 \cancel{\text{mL}} \left(\frac{1 \text{ fl oz}}{29.6 \cancel{\text{mL}}}\right) = 4.1 \text{ fl oz}$$

Therefore, the containers hold approximately 4.1 fl oz of lubricant.

PRACTICE PROBLEM 10: An agricultural research company allots 20 fl oz of water a week for each of its smaller plants. How many mL is this?

EXAMPLE 11: A hospital orders antiseptic liquid in 2-gallon containers. Find the number of liters in each container.

Solution:
Since we don't have a direct conversion from gallons to liters, we will first change gallons to quarts (4 qt = 1 gal) and then change quarts to liters (1 qt = 0.9 L). Note the cancellation that occurs in the following multiplication.

$$2 \cancel{\text{gal}} \left(\frac{4 \cancel{\text{qt}}}{1 \cancel{\text{gal}}}\right)\left(\frac{0.9 \text{ L}}{1 \cancel{\text{qt}}}\right) = 7.2 \text{ L}$$

Thus, each container has approximately 7.2 L of liquid.

PRACTICE PROBLEM 11: A commercial doughnut recipe requires 210 mL of liquid. Calculate the number of tablespoons needed to the nearest tenth.

ANSWERS FOR PRACTICE PROBLEMS

1. 24 1-quart containers **2.** $\frac{7}{8}$ pt **3.** 85 L **4.** 1.2 cL

5. (b) **6.** (a) **7.** (b) **8.** (a) **9.** (a) or (b)

10. 592 mL **11.** 14.3 tablespoons

EXERCISE SET 9.3A

Skill Development

Determine the following conversions.

1. 20 qt = __________ gal
2. 7 pt = __________ c
3. 1.5 tbsp = __________ tsp
4. 4 fl oz = __________ c
5. $2\frac{1}{2}$ qt = __________ gal
6. $5\frac{2}{3}$ qt = __________ pt
7. 20.5 tsp = __________ tbsp
8. 3 gal = __________ pt
9. 240 fl oz = __________ qt
10. 4,800 tsp = __________ gal
11. 8.25 qt = __________ tsp
12. 5.63 cL = __________ dL
13. 5 kL = __________ L
14. 18 dL = __________ hL
15. 70.2 mL = __________ cL
16. 9 daL = __________ mL
17. 200 cL = __________ daL
18. 18 tsp = __________ mL
19. 74 mL = __________ fl oz
20. 18 L = __________ qt
21. 8 fl oz = __________ mL

Applications

22. A utility company billed a business for 160 gallons of fuel oil. How many quarts is this?

23. A pharmacist orders 320 fl oz of a solution. Determine the number of pints she has ordered.

24. A vending machine is designed to dispense coffee in 240 mL amounts. Determine the amount in liters.

25. A cooling unit has a capacity of 0.5 kL. What is the capacity in dekaliters?

For Some Extra Thought

26. A winery has produced 32,000 gallons of wine. How many 3L bottles will this fill?

27. Which is the better buy, 4 oz of machine oil for $1.15 or 20 cL for $1.95?

28. The contents of 3 bottles of an antiseptic solution are to be combined into liter bottles. How many liter bottles are needed if the contents are 330 cL, 2,500 mL, and 52 dL?

Review Problems for Maintaining Your Skills

29. Round 12,687,102 to the nearest million.

30. $3\frac{2}{5} + 1\frac{1}{4} =$

31. $\frac{5}{6} \times \frac{1}{5} =$

32. Divide 28.1 by 1.06 and round the quotient to the nearest hundredth.

33. Find 6.8% of $750.

34. 4% of what amount equals $45.80?

35. $(-3)(-7.2)(-1) =$

36. Solve: $3w - 5 = 5w - 3$

37. $(-5)^{-3}(-5)^{7} =$

38. $\frac{3^2}{3^{-2}} =$

39. Write 41,200,000 in scientific notation.

40. Write 1.6×10^5 as a whole number.

41. Divide and express your result in scientific notation.
$\frac{8.4 \times 10^5}{4.2 \times 10^3} =$

42. Find the value of $-\sqrt{144}$.

43. Use a calculator to approximate $\sqrt{140}$ to the nearest tenth.

44. Determine the following conversions.
a) 15 dam = __________ hm **b)** 24.6 in = __________ ft

45. Determine the following conversions.
a) 30 km = __________ m **b)** 7 yd = __________ m

EXERCISE SET 9.3B

Skill Development

Determine the following conversions.

1. 24 fl oz = __________ c

2. 9 qt = __________ pt

3. 3.6 c = __________ fl oz

4. 10 tsp = __________ tbsp

5. $3\frac{1}{2}$ c = __________ pt

6. $8\frac{3}{5}$ gal = __________ qt

7. 8.02 pt = __________ qt

8. 5 qt = __________ c

9. 640 tsp = __________ qt

10. 320 tbsp = __________ gal

11. 2.35 gal = __________ c

12. 32.9 daL = __________ hL

13. 6 L = __________ mL

14. 24 cL = __________ daL

15. 8.04 dL = __________ L

16. 2 hL = __________ cL

17. 350 mL = __________ kL

18. 4 qt = __________ L

19. 81 L - __________ qt

20. 49 mL = __________ tsp

21. 20 fl oz = __________ mL

Applications

22. A can of spray paint contains 1.5 pts. Find the number of fluid ounces the can contains.

23. A cleaning agency used 22 1-quart containers of detergent. How many gallons is this?

24. A doctor prescribes 24 cL of medicine to be given daily to a patient. Find the dosage in deciliters.

25. A salesperson claimed expenses for 18 kL of gasoline. Determine the number of hectoliters claimed.

For Some Extra Thought

26. The participants at an ice cream eating contest consumed the following quantities of ice cream: 3 pts, 2 qts, $2\frac{1}{2}$ qts, 0.9 gal, 1.25 gal, and 5 pts. How many total gallons of ice cream were consumed?

27. A soda company produced 5 kL of soda. How many 2-qt containers can it fill?

28. Which is the better buy, 0.7 oz of cinnamon at \$1.39 or 2 cL at \$1.25?

Review Problems for Maintaining Your Skills

29. Round 13,608 to the nearest ten.

30. $2\frac{4}{5} + 3\frac{2}{5} =$

31. $\frac{9}{10} \times \frac{2}{3} =$

32. Divide 13 by 1.7 and round the quotient to the nearest tenth.

33. Find 150% of 34.

34. 12% of what amount equals \$18?

35. $(4.2)(-5)(-5) =$

36. Solve: $4(3 - x) = x + 7$

37. $(-2)^{-7}(-2)^{10} =$

38. $\frac{5^{10}}{5^7} =$

39. Write 0.000000002 in scientific notation.

40. Write 6.3×10^{-5} as a decimal.

41. Multiply and express your result in scientific notation.
$(4.3 \times 10^{12}) \times (5.1 \times 10^{-9}) =$

42. Find the value of $\sqrt{121}$.

43. Use a calculator to approximate $\sqrt{89}$ to the nearest hundredth.

44. Determine the following conversions.

a) 24.3 ft = __________ yd b) 2400 yd = __________ in

45. Determine the following conversions.

a) 12.6 in = __________ cm b) 87 mi = __________ km

9.4 WEIGHT MEASUREMENTS

Preview Test

1. A supermarket purchases plastic bags having a strength of 10 pounds. How many ounces can each bag hold?
2. An electrical component manufacturer orders a 1,200-lb delivery of ohmmeters. Determine the weight of the order in tons.
3. A stone mason lays 1,000 stones with an average weight of 32 ounces each. Find the weight of the stones in tons.
4. A frozen dinner has a weight of 250 g. Find the weight in kilograms.
5. A solution requires 5 dg of terramycin. What is the weight in centigrams?
6. Select the metric unit(s) appropriate to measure the weight of an apple.

   a) kg **b)** g **c)** dg
7. Select the metric unit(s) appropriate to measure the weight of a construction crane.

   a) kg **b)** g **c)** mg
8. An airline estimates the average weight of passenger luggage to be 25 pounds. What is the weight in kilograms? (1 lb = 0.45 kg)
9. If a single egg has a weight of 50 g, determine the number of pounds of a dozen eggs to the nearest tenth.

So far we've dealt with such matters as the distance from Boston to New York (length measurements) and the comparison of a quart of milk and a liter of milk (liquid volume measurements). We can now consider weight measurements. The most common weight measurements in the U.S. Customary system are listed in the following chart.

Weight Measurements (U.S. Customary)

1 pound (lb) = 16 ounces (oz)

2000 pounds = 1 ton (t)

As with previous conversions that we've done in the U.S. Customary system, we can use ratios to convert weight measurements. Study these two conversion examples and then try the practice problems.

EXAMPLE 1: An alloy contains 83.2 ounces of aluminum. How many pounds is this?

Solution:

Since there are 16 ounces for each pound, we will use the ratio $\frac{1 \text{ lb}}{16 \text{ oz}}$ so that the oz will

ANSWERS FOR PREVIEW TEST

1. 160 oz **2.** 0.6 t **3.** 1 t **4.** 0.25 kg **5.** 50 cg
6. (b) or (c) **7.** (a) **8.** 11.25 kg **9.** 1.3 lb

cancel out. Multiplying and cancelling, we get

$$83.2\,\cancel{oz}\left(\frac{1\text{ lb}}{16\,\cancel{oz}}\right) = 5.2\text{ lb}$$

The alloy contains 5.2 pounds of aluminum.

EXAMPLE 2: An imprint machine weighs $\frac{1}{40}$ ton. How many ounces is this?

Solution:
First convert the tons to pounds (2000 lb = 1 t), then convert the pounds to ounces (16 oz = 1 lb).

$$\frac{1}{40}\cancel{t}\left(\frac{2000\,\cancel{lb}}{1\,\cancel{t}}\right)\left(\frac{16\text{ oz}}{1\,\cancel{lb}}\right) = 800\text{ oz}$$

The machine weighs 800 ounces.

PRACTICE PROBLEM 1: An automobile weighs $\frac{3}{4}$ tons. What is its weight in pounds?

PRACTICE PROBLEM 2: A trucking firm needs to deliver 7,000 18-oz tins of sardines. What minimum weight (in tons) should the assigned truck be able to transport?

The basic weight measurement in the metric system is the gram, which is the weight of about two paper clips. The other metric weight measurements are defined as we did with length and liquid volume measurements.

Weight Measurements (Metric)

$$\begin{aligned}
1\text{ kilogram (kg)} &= 1000\text{ grams (g)}\\
1\text{ hectogram (hg)} &= 100\text{ grams}\\
1\text{ dekagram (dag)} &= 10\text{ grams}\\
1\text{ decigram (dg)} &= \frac{1}{10}\text{ or }0.1\text{ gram}\\
1\text{ centigram (cg)} &= \frac{1}{100}\text{ or }0.01\text{ gram}\\
1\text{ milligram (mg)} &= \frac{1}{1000}\text{ or }0.001\text{ gram}
\end{aligned}$$

Let's take a look at some examples of converting metric weight measurements.

EXAMPLE 3: The average weight of an adult female is approximately 63,000 g. What is the weight in kilograms?

Solution:
Set up the chart of weight measurements and count the number of blocks from grams to kilograms.

kilo- hecto- deka- gram deci- centi- milli-

3 blocks to the left

Now, move the decimal point for 63,000 three places to the left.

63,000.

3 places to the left

The average weight of an adult female is 63 kg.

EXAMPLE 4: A box of soap detergent has a weight of 4.8 dag. What is the weight of the detergent in decigrams?

Solution:
Set up the chart and count the number of blocks from dekagrams to decigrams.

kilo- hecto- deka- gram deci- centi- milli-

2 places to the right

Now move the decimal point for 4.8 two places to the right.

4.80.

two places to the right

The detergent has a weight of 480 dg.

PRACTICE PROBLEM 3: A bolt has a weight of 2.5 dag. What is the weight of the bolt in grams?

PRACTICE PROBLEM 4: The weight of a cold tablet is 0.5 g. What is the weight of the tablet in milligrams?

Now that we've practiced conversions within systems, let's work on conversions from one system to another. Some estimates for conversions between the U.S. Customary system weight and metric system weight are

$$1 \text{ oz} = 30 \text{ g}$$

$$1 \text{ lb} = \frac{1}{2} \text{ kg}$$

Notice that ounces and grams are comparable, as are pounds and kilograms. Keep this in mind as you read these examples.

EXAMPLE 5: Select the metric unit(s) appropriate to measure the weight of a VCR.

a) kg **b)** g **c)** mg

Solution:
Since a VCR is likely to be measured in pounds, the kilogram is an appropriate measure. The gram and kilogram measurements are too small.

EXAMPLE 6: Select the metric unit(s) appropriate to measure the weight of a slice of cheese.

a) kg **b)** g **c)** dg

Solution:
A slice of cheese will be very light, so either the gram or the decigram is appropriate. The kilogram is a larger measurement than is needed.

PRACTICE PROBLEM 5: Select the metric unit(s) appropriate to measure the weight of a vitamin tablet.

a) kg **b)** dag **c)** g

PRACTICE PROBLEM 6: Select the metric unit(s) appropriate to measure the weight of a desktop personal computer.

a) kg **b)** hg **c)** g

Now let's use the estimates to approximate weights of objects.

EXAMPLE 7: Choose the value(s) that best estimate(s) the weight of a 10-lb bag of potatoes.

a) 5 kg **b)** 5 g **c)** 150 g

Solution:
Since a pound is approximately 1/2 kg, (a) 5 kg is the best estimate.

EXAMPLE 8: Choose the value(s) that best estimate(s) the weight of a camera battery.

a) 3 kg **b)** 30 g **c)** 300 dg

Solution:
A camera battery is very light, perhaps weighing an ounce. Answer (b) 30 g is a good estimate. Since 300 dg is equivalent, it is also an appropriate estimate. Answer (a) 3 kg is too large a weight to estimate the camera battery.

PRACTICE PROBLEM 7: Choose the value(s) that best estimate(s) the weight of a box of crackers.

a) 10 kg **b)** 300 g **c)** 3 mg

PRACTICE PROBLEM 8: Choose the value(s) that best estimate(s) the weight of a dictionary.

a) 1 kg **b)** 1 g **c)** 10 cg

Let's see how ratios help us to obtain more exact measurement conversions.

U.S. Customary/Metric Weight Conversions

$$1 \text{ oz} = 28.4 \text{ g}$$

$$1 \text{ lb} = 0.45 \text{ kg}$$

EXAMPLE 9: A grocery warehouse has determined that the best-selling box of crackers is the 454 g size. However, the brand that the warehouse wants to purchase is sold in ounces, not grams. To the nearest whole ounce, what weight of box is needed?

Solution:

Since 1 oz = 28.4 g, set up the ratio $\frac{1 \text{ oz}}{28.4 \text{ g}}$ so that the grams will cancel out. Multiplying and cancelling, we get

$$454 \not{g} \left(\frac{1 \text{ oz}}{28.4 \not{g}} \right) = 15.9 \text{ oz} \cong 16 \text{ oz}$$

The warehouse should purchase boxes that weigh approximately 16 oz.

EXAMPLE 10: The motor for an industrial use pulley system weighs $\frac{3}{4}$ tons. What is its weight in kilograms?

Solution:
First convert the tons to pounds (1 t = 2000 lb), then convert the pounds to kilograms (1 lb = 0.45 kg). Notice that the ratios are set up in the following calculation so that the appropriate units will cancel out.

$$\frac{3}{4} \not{t} \left(\frac{2000 \not{lb}}{1 \not{t}} \right) \left(\frac{0.45 \text{ kg}}{1 \not{lb}} \right) = 675 \text{ kg}$$

The motor weighs 675 kg.

In Example 9, the conversion factors were divided, and in Example 10 they were multiplied. How do you know which operation to use? If you set up your ratio so that the appropriate units will cancel, the ratio will dictate the operation. Keep this in mind as you do these Practice Problems.

PRACTICE PROBLEM 9: An IV solution requires 50 ounces of sucrose per solution. How many grams of sucrose is needed?

PRACTICE PROBLEM 10: Determine the tonnage of an 1800-kg vehicle.

ANSWERS FOR PRACTICE PROBLEMS

1. 1500 lb	**2.** 4 t	**3.** 25 g	**4.** 500 mg	**5.** (c)
6. (a) or (b)	**7.** (b)	**8.** (a)	**9.** 1420 g	**10.** 2 t

EXERCISE SET 9.4A

Skill Development

Determine the following conversions.

1. 12 lb = __________ oz
2. 6000 lb = __________ t
3. 12.8 oz = __________ lb
4. 3.2 t = __________ lb
5. $8\frac{3}{4}$ lb = __________ oz
6. 24 oz = __________ lb
7. 32,000 oz = __________ t
8. 2.1 t = __________ oz
9. 1.2 g = __________ dg
10. 6 hg = __________ kg
11. 12 cg = __________ dag
12. 5 dg = __________ mg
13. 15 oz = __________ g
14. 100 lb = __________ kg
15. 710 g = __________ oz
16. 22.5 kg = __________ lb

Applications

17. A pocketbook weighs 80 oz. How many pounds is this?

18. A trucking company has been commissioned to move three 500-pound milling machines. Find the total weight in tons that the company must move.

19. A construction company orders cinderblock in 900-gram weights. What is the weight in kilograms?

20. A package weighing 255 grams contains 12 slices of cheese. Determine the weight in milligrams of one slice of cheese.

For Some Extra Thought

21. A Pekinese dog weighs 2.7 kg. A German shepherd dog weighs 50 pounds. How much heavier is the shepherd than the Pekinese?

22. Which is the better buy, 500 lb of grain for $425 or one ton of grain for $1600?

23. A moving van is loaded with the following items:

one sofa weighing 300 pounds
two tables weighing 54 kg
one wardrobe weighing 2250 g

What total weight, to the nearest pound, will the van be transporting?

Review Problems for Maintaining Your Skills

24. Write 2,075,300 in words.

25. $\frac{4}{5} - \frac{1}{3} =$

26. $3\frac{2}{3} \div 7\frac{1}{3} =$

27. $24.8 \times 1.26 =$

28. To the nearest tenth of a percent, 29 is what percent of 43?

29. 4% of what amount equals \$29.60?

30. Solve: $5 - 6x = 7 + 2(x + 3)$

31. $2^4 \cdot 2^2 =$

32. $\frac{5^8}{5^3} =$

33. Write 0.00000403 in scientific notation.

34. Write 4.8×10^5 as a whole number.

35. Multiply and express your result in scientific notation.
$(3.2 \times 10^{-6}) \times (1.5 \times 10^{-2}) =$

36. Use a calculator to approximate $\sqrt{147}$ to the nearest tenth.

Determine the following conversions.

37. 24.6 in = __________ ft

38. 0.06 m = __________ dam

39. 60.96 cm = __________ in

40. 7.4 m = __________ km

41. 40 qt = __________ gal

42. 6.72 cL = __________ dL

EXERCISE SET 9.4B

Skill Development

Determine the following conversions.

1. 3 t = __________ lb
2. 48 oz = __________ lb
3. 1450.4 lb = __________ t
4. 1.6 lb = __________ oz
5. $2\frac{1}{4}$ t = __________ lb
6. 30 oz = __________ lb
7. 16,000 oz = __________ t
8. 1.6 t = __________ oz
9. 7.02 hg = __________ dag
10. 8 cg = __________ dg
11. 23 mg = __________ g
12. 1 kg = __________ g
13. 20 lb = __________ kg
14. 25 oz = __________ g
15. 6.75 kg = __________ lb
16. 426 g = __________ oz

Applications

17. A compact car weighs 2200 pounds. How many tons does it weigh?

18. A compact computer weighs 8 pounds. Find its weight in ounces.

19. An athlete begins an exercise routine to gain 0.3 kg per week. What weight gain in grams can be expected after two weeks?

20. A patient is required to take a cold capsule containing 500 mg of medication. Determine the centigrams of medication.

For Some Extra Thought

21. A manual lathe weighs 350 pounds. A computer lathe weighs 225 kg. Which machine is heavier and by how much?

22. An I-beam weighs 35 pounds per foot. How much, to the nearest pound, would a 5-meter beam weigh?

23. Which is the better buy, 13 oz of Brand A oatmeal for \$1.89 or 380 g of Brand B oatmeal for \$2.00?

Review Problems for Maintaining Your Skills

24. Write 32,450,000,000 in words.

25. $\frac{13}{15} - \frac{1}{10} =$ **26.** $6\frac{3}{4} \div 2\frac{1}{2} =$

27. $10{,}000 \times 0.00456 =$

28. To the nearest hundredth of a percent, 2 is what percent of 7?

29. 1.5% of what amount equals \$18.75?

30. Solve: $-6x + 9 = -5x - 17$

31. $(-4)^{-5}(-4)^{7} =$ **32.** $\frac{3^6}{3^4} =$

33. Write 120,000 in scientific notation.

34. Write 2.7×10^3 as a whole number.

35. Multiply and express your result in scientific notation.
$(3.4 \times 10^{12}) \times (6.7 \times 10^{-8}) =$

36. Use a calculator to approximate $\sqrt{200}$ to the nearest hundredth.

Determine the following conversions.

37. 39 yds = __________ ft **38.** 3.6 cm = __________ mm

39. 4 mi = __________ km **40.** 7.4 ft = __________ cm

41. 6 gal = __________ pt **42.** 4 daL = __________ mL

PRACTICE TEST 9A

Determine the following conversions.

1. 3 mi = __________ ft
2. 540 in = __________ yd
3. 56.4 mm = __________ cm
4. 0.04 m = __________ dam
5. 9 ft = __________ m
6. 15 mi = __________ km
7. 60.96 cm = __________ ft
8. 24 in = __________ cm
9. 4.5 c = __________ pt
10. 4.35 gal = __________ tbsp
11. 450 mL = __________ kL
12. 6 qt = __________ L
13. 15 lb = __________ oz
14. 2.4 g = __________ dg
15. 3.4 t = __________ oz
16. 200 lb = __________ kg

PRACTICE TEST 9B

Determine the following conversions.

1. 15.6 in = ________ ft
2. 10 mi = ________ yd
3. 4 km = ________ dm
4. 24 dam = ________ hm
5. 6.75 ft = ________ m
6. 3.81 cm = ________ ft
7. 1.4 mi = ________ km
8. 12 ft = ________ cm
9. 6.24 hL = ________ L
10. 10.4 qt = ________ pt
11. 840 tsp = ________ qt
12. 12 cL = ________ daL
13. 36 kg = ________ lb
14. 6 dg = ________ mg
15. 14.6 oz = ________ lb
16. 3,600 lb = ________ t

CUMULATIVE REVIEW EXERCISES FOR CHAPTERS 1 THROUGH 9

1. Write 12,643,000 in words.

2. Round 45,616 to the nearest hundred.

In 3 through 6, reduce each fraction to its lowest terms and express any result that is an improper fraction as a mixed number.

3. $3\frac{1}{3} \times 2\frac{2}{5} =$

4. $\frac{11}{12} \div \frac{2}{12} =$

5. $\frac{5}{6} + \frac{1}{2} =$

6. $5\frac{3}{4} - 2\frac{1}{3} =$

7. Write 5.012 in words.

8. Round 1.4506 to the nearest thousandth.
9. Round \$3.12455 to the nearest cent.
10. Divide 12.9 by 1.04 and round the quotient to the nearest hundredth.

11. $4.5 \times 6.7 - 1.45 =$
12. Convert 225% to a decimal.
13. Convert 0.00487 to a percent.

14. Convert 26% to a fraction reduced to its lowest terms.

15. 14 is what percent of 50?
16. Find 2.6% of \$8,000.
17. 5% of what amount equals \$64?
18. If a price decreases from \$200 to \$174, find the percent decrease.

19. Solve the proportion: $\frac{4}{5} = \frac{2}{R}$
20. $-13 - 20 + 40 =$
21. $(-21) - (+30) =$
22. $(+2.6) \times (-7.6) =$
23. $(-80) \div (-0.5) =$
24. Write in words: $5x - 7w + 2$
25. Solve: $7x - 5 = 9x + 19$
26. Solve: $4(3x - 7) = x - 17$
27. Write $(-1) \times (-1) \times (-1)$ in exponential form.
28. Write 5^6 as a product of factors.
29. Evaluate each term or expression.

    a) 6^2 **b)** $(-2)^5$ **c)** $8^2 - 7^2$

30. Find the value of each.

a) $3^4 \cdot 5^2$ **b)** $\left(\frac{2}{5}\right)^3$

31. Write each number in scientific notation.

a) 45,000,000 **b)** 0.000823

32. Write as a whole number or decimal.

a) 4.29×10^{-3} **b)** 3.44×10^5

33. Express the result in scientific notation.
$(6.3 \times 10^{12}) \times (2.2 \times 10^{-7})$

34. Find the square roots of 196.

35. Find the value of **a)** $-\sqrt{121}$ **b)** $\sqrt{225}$

36. Find two consecutive integers that bound $\sqrt{39}$.

37. Use a calculator to approximate each square root to the nearest tenth.

a) $\sqrt{43}$ **b)** $\sqrt{150}$ **c)** $\sqrt{77}$

In 38 through 43 determine each conversion.

38. 180 in = __________ ft **39.** 46 hm = __________ m

40. 7.3 mi = __________ km **41.** 666 tsp = __________ qt

42. 72 kg = __________ lb **43.** 740 mg = __________ g

CHAPTER 10

GEOMETRY

10.1 SOME COMMON GEOMETRIC FORMS AND PERIMETER

10.2 AREA

10.3 CIRCLES

10.4 VOLUME

How do you determine the length of a fence to enclose a back yard or the quantity of stripping to weatherstrip the windows in a house? To find these lengths, you have to know how to find the **perimeter** of the yard or windows.

How do you determine the number of square feet (1 ft by 1 ft) of tile to cover a basement floor or the number of square yards of carpet to cover a classroom floor? To find these measurements you have to know how to find the **area** of the basement or classroom floor.

When you purchase an air-conditioner for a room you have to know the size of the room in cubic units (1 ft by 1 ft by 1 ft). The size of a room in cubic units is its **volume**.

In order to learn about perimeter, area, and volume it is important to become familiar with some common geometric shapes (squares, rectangles, triangles, circles, and so on). Therefore, geometric shapes will be the focus of our attention in the first part of section 1 of this chapter.

10.1 SOME COMMON GEOMETRIC FORMS AND PERIMETER

Preview Test

1. The number of degrees in a complete revolution is ______.
2. A right angle is ______ degrees.
3. One-third of a revolution is ______ degrees.
4. Label the following using the appropriate terms, either straight angle or right angle.

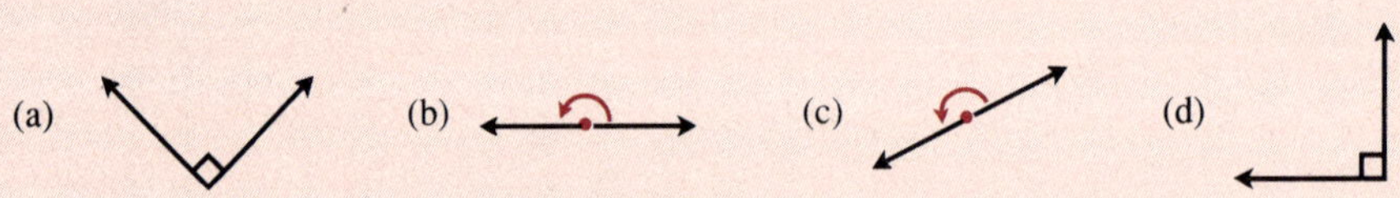

5. Label the following using all appropriate terms—perpendicular lines, parallel lines, intersecting lines, right angles.

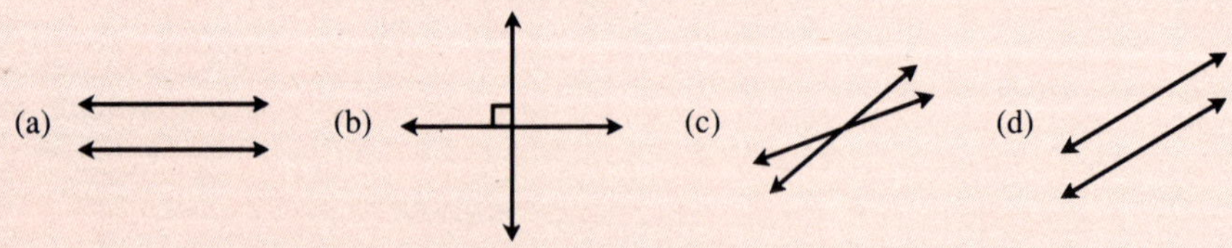

6. Label the following using all appropriate terms—triangle, rectangle, square, parallelogram, trapezoid, circle.

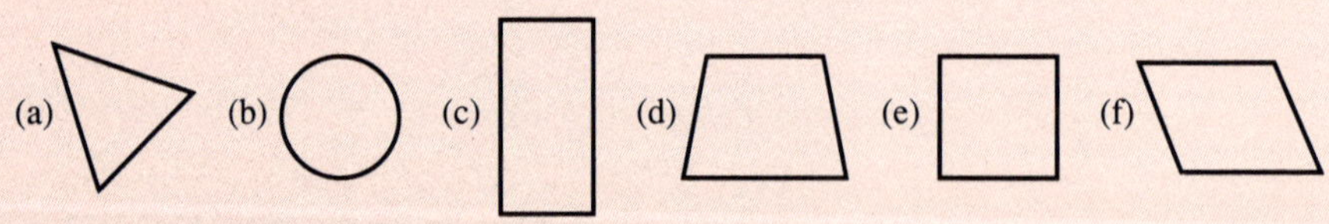

7. Metal edging is to be installed around the plate diagrammed below. What length of edging is needed for this process?

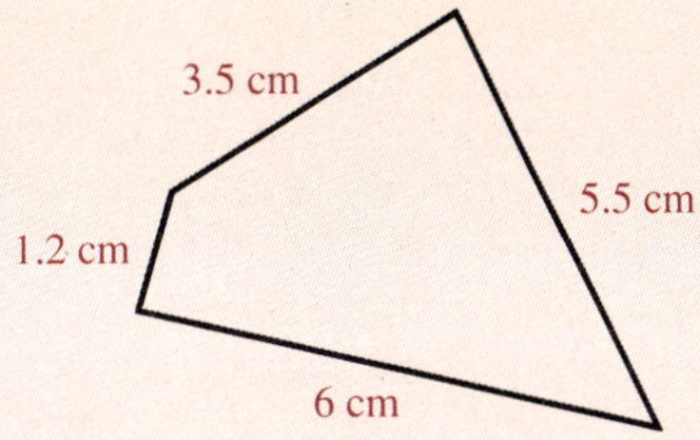

8. The baseboard around the edges of a triangular shaped room is to be finished off with quarter-round moulding. What quantity of moulding is needed?

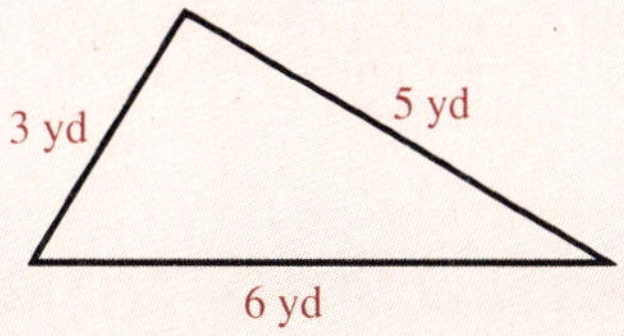

9. Cloth covers for the headrest of a certain model car require a $1\frac{2}{3}$ ft square piece of material. Using the formula for the perimeter of a square, determine the length of binding needed to finish the material for one headrest.

10. What's the minimum distance a security person travels when patroling the exterior of a rectangular warehouse with a length of 120 meters and a width of 35 meters?

Let's begin with some vocabulary that will be helpful in describing different figures.

An **angle** is formed by two rays (half lines) that have a common endpoint (called the **vertex**).

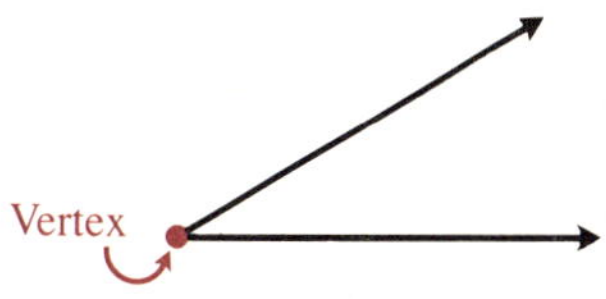

ANSWERS FOR PREVIEW TEST

1. 360 **2.** 90 **3.** 120

4. a) right angle **b)** straight angle **c)** straight angle **d)** right angle

5. a) parallel lines **b)** perpendicular lines, intersecting lines, right angles
c) intersecting lines **d)** parallel lines

6. a) triangle **b)** circle **c)** rectangle, parallelogram
d) trapezoid **e)** square, rectangle, parallelogram **f)** parallelogram

7. 16.2 cm **8.** 14 yd **9.** $6\frac{2}{3}$ ft **10.** 310 m

Angles can be measured in a standard unit called **degrees**, where one complete counterclockwise revolution of a ray back onto its initial position is measured as 360 degrees. Note that the notation for degrees is a superscript o. So, 360 degrees is written as 360°.

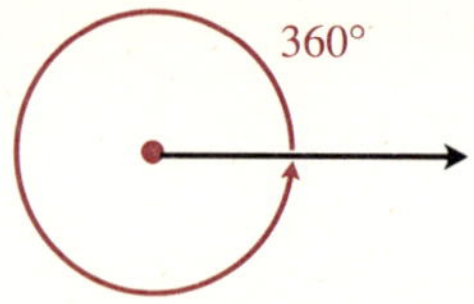

Measures of some other angles can be determined using 360° for one revolution.

EXAMPLE 1: Determine the measure of an angle that is half of a revolution.

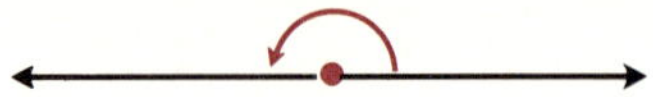

Solution:

Because "half" means 1/2 of (or times) 360°, we only have to multiply 360 by $\frac{1}{2}$.

The measure of half a revolution $= \frac{1}{2}(360^\circ) = 180^\circ$.

PRACTICE PROBLEM 1: Determine the angle that is one-quarter of a revolution.

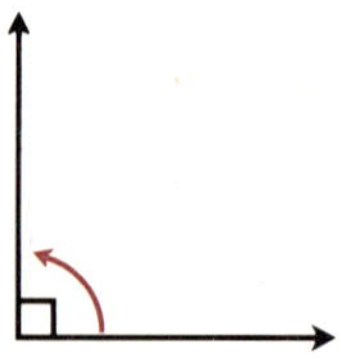

Some common angles have names. The angle measuring 180° is named a **straight angle**.

The angle measuring 90° is named a **right angle**. Right angles are identified with a special "box" symbol.

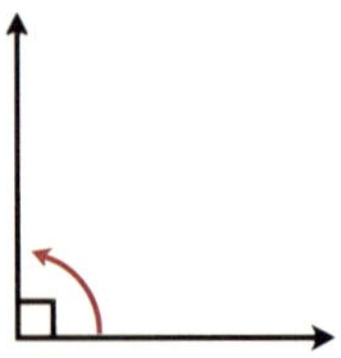

EXAMPLE 2: Label the following using the appropriate term, straight angle or right angle.

(a) 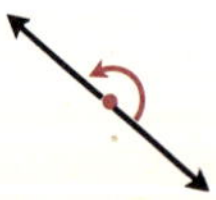 (b)  (c) 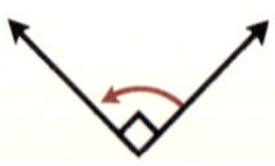 (d) 

Solution:
By observation, the straight angles (half a revolution) are (a) and (d). The right angles (quarter of a revolution) are (b) and (c).

PRACTICE PROBLEM 2: Label the following using the appropriate term, straight angle or right angle.

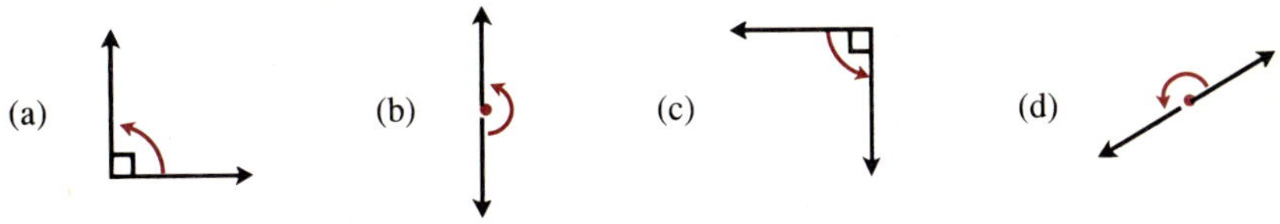

Just as we have certain names for some angles, we have certain names for some types of lines. When two lines cross one another they are called **intersecting lines**. Notice that the distance between the lines decreases up to the point where they cross, called the point of intersection.

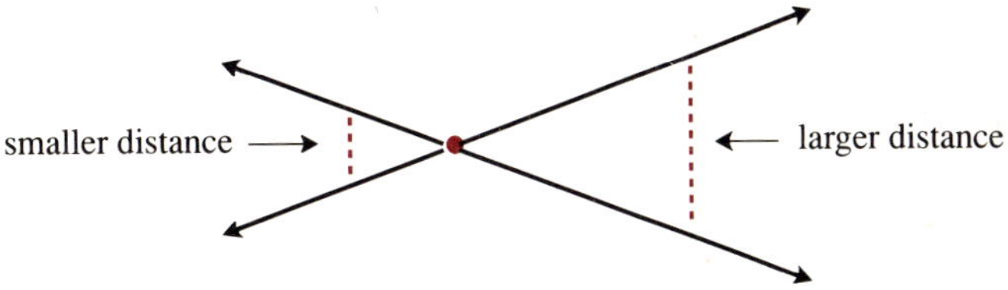

If the distance between the lines remains the same (never changes) the lines will never intersect. Lines that never intersect are called **parallel lines**.

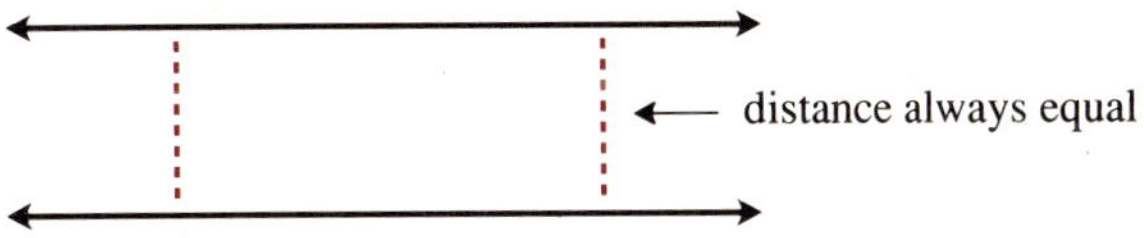

If the angles between the lines are right angles, the lines are called **perpendicular**.

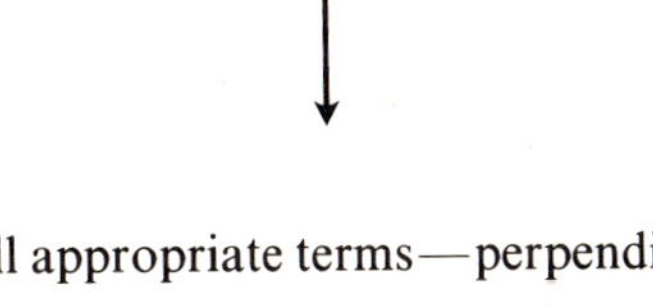

EXAMPLE 3: Label the following using all appropriate terms—perpendicular, parallel, intersecting, right angles.

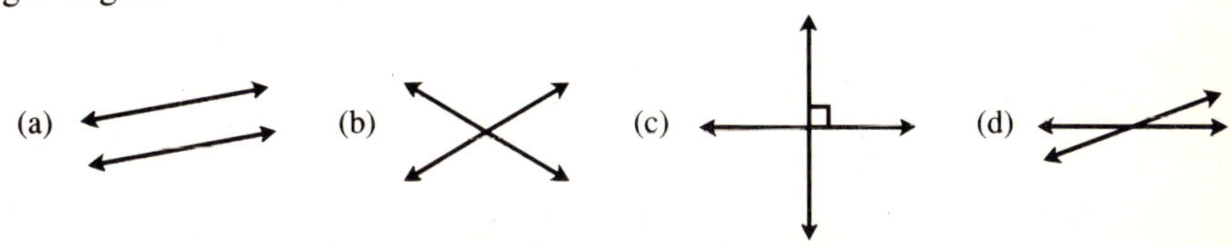

Solution:
The easiest lines to identify are the intersecting ones in (b), (c), and (d). The lines in (a) don't intersect, so they are parallel. The lines in (c) form right angles, so they are perpendicular.

PRACTICE PROBLEM 3: Label the following using all appropriate terms—perpendicular, parallel, intersecting, right angles.

(a)  (b)  (c) 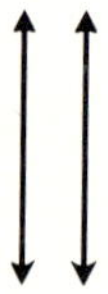 (d) 

The terms discussed so far can be used to describe various geometric figures. A **parallelogram** has four sides with opposite sides parallel.

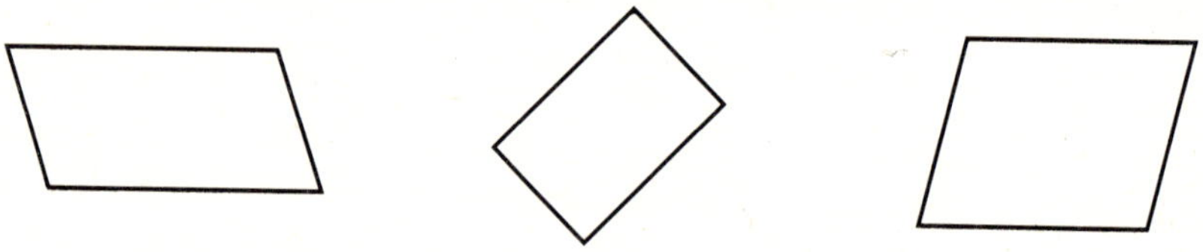

A **rectangle** is a parallelogram that has four perpendicular sides. Rooms are often shaped as rectangles.

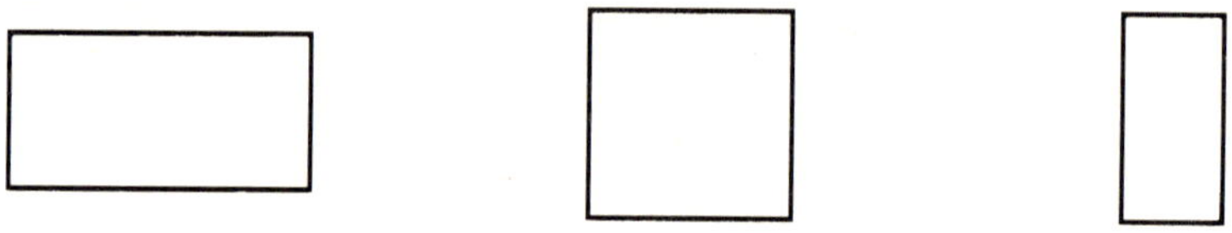

A **square** is a rectangle with four equal sides. The faces of a die are squares.

From the definitions of a parallelogram, rectangle, and square you can see that

1. there are two types of parallelograms, those that are rectangles and those that are not rectangles;

2. there are two types of rectangles, those that are squares and those that are not squares.

A **trapezoid** is a four-sided figure with one pair of sides parallel but not the same length. The front window of a car is approximately the shape of a trapezoid.

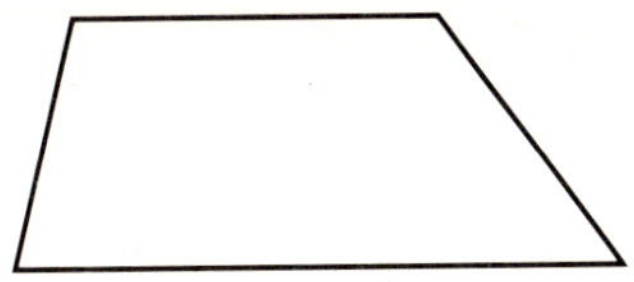

A **triangle** is a three-sided figure. The front of a tent may be shaped like a triangle.

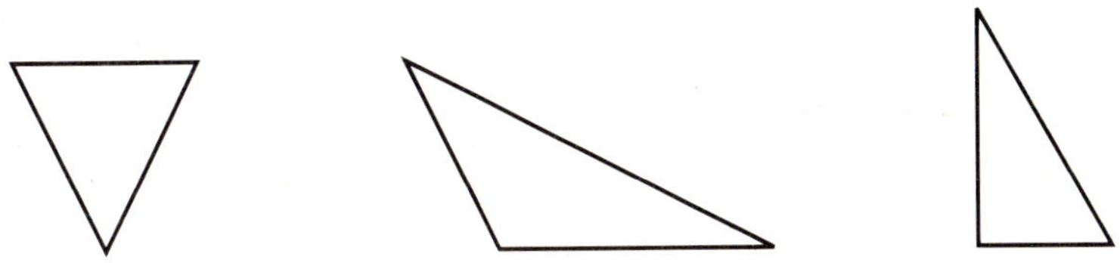

A **circle** has no sides. All the points on a circle are the same distance, called the radius, from a fixed point called its center.

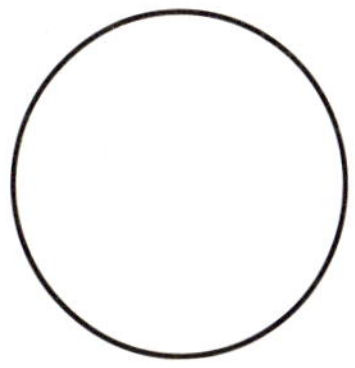

EXAMPLE 4: Label these figures using all the appropriate terms—triangle, rectangle, square, parallelogram, trapezoid, circle.

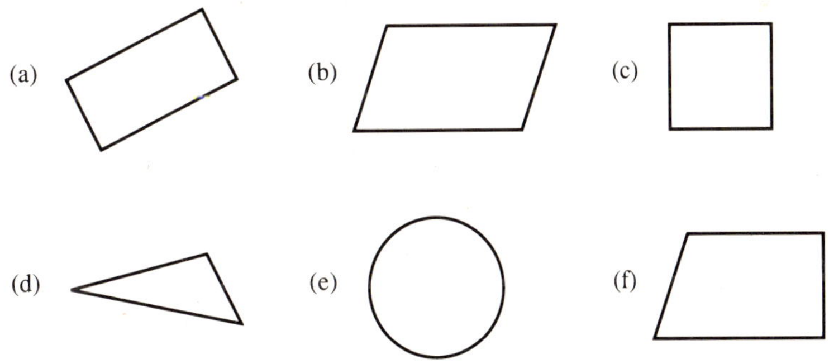

Solution:

a) parallelogram and rectangle
b) parallelogram
c) square, rectangle, and parallelogram
d) triangle
e) circle
f) trapezoid

PRACTICE PROBLEM 4: Label these figures using all the appropriate terms—triangle, rectangle, square, parallelogram, circle, trapezoid.

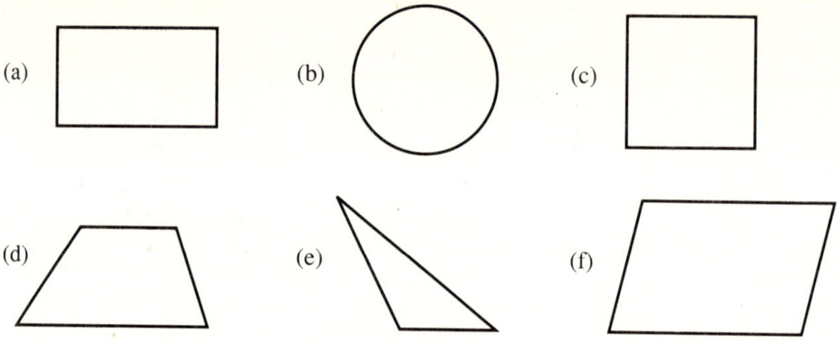

Now that you're familiar with some common geometric figures, we can discuss an important measurement associated with them. The **perimeter** of a figure is the distance around it. How do you determine the perimeter? Add the lengths of the sides.

PERIMETER = the sum of the lengths of the sides

EXAMPLE 5: Marks wants to build a fence enclosing his back yard. How many meters of fence will he need?

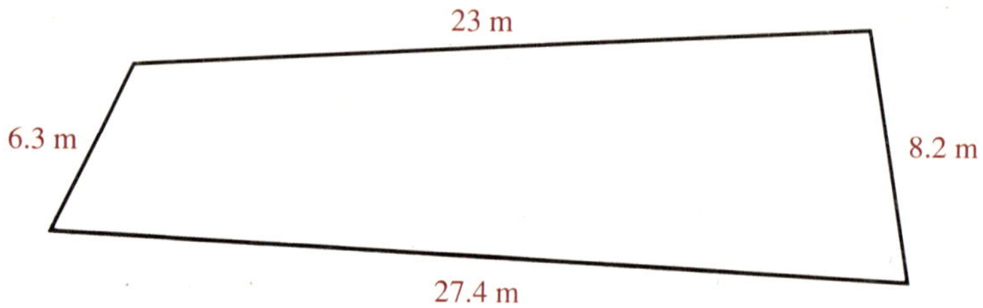

Solution:
To determine the length of fence, we will have to determine the perimeter of the yard. The lengths of the sides are 6.3 m, 23 m, 8.2 m, and 27.4 m. Adding the lengths to get the perimeter,

$$\begin{aligned}\text{PERIMETER} &= \text{sum of the lengths of the sides}\\ &= 6.3\text{ m} + 23\text{ m} + 8.2\text{ m} + 27.4\text{ m} = 64.9\text{ m}\end{aligned}$$

Mark will need 64.9 m of fence.

PRACTICE PROBLEM 5: A carpet, illustrated below, needs binding attached to its edges. How many feet of binding are needed?

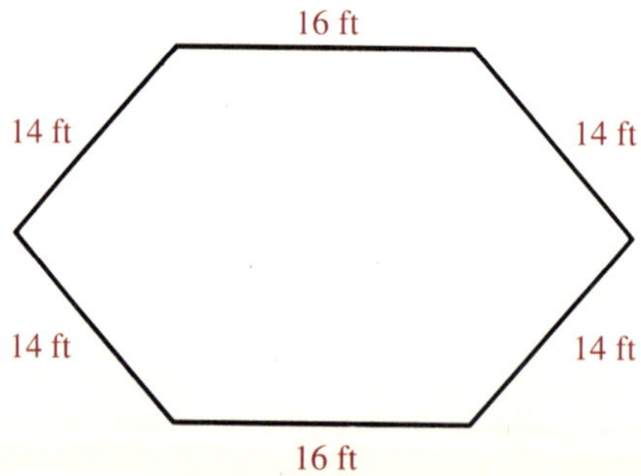

EXAMPLE 6: Packing foam is to be wrapped around the following steel gusset for mailing purposes. What length of foam is needed?

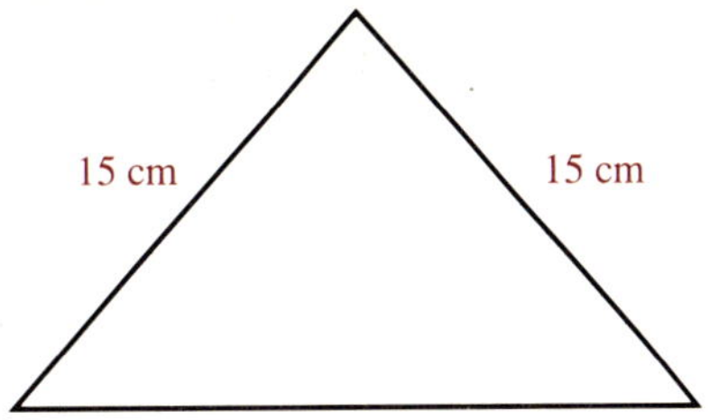

Solution:
Notice that the gusset has the shape of a triangle. The sides have lengths of 15 cm, 15 cm, and 20 cm. Adding to get the perimeter,

$$\begin{aligned}\text{PERIMETER} &= \text{sum of lengths of sides}\\ &= 15\text{ cm} + 15\text{ cm} + 20\text{ cm} = 50\text{ cm}\end{aligned}$$

50 cm of foam are needed to enclose the gusset.

PRACTICE PROBLEM 6: A triangular shaped lot is to be enclosed with cement curbing. What length of curbing is needed?

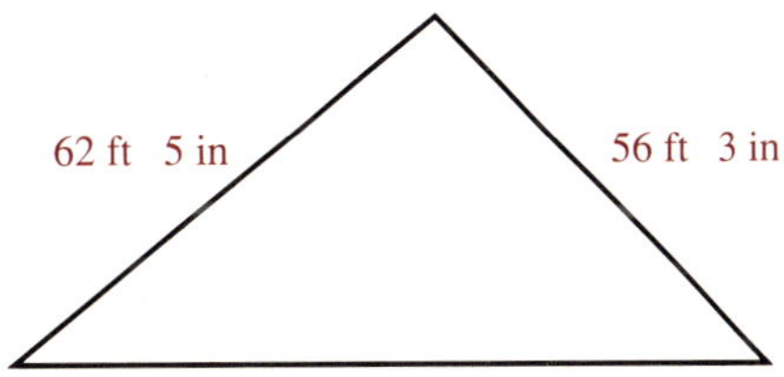

EXAMPLE 7: The edge of a work table is to be enclosed with metal trim. What length of trim is needed?

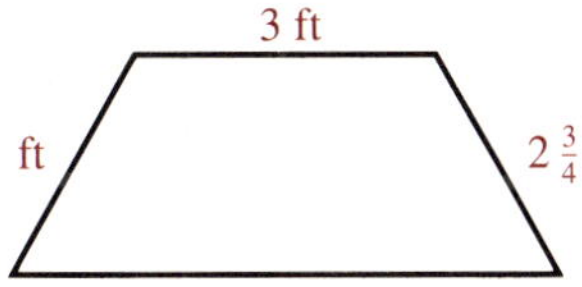

Solution:
You might recognize the shape of the table as a trapezoid. Adding the lengths of the sides to get the perimeter,

$$\begin{aligned}\text{PERIMETER} &= \text{sum of lengths of sides}\\ &= 2\frac{3}{4}\text{ ft} + 5\frac{1}{2}\text{ ft} + 2\frac{3}{4}\text{ ft} + 3\text{ ft} = 14\text{ ft}\end{aligned}$$

14 ft of trim are needed to enclose the table.

PRACTICE PROBLEM 7: A store owner wants to install fluorescent lighting around her store sign. How much lighting will she need?

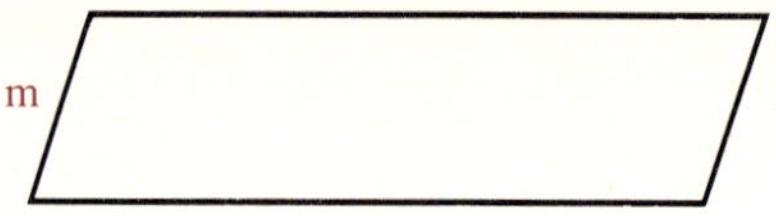

The procedure for finding the perimeters of some figures is made easier because of their shapes. Since a square has four sides of equal length, its perimeter is four times the length of a side. Writing this as a formula, where P represents the perimeter and S represents the length of a side, we have

$$P = 4S$$

EXAMPLE 8: Bud has decided to weatherstrip his two den windows. If the windows are square with sides 16 in long, what length of weatherstripping is needed?

Solution:
It's helpful to first sketch the square windows.

Now apply the formula for the perimeter of a square.

Write the formula:	$P = 4S$
Substitute 16 in for S:	$P = 4(16\text{ in})$
Multiply:	$P = 64\text{ in}$

Therefore, 64 in of weatherstripping are needed for each window. Since there are two windows, this length must be doubled. So, a total of 2(64 in) or 128 in of weatherstripping will be used.

PRACTICE PROBLEM 8: Katy is going to enclose the 3-ft-square atrium in her living room with treated wood logs. Using the formula for the perimeter of a square, determine the length of logs she will need. Allow 1 ft for total waste.

Just as the formula for the perimeter of a square is simple because of its shape, so is the formula for the perimeter of a rectangle. Recall that the opposing sides of a rectangle have the same length. Using L and W to represent the length and width, the formula for the perimeter, P, of a rectangle is

$$P = 2L + 2W$$

EXAMPLE 9: Telephone wire is being installed around the edge of a rectangular office with length 20 meters and width 15 meters. What length of wire is needed?

Solution:
Diagram the office first.

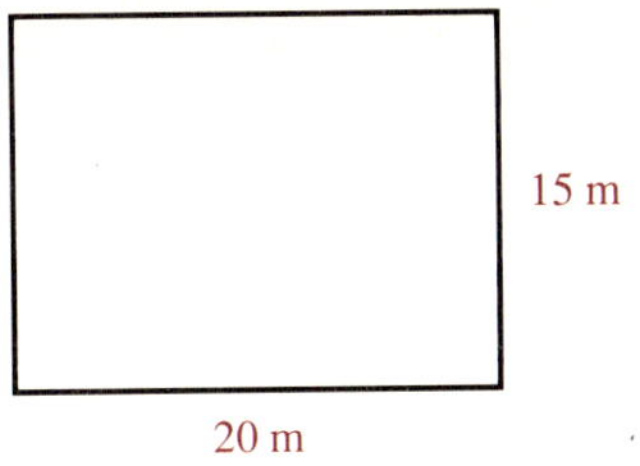

Now apply the formula for the perimeter of a rectangle, substituting 20 m for the length and 15 m for the width.

Write the formula:	$P = 2L + 2W$
Substitute values:	$P = 2(20 \text{ m}) + 2(15 \text{ m})$
Multiply and add:	$P = 40 \text{ m} + 30 \text{ m} = 70 \text{ m}$

Therefore 70 m of wire are required.

Be Careful About This

A common error when calculating the perimeter of a rectangle is to add just one length and width together. Don't forget to double the length and width before adding.

PRACTICE PROBLEM 9: A car dealer's showroom is rectangular with length 85 ft and width 35 ft. A decorative horizontal stripe is painted on the walls. How long is the stripe if it goes around the showroom?

ANSWERS FOR PRACTICE PROBLEMS

1. 90° **2. a)** right **b)** straight **c)** right **d)** straight
3. a) intersecting, perpendicular, right angles **b)** parallel
c) parallel **d)** intersecting
4. a) rectangle, parallelogram **b)** circle
c) square, rectangle, parallelogram **d)** trapezoid
e) triangle **f)** parallelogram
5. 88 ft **6.** 203 ft, 3 in **7.** 9 m **8.** 13 ft **9.** 240 ft

EXERCISE SET 10.1A

Skill Development

1. Label the following using the appropriate term, straight angle or right angle.

(a)  (b)  (c) 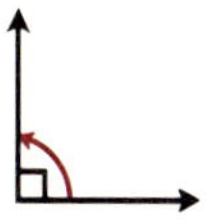 (d) 

2. Label the following using all the appropriate terms—perpendicular, parallel, intersecting, right angles.

(a) 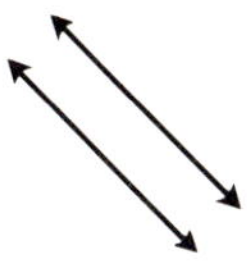 (b)  (c)  (d) 

3. Label these figures using all the appropriate terms—triangle, rectangle, square, parallelogram, trapezoid, circle.

(a) 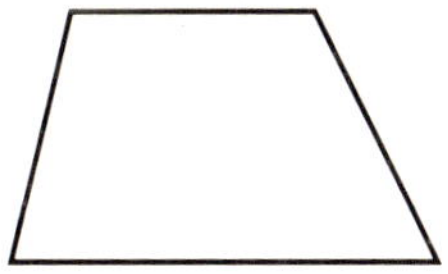 (b) 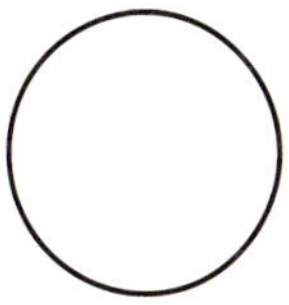 (c) 

(d) 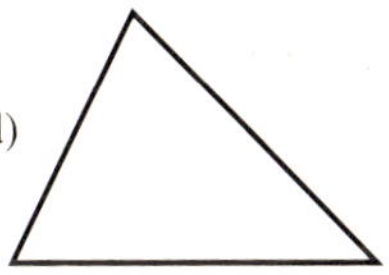 (e)  (f) 

4. Find the perimeter of a square whose side measures 2.5 cm.

5. Find the perimeter of a rectangle whose length and width are 3.7 m and 2.4 m, respectively.

6. Find the perimeter of a triangle whose sides measure 4 ft, 3.7 ft, and 2.9 ft.

7. Find the perimeter of a five-sided figure whose sides measure 2 cm, 3 cm, 5 cm, 7 cm, and 8 cm.

Applications

8. The Barry family wants to fence in their land. The property is rectangular and has a length of 123 ft and a width of 87 ft. Find the length of the fence.

9. A greeting card company has designed an anniversary card with a gold border. If the length and width of the card are 17.2 cm and 12.3 cm, find the length of the border.

10. A sales counter for the jewelry department of a retail store has the shape shown below. Find the perimeter of the counter.

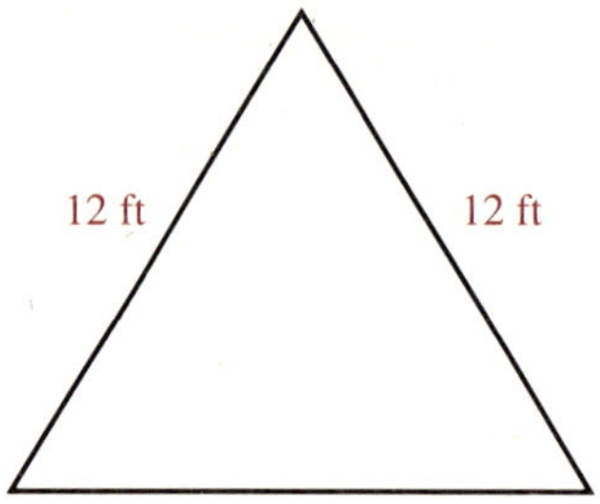

11. A foreign coin has the shape shown below. Find the perimeter of the coin.

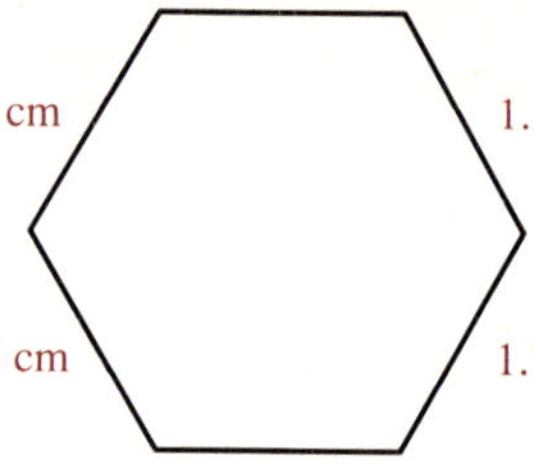

For Some Extra Thought

12. If one side of a triangle measures 12 in, can the other two sides measure 5 in and 6 in? Explain.

13. If a rectangle that has a length of 6 cm and width of 4 cm is reshaped so the perimeter remains the same but its shape is a square, what will each side measure?

Review Problems for Maintaining Your Skills

14. Round 1,608,722 to the nearest ten thousand.

15. $3\frac{1}{4} + 5\frac{2}{5} =$

16. $\frac{3}{10} \times \frac{5}{6} =$

17. Divide 12.04 by 7.3 and round the quotient to the nearest hundredth.

18. Find 17.25% of \$25,000.

19. Solve: $4(5 - 2x) + 10x = 30$

20. $5^4 \cdot 5^{-1} =$

21. Divide and express your result in scientific notation.

$$\frac{6.4 \times 10^4}{1.6 \times 10^{-5}} =$$

22. Use a calculator to approximate $\sqrt{93}$ to the nearest hundredth.

Determine the following conversions.

23. 72 yds = __________ ft

24. 300 mi = __________ km

25. 6 fl oz = __________ c

26. 27 cL = __________ dL

27. 5 t = __________ lb

28. 2 kg = __________ g

EXERCISE SET 10.1B

Skill Development

1. Label the following using the appropriate term, straight angle or right angle.

(a)  (b)  (c)  (d) 

2. Label the following using all the appropriate terms—perpendicular, parallel, intersecting, right angles.

(a)  (b)  (c)  (d) 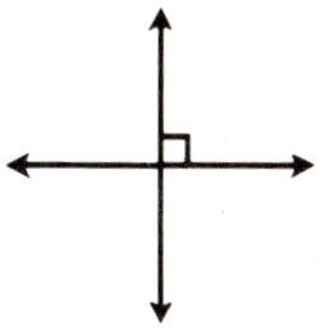

3. Label these figures using all the appropriate terms—triangle, rectangle, square, parallelogram, trapezoid, circle.

(a)  (b) 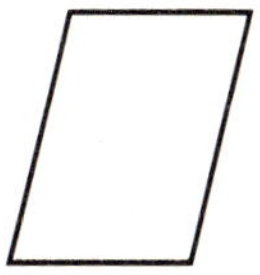 (c) 

(d)  (e) 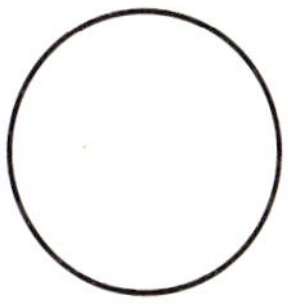 (f) 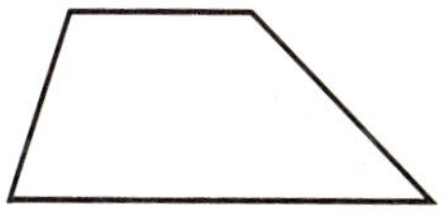

4. Find the perimeter of a square whose side measures 6.7 yd.

5. Find the perimeter of a rectangle whose length and width are 34.8 cm and 17.6 cm, respectively.

6. Find the perimeter of a triangle whose sides measure 5 cm, 6.4 cm, and 7.3 cm.

7. Find the perimeter of a six-sided figure if each side measures 17.5 in.

Applications

8. Moulding is to be placed around the ceiling of a classroom whose dimensions are 26 ft by 22 ft. Find the length of moulding needed for the job.

9. If it is 90 ft between bases, find the minimum distance a batter must run when he or she hits a home run.

10. A piece of jewelry has the shape shown below. Find its perimeter,

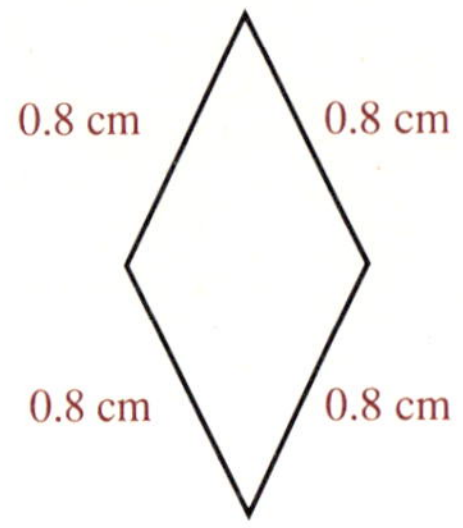

11. A parcel of farm land has the dimensions shown below. How many meters of fence are needed to fence in the land?

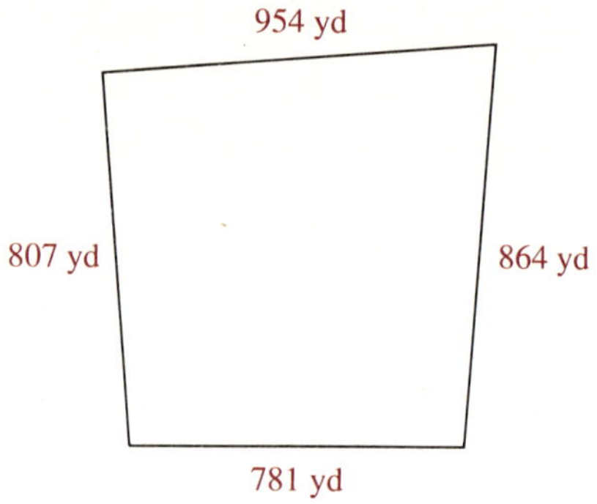

For Some Extra Thought

12. If the distance from one vertex of a square to the opposite vertex is 20 cm, can the perimeter be 80 cm? Explain.

13. If a rectangle has a perimeter of 124 in, and its width is 30 in, is the rectangle a square? Explain.

Review Problems for Maintaining Your Skills

14. Round 156,378 to the nearest thousand.

15. $5\frac{4}{5} + 7\frac{1}{5} =$

16. $\frac{15}{16} \times \frac{2}{11} =$

17. Divide 23.578 by 1.71 and round the quotient to the nearest thousandth.

18. Find 250% of 4.

19. Solve: $3(x - 7) - (4 - x) = 3$

20. $(-4)^{-3}(-4)^{6} =$

21. Divide and express your result in scientific notation.

$$\frac{8.0 \times 10^{-4}}{1.6 \times 10^{-7}} =$$

22. Use a calculator to approximate $\sqrt{125}$ to the nearest tenth.

Determine the following conversions.

23. 480 mm = __________ cm

24. 13 mi = __________ yd

25. 3.25 in = __________ cm

26. 5 kL = __________ L

27. 60 lb = __________ kg

28. 30,000 oz = __________ t

10.2 AREA

Preview Test

1. Pat is decorating a 15 ft by 15 ft wall with 12-inch-square mirror tiles. How many tiles will she need?
2. The Fixit Company is estimating the cost of a potential job. The job involves constructing a shed with end walls of size 10 ft by 12 ft and side walls of size 10 ft by 17 ft. If the wallboard costs \$0.16 per square foot, find the cost of the wallboard.
3. One end of a jungle gym set for a playground is sketched as follows. Determine the amount of hardwood needed to construct both ends.

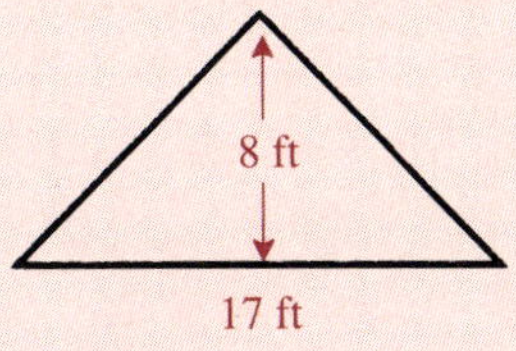

4. An advertisement claims that one gallon of a certain brand of exterior paint covers 12 sq m. If 4 gallons of paint are needed to paint the side of a hip roof drawn below, how accurate is the advertisement?

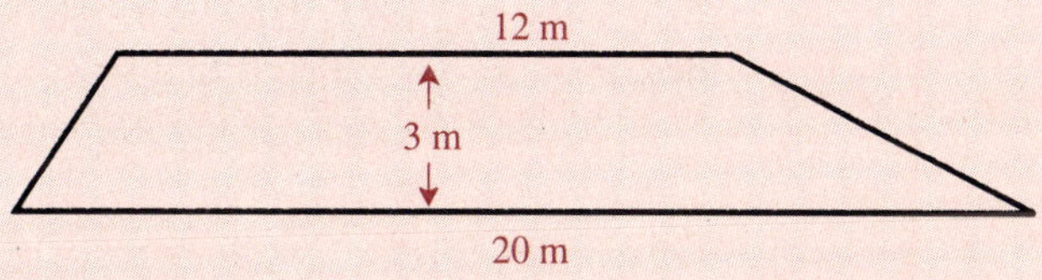

5. The top of a formica laboratory bench is constructed in the shape sketched below. What amount of formica is needed to make the top of the bench?

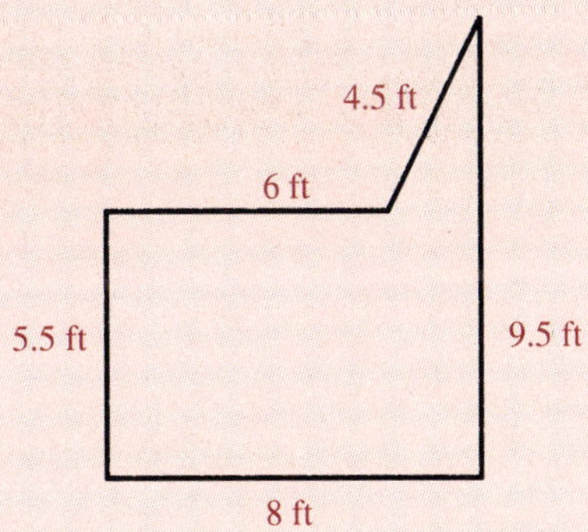

ANSWERS FOR PREVIEW TEST

1. 225 **2.** \$92.80 **3.** 136 sq ft

4. Area of roof = 48 sq m requires 48/12 = 4 gal of paint. The ad is correct.

5. 48 sq ft **6.** 68 sq in

6. Below is the drawing of the front of a television console. Determine the area of plastic needed to cover the front of the console and provide an opening for the television screen.

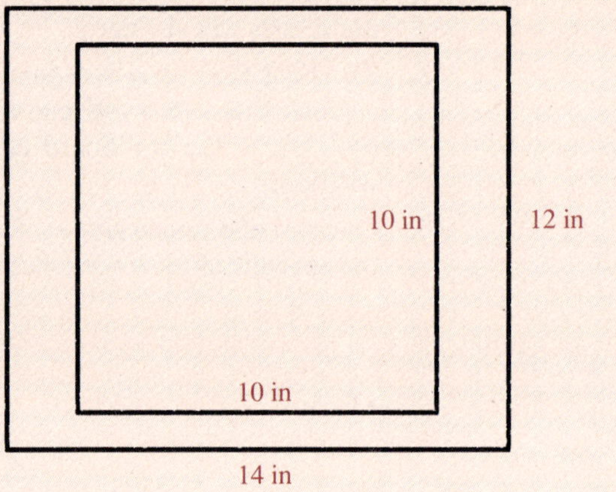

In the last section we used the perimeter of figures to solve problems such as finding the length of weatherstripping for a window. However, perimeter won't help us solve problems such as determining the amount of carpet to cover a floor. To deal with these applications, we need to discuss the **area** of figures.

AREA is the measure of the number of square units, such as the square inch or square centimeter, that can be placed inside a figure.

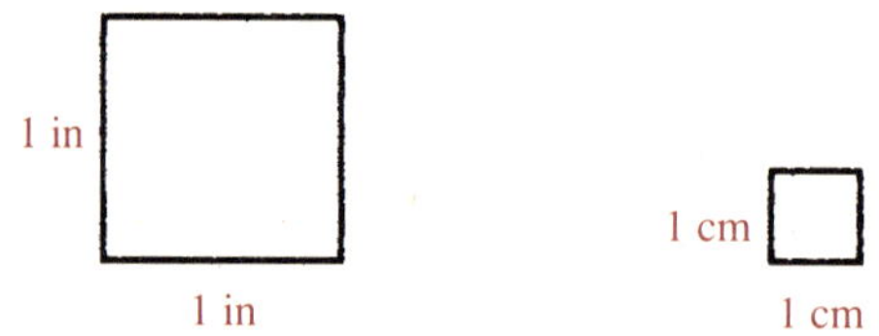

In the figure shown here there are twelve 1-square-centimeter squares that are contained in a rectangle. The area of the rectangle is then reported as 12 square centimeters.

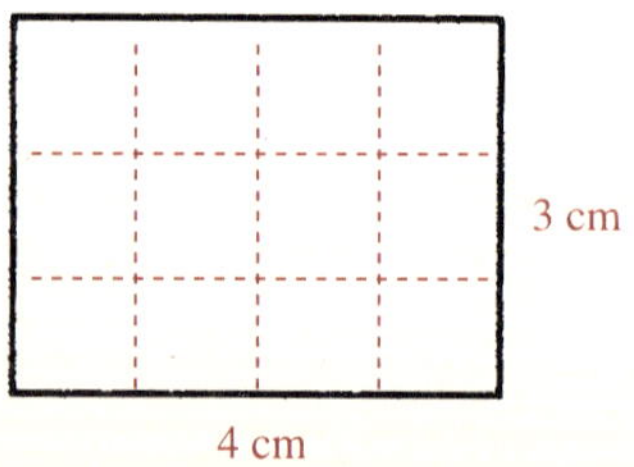

Be Careful About This

It's important to distinguish between area and perimeter. Perimeter is the distance *around* a figure and area is the measure of the region *within* the figure. Perimeter is measured in units such as feet, centimeters, and so on. Area is measured in units such as square feet, square centimeters, and so on.

In the figure whose area is 12 square centimeters the area was found by counting squares. Since there were 3 rows of 4 squares, it makes sense to find the area by multiplying 4 cm times 3 cm; that is, the length times the width. This suggests the following formulas for area.

The area of a rectangle is its length times its width. If A represents the area, L the length, and W the width, the formula is

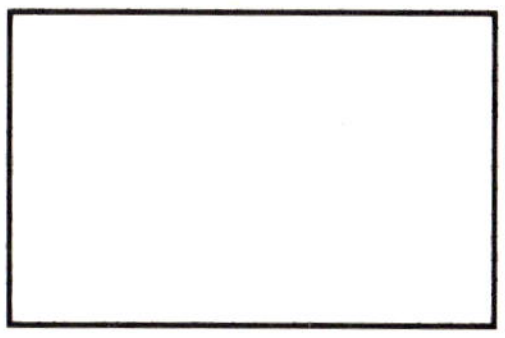

$$A = LW$$

Since the length and width of a square are equal, the formula for the area of a square, where the side is represented by S is

$$A = S^2$$

EXAMPLE 1: Lynn Cortez wants to wallpaper one wall of her bedroom. How many square feet will need to be covered?

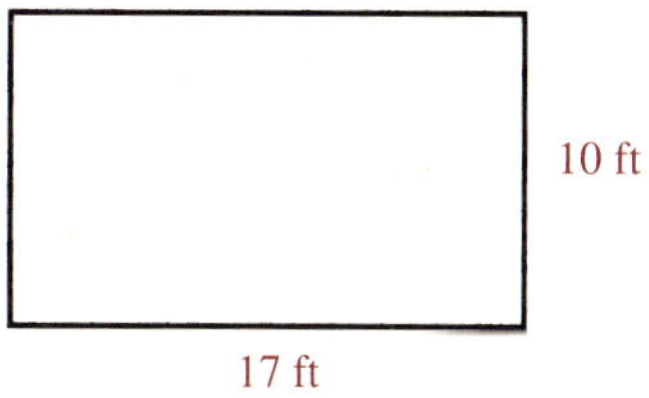

Solution:
Lynn's wall is in the shape of a rectangle. Using the appropriate formula we have

$$A = LW = (17\text{ ft})(10\text{ ft}) = 170\text{ sq ft}$$

Lynn will be papering 170 square feet.

PRACTICE PROBLEM 1: Find the area of a brick walkway that is $12\frac{1}{5}$ feet long and $1\frac{2}{3}$ feet wide.

EXAMPLE 2: Determine the square meters of aluminum that are needed to make a square plate with sides 0.3 meters long.

Solution:
The plate looks like this.

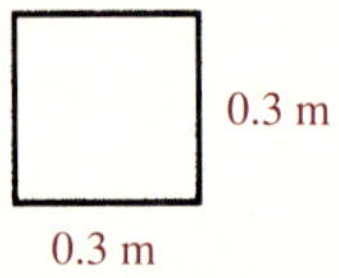

Now apply the formula for the area of a square.

$$\begin{aligned} A &= S^2 \\ &= (0.3 \text{ m})(0.3 \text{ m}) \\ &= 0.09 \text{ sq m} \end{aligned}$$

The machinist will use 0.09 sq m of aluminum.

PRACTICE PROBLEM 2: A store owner wants to mount the first dollar he earned in a frame that is 13 inches square. How many square inches of backing does he need?

So much for the areas of squares and rectangles. Now on to other shapes and formulas.

SHAPE	NAME	FORMULA
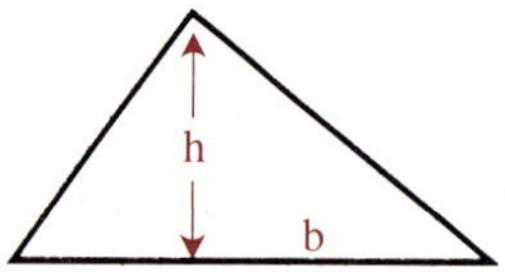	Triangle	$A = \frac{1}{2}bh$
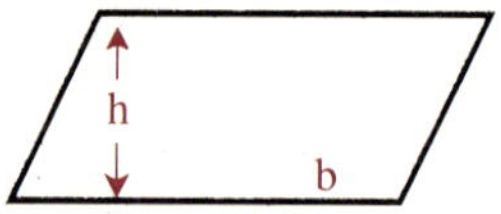	Parallelogram	$A = bh$
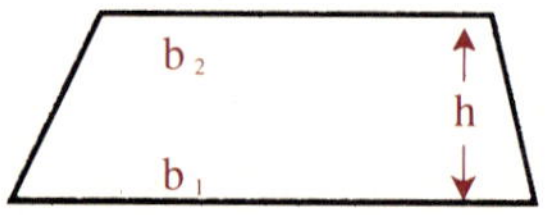	Trapezoid	$A = \frac{1}{2}h(b_1 + b_2)$

The formula for the area of a parallelogram is based on the formula for the area of a rectangle. For example, a parallelogram can be converted to a rectangle by cutting a right triangle from the left side and moving it to the right side. (See the illustration below). The base, b, is then the length of the rectangle and the height, h, is the width of the rectangle.

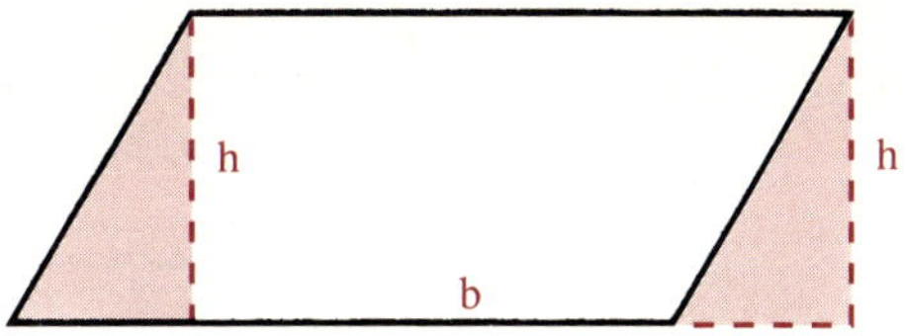

Before we tackle some examples, it might be helpful to set up a strategy.

1. Carefully read the problem to determine what shape is involved.
2. Select the appropriate formula.
3. Substitute the values into the formula and evaluate.
4. Label the answer with the correct units.
5. Check to see if the answer makes sense.

EXAMPLE 3: How many square feet of siding are needed to cover the gable end of a house sketched below.

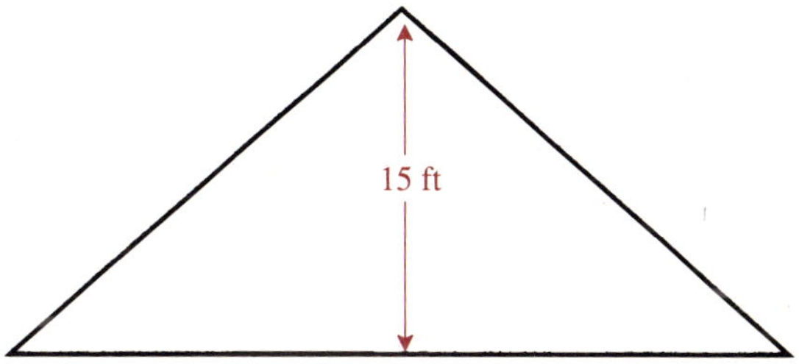

Solution:
The gable is in the shape of a triangle with base 35 ft and height 15 ft. Applying the formula for the area of a triangle,

$$A = \frac{1}{2}bh = \frac{1}{2}(35 \text{ ft})(15 \text{ ft}) = 262.5 \text{ sq ft}$$

262.5 square feet of siding is needed.

PRACTICE PROBLEM 3: The end flap of a canvas tent is diagrammed below. Determine the amount of material needed to make the flap.

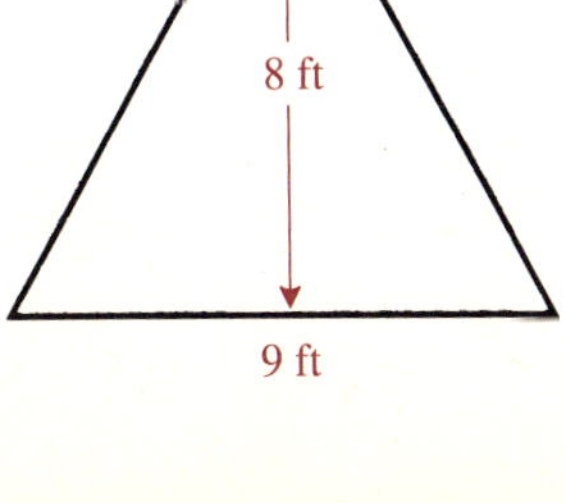

EXAMPLE 4: Bob Mayer orders a cover for his swimming pool. Find the square meters of material needed if the pool is shaped like a parallelogram with height 6.2 meters and base 14 meters.

Solution:
The sketch of the pool is shown here.

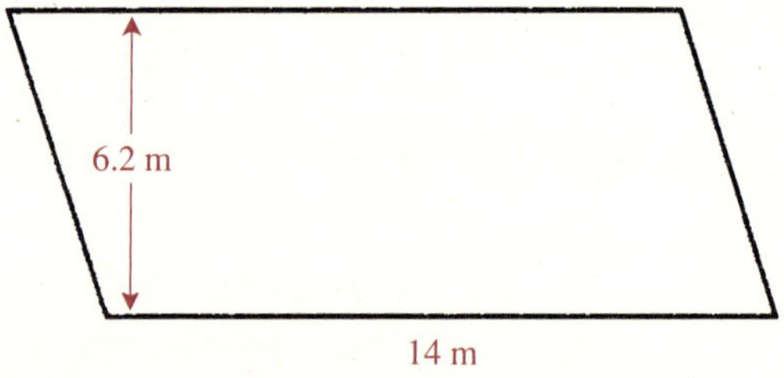

Using the area formula for a parallelogram,
$A = bh = (14\text{ m})(6.2\text{ m}) = 86.8\text{ sq m}$
The cover will need 86.8 square meters of materials.

PRACTICE PROBLEM 4: Mary Maklin is framing an abstract picture with the following shape. What quantity of glass does she need to cover the front?

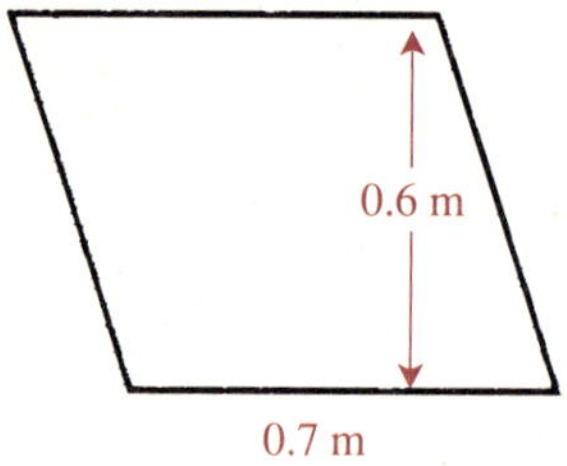

EXAMPLE 5: The top of a laboratory table is to be covered with a specially treated chemical. Determine the area to be covered.

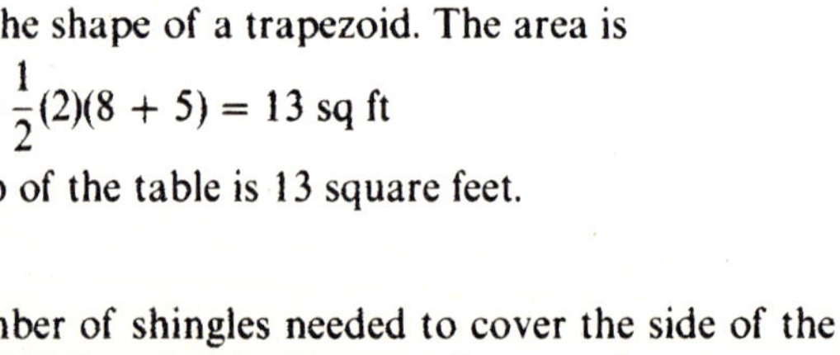

Solution:
The table top has the shape of a trapezoid. The area is
$A = \frac{1}{2}h(b_1 + b_2) = \frac{1}{2}(2)(8 + 5) = 13\text{ sq ft}$
The area of the top of the table is 13 square feet.

PRACTICE PROBLEM 5: Determine the number of shingles needed to cover the side of the hip roof sketched at the top of page 507 if each shingle covers 2 square feet.

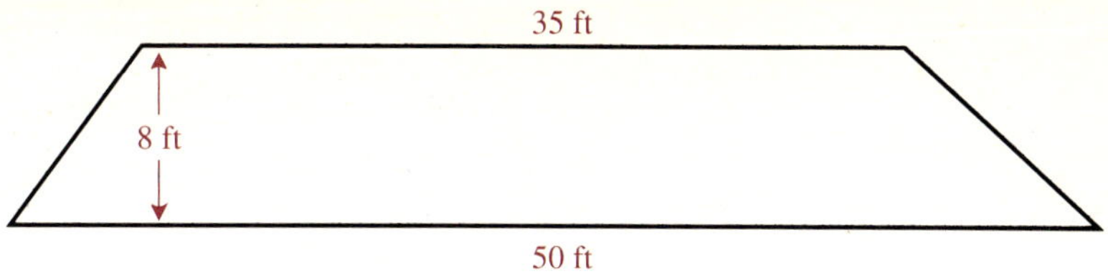

Some applications involve more than one shape or more than one measurement. If the figure you're working with does not fit a known shape, see if you can cut it up into identifiable shapes.

EXAMPLE 6: The side of a house is in the shape shown below. Find its area.

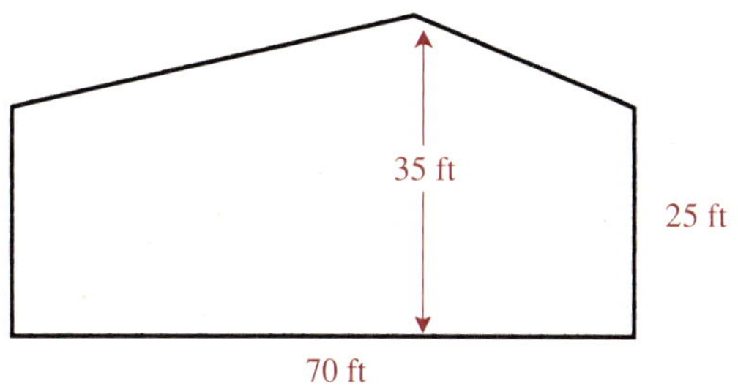

Solution:
As shown below, you consider the side of the house as a combination of a rectangle and a triangle.

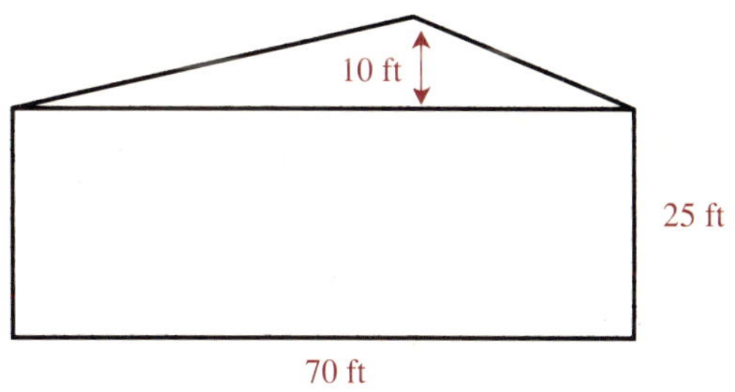

The area of the side of the house is the sum of the areas of the triangle and rectangle.

Triangle: $A = \frac{1}{2}bh = \frac{1}{2}(70\text{ ft})(10\text{ ft}) = 350\text{ sq ft}$

Rectangle: $A = LW = (70\text{ ft})(25\text{ ft}) = 1750\text{ sq ft}$

The total area is $350\text{ sq ft} + 1750\text{ sq ft} = 2100\text{ sq ft}$

PRACTICE PROBLEM 6: The window for a contemporary professional building is designed as sketched below. Find its area.

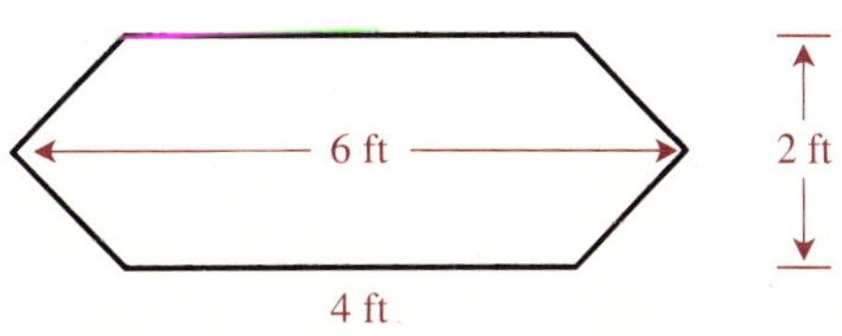

ANSWERS FOR PRACTICE PROBLEMS

1. $20\frac{1}{3}$ sq ft **2.** 169 sq in **3.** 36 sq ft **4.** 0.42 sq m

5. 170 shingles **6.** 10 sq ft

EXERCISE SET 10.2A

Skill Development

Find the area of each square.

1. $S = 5$ cm **2.** $S = 7.2$ m **3.** $S = 4.67$ ft

4. $S = 100$ in **5.** $S = \frac{3}{4}$ mi **6.** $S = 2\frac{1}{2}$ cm

Find the area of each rectangle.

7. $L = 7.4$ in, $W = 3.5$ in **8.** $L = 9.2$ ft, $W = 4.1$ ft

9. $L = 12$ mi, $W = 2.5$ mi **10.** $L = 14.6$ m, $W = 4.25$ m

Find the area of each triangle.

11. $b = 5$ cm, $h = 3$ cm **12.** $b = 8.7$ km, $h = 2.3$ km

13. $b = 56.4$ mm, $h = 24.8$ mm **14.** $b = 6.03$ m, $h = 2.15$ m

Find the area of each parallelogram.

15. $b = 4.6$ yd, $h = 2.5$ yd **16.** $b = 10.5$ in, $h = 2.6$ in

17. $b = \frac{7}{8}$ ft, $h = \frac{1}{2}$ ft **18.** $b = 5\frac{3}{4}$ mi, $h = 2\frac{1}{4}$ mi

Find the area of each trapezoid.

19. $b_1 = 5$ cm, $b_2 = 4$ cm, $h = 2$ cm

20. $b_1 = 4.6$ ft, $b_2 = 1.9$ ft, $h = 0.6$ ft

21. $b_1 = 10.75$ cm, $b_2 = 9.64$ cm, $h = 5$ cm

22. $b_1 = 125$ mm, $b_2 = 80$ mm, $h = 60$ mm

Applications

23. A wheat field is rectangular and has a length of 600 m and a width of 400 m. Find its area.

24. A desk top has the shape of a trapezoid. The bases measure 6.2 ft and 5.8 ft. If the height measures 3.4 ft, find its area.

25. Find the area of the figure shown below.

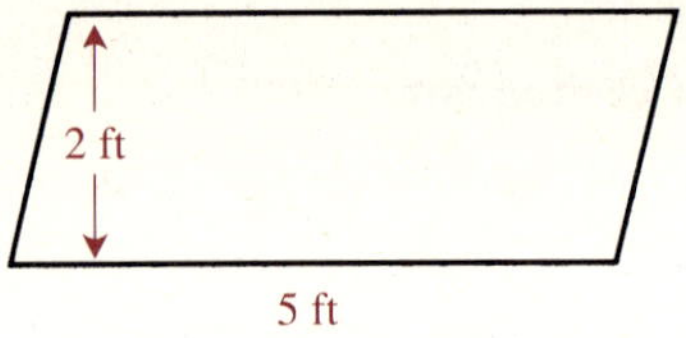

26. Find the area of the figure shown below.

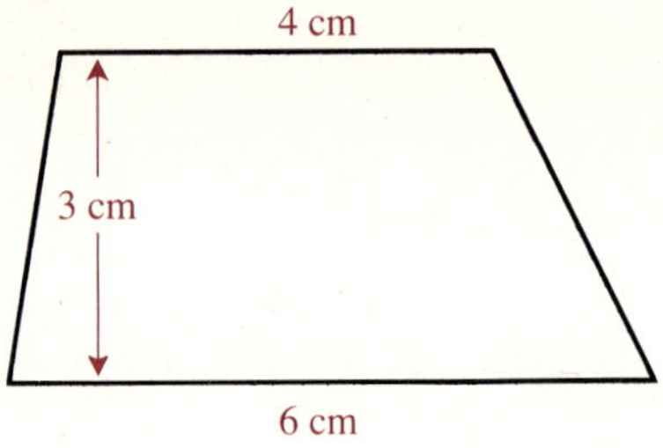

27. Find the area of the figure shown below.

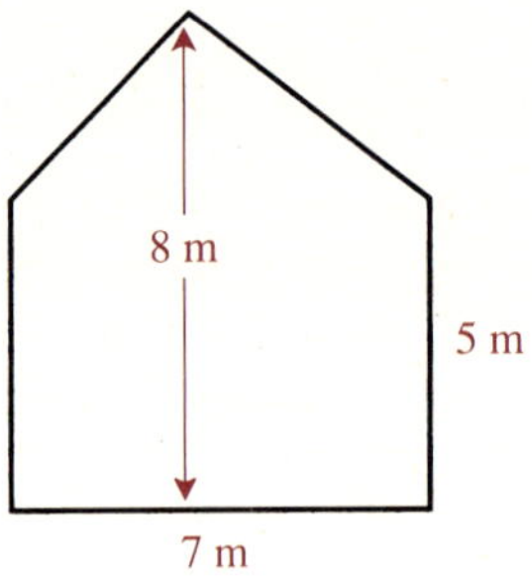

28. Find the area of the figure shown below.

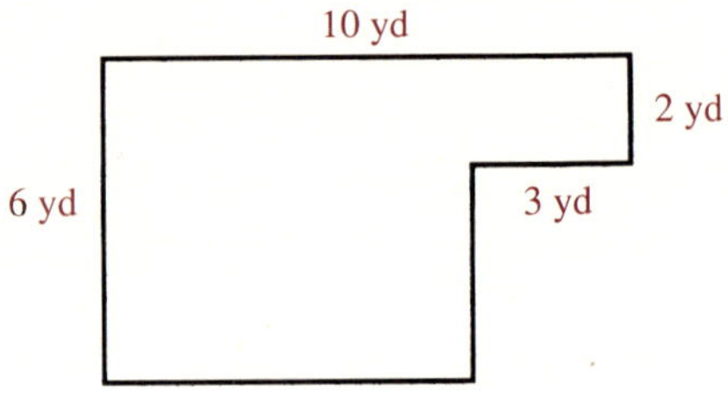

For Some Extra Thought

29. If the area of a rectangle is 24.6 square centimeters and its width is 4 cm, find its length.

30. The base of a triangle measures 16 ft and its height measures 4 yd. Find its area.

Review Problems for Maintaining Your Skills

31. Write 2,105,007 in words.

32. $4\frac{1}{2} - 2\frac{3}{5} =$

33. $\frac{14}{15} \div \frac{2}{15} =$

34. $2.0041 \times 10{,}000 =$

35. To the nearest tenth of a percent, 17 is what percent of 29?

36. Solve: $12 - 2x = 3x + 42$

37. $(-6)^{-2}(-6)^5 =$

38. Convert 0.00000056 to scientific notation.

Determine the following conversions.

39. 39 ft = ________ yd

40. 0.03 mi = ________ ft

41. 50 in = ________ cm

42. 6 t = ________ lb

43. Find the perimeter of a rectangle whose length and width are 48.3 m and 36.8 m, respectively.

44. Find the perimeter of a square whose side measures 1 mi.

EXERCISE SET 10.2B

Skill Development

Find the area of each square.

1. $S = 12$ in
2. $S = 67.4$ mm
3. $S = 8.84$ yd
4. $S = 56$ km
5. $S = \frac{3}{8}$ ft
6. $S = 6\frac{1}{2}$ cm

Find the area of each rectangle.

7. $L = 18.4$ cm, $W = 8.4$ cm
8. $L = 12$ mi, $W = 8$ mi
9. $L = 0.07$ km, $W = 0.06$ km
10. $L = 4.65$ m, $W = 3$ m

Find the area of each triangle.

11. $b = 86$ mm, $h = 64$ mm
12. $b = 2.6$ dm, $h = 1.2$ dm
13. $b = 26$ in, $h = 10$ in
14. $b = 3.9$ m, $h = 2.1$ m

Find the area of each parallelogram.

15. $b = 5.3$ mi, $h = 4.1$ mi
16. $b = 4.8$ in, $h = 3.3$ in
17. $b = \frac{2}{5}$ ft, $h = \frac{4}{5}$ ft
18. $b = 4\frac{2}{3}$ in, $h = 2\frac{1}{4}$ in

Find the area of each trapezoid.

19. $b_1 = 12$ mm, $b_2 = 9$ mm, $h = 4$ mm
20. $b_1 = 2.5$ mi, $b_2 = 1.4$ mi, $h = 1.2$ mi
21. $b_1 = 3.25$ yd, $b_2 = 2.64$ yd, $h = 1.46$ yd
22. $b_1 = 24.6$ dm, $b_2 = 22.1$ dm, $h = 10.4$ dm

Applications

23. A modern painting has the shape of a parallelogram whose base and height are 63 in and 42 in, respectively. Find the area of the painting.

24. An attic window has the shape of a trapezoid. The bases measure 124 cm and 92 cm. The height measures 80 cm. Find the area of the window.

25. Find the area of the figure shown below.

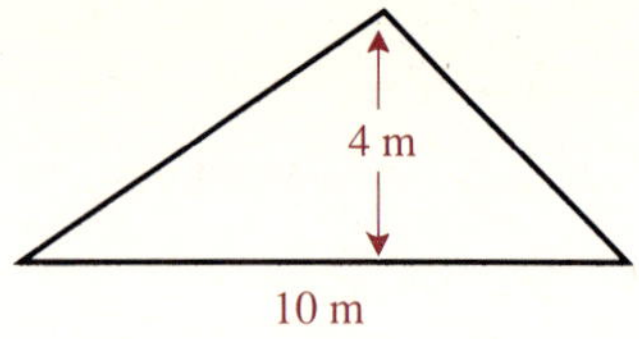

26. Find the area of the figure shown below.

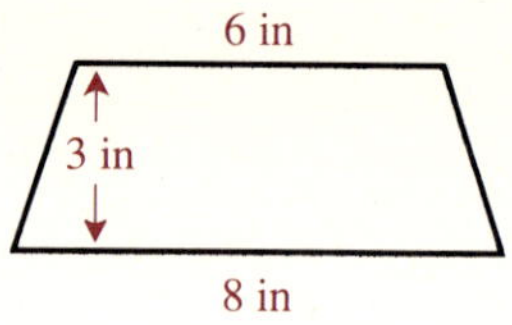

27. Find the area of the figure shown below.

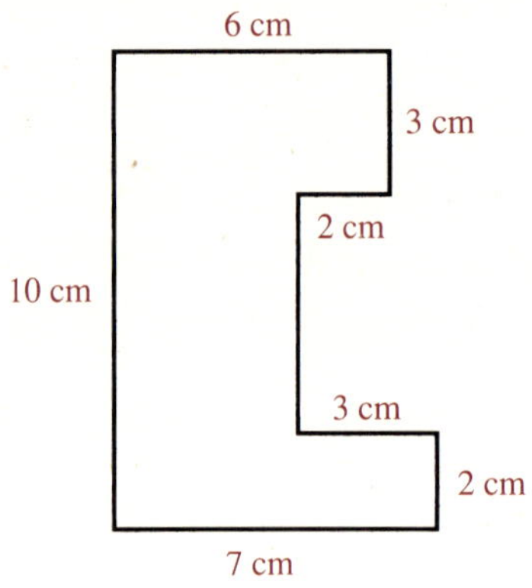

28. Find the area of the figure shown below.

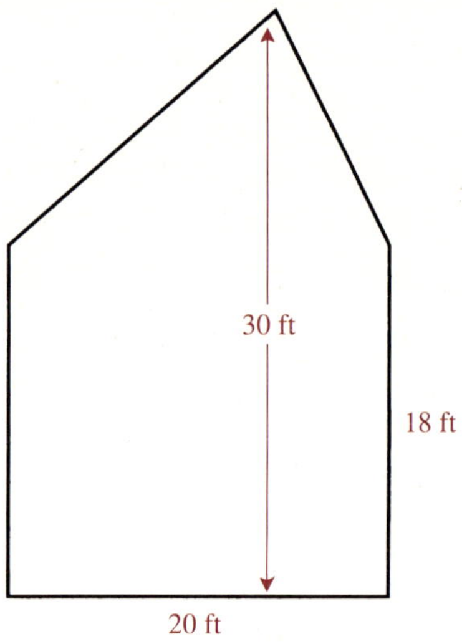

For Some Extra Thought

29. If the area of a triangle is 36 square meters and its height is 9 meters, find its base.

30. The base of a parallelogram measures 73.4 cm and its height measures 14.5 dm. Find its area.

Review Problems for Maintaining Your Skills

31. Write 34,050 in words.

32. $10\frac{1}{7} - 8 =$

33. $\frac{4}{7} \div \frac{1}{5} =$

34. $0.005 \times 0.0072 =$

35. To the nearest hundredth of a percent, 200 is what percent of 950?

36. Solve: $25 + x = 3(5 - x)$

37. $10^2 \cdot 10^3 =$

38. Convert 2,000,000,000 to scientific notation.

Determine the following conversions.

39. 300 yd = _________ ft **40.** 4.2 mi = _________ km

41. 8 fl oz = _________ c **42.** 6 kg = _________ g

43. Find the perimeter of a rectangle whose length and width are 63.5 in and 45.8 in, respectively.

44. Find the perimeter of a square whose side measures 24.6 cm.

10.3 CIRCLES

Preview Test

1. A circle is the set of all __________ at the same __________ (called the __________) from a single point (called the __________).
2. The diameter of a steering wheel with radius 7 inches is __________ .
3. What is the radius of a wheel in a pulley system that has a diameter of 2.34 meters?
4. The name of the ratio of the circumference of a circle to its diameter is __________. The approximate value of this ratio rounded to the nearest hundredth is __________ .
5. Insulating material is to be placed around a wire with a diameter of 0.16 inch. What width of material is needed to wrap around the circumference of the wire? Round the result to the nearest hundredth of an inch.
6. Lee Huang wants to install a tile border around her hot tub. If her tub has a radius of 1 meter, find the length to be tiled to the nearest tenth of a meter.
7. To the nearest square mile, what is the receiving area of a television station that can transmit a beam 50 miles in any direction?
8. Kitchen Coverups is designing flame retardant covers for electric stove burners. To the nearest square centimeter, what amount of material is needed for a burner that has a diameter of 16 cm?

You may have noticed that in this chapter we have not yet dealt with the perimeter or area of a circle. Let's begin with some definitions.

The **radius** of a circle is the distance from a point on the circle to its center. Of interest is that the radius of a circle is the same regardless of the point used on the circle. This, in fact, leads to the definition of a circle.

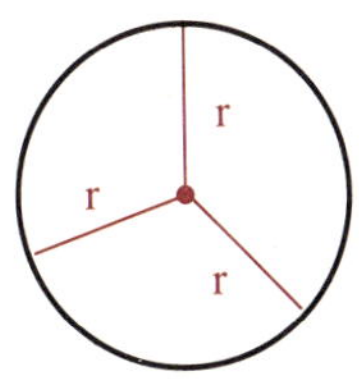

A **circle** is the set of all points in a plane of equal distance (the radius) from a single point (the center).

The radius of a circle is very closely related to its **diameter**. The diameter of a circle is the distance from one side of a circle, through the center, to the other side. It is not difficult to see that the diameter, d, is twice the radius, r. That is, $d = 2r$.

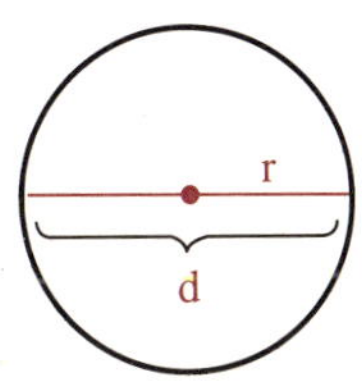

ANSWERS FOR PREVIEW TEST

1. points, distance, radius, center **2.** 14 in **3.** 1.17 m **4.** pi, 3.14
5. 0.50 in **6.** 6.3 m **7.** 7850 sq mi **8.** 201 sq cm

Since some of the formulas for circles involve the diameter and some involve the radius, it is important to be able to find the radius given the diameter, and vice versa.

EXAMPLE 1: Determine the diameter of a circular storage tank with a radius of 6 meters.

Solution:
Substitute 6 m for r in the formula $d = 2r$.

$$d = 2r = 2(6 \text{ m}) = 12 \text{ m}$$

Therefore, the diameter of the tank is 12 meters.

PRACTICE PROBLEM 1: Find the diameter of a circular cable with a radius of $2\frac{5}{8}$ inches.

EXAMPLE 2: What is the radius of a pipe with a 0.03 ft diameter?

Solution:
Substitute 0.03 ft for d in the formula $d = 2r$.

$$d = 2r$$
$$0.03 \text{ ft} = 2r$$

We must now solve this equation for r by dividing both sides by 2.

$$\frac{0.03 \text{ ft}}{2} = \frac{2r}{2}$$
$$0.015 \text{ ft} = r$$

The radius of the pipe is 0.015 feet.

PRACTICE PROBLEM 2: What is the radius of a chimney pipe with a 0.67 m diameter?

The circumference of a circle is the distance around the circle. Sound like perimeter? Well, that's what it is! An important characteristic of circles is that the ratio of the circumference to the diameter is the same regardless of the size of the circle. The value of this ratio is about 3.14. Since this value occurs so frequently when working with circles, it is given a name "pi" and is written with the Greek symbol for pi which is "π".

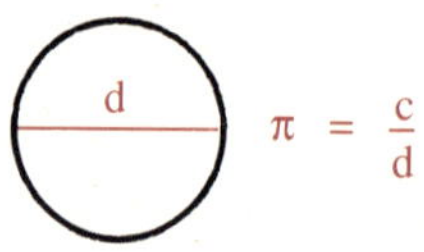

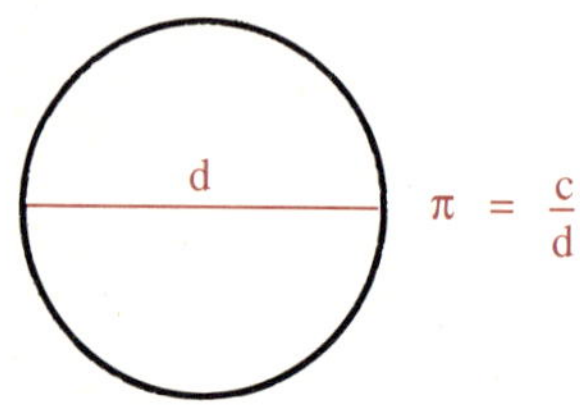

If we solve $\frac{C}{d} = \pi$ for C, by multiplying both sides of the equation by d, we have the formula for the circumference of a circle.

$$C = \pi d \qquad \text{where } \pi \text{ is approximately } 3.14$$

EXAMPLE 3: A gasket is to be placed around a pipe with an outside diameter of 110 mm. What size gasket is needed?

Solution:
Substitute 110 mm for d and 3.14 for π in the formula $C = \pi d$.

$$C = \pi d = (3.14)(110 \text{ mm}) = 345.4 \text{ mm}$$

345.4 mm of gasket will be needed.

PRACTICE PROBLEM 3: A circular opening with a diameter of 1.22 m is to be cut in a piece of metal to allow for the installation of a flywheel. Find the circumference of this opening to the nearest hundredth of a meter.

EXAMPLE 4: A circular speaker is to be bound with a rubber strip. If the speaker has a 4 in radius, find the length of the rubber strip to be used.

Solution:
Be careful with this example—we are given the radius and need to find the diameter for the formula. So, first substitute 4 in for r in the formula $d = 2r$.

$$d = 2r = 2(4 \text{ in}) = 8 \text{ in}$$

Now use this value in the formula $C = \pi d$, with 3.14 for π.

$$C = \pi d = (3.14)(8 \text{ in}) = 25.12 \text{ in.}$$

The length of the rubber strip must be 25.12 in.

PRACTICE PROBLEM 4: Larry Carlin is constructing a circular driveway that will be planted with hedges on the inner edge. If the radius measures 14 ft to the inner edge, what length of ground, to the nearest tenth of a foot, will be planted with hedges?

Now that we've dealt with the circumference of a circle we can consider the **area** of a circle. The formula for the area, A, of a circle with radius r is

$$A = \pi r^2$$

Notice that the area formula involves the radius of the circle, not the diameter as with the circumference formula.

Be Careful About This

When working with the area formula the rule for order of operations directs us to evaluate the exponential factor first (that is, square the radius first), then multiply by the value for π (3.14). Make sure that you do not multiply the radius by 3.14 first and then square.

EXAMPLE 5: A filter is to be placed over the opening of a circular dryer hose. If the hose has a radius of 3 in, what area of filter is needed?

Solution:
Substitute 3 in for r and 3.14 for π in the area formula.

$$\begin{aligned} A = \pi r &= (3.14)(3 \text{ in})^2 \\ &= (3.14)(9 \text{ sq in}) \\ &= 28.26 \text{ sq in} \end{aligned}$$

The filter will have an area of 28.26 sq in.

PRACTICE PROBLEM 5: What area of lawn can be watered with a sprinkler that can spray water 4 yards in any direction?

> **Be Careful About This**
>
> Watch your measurement units when solving problems involving circumference or area. The circumference is a length and has units such as inches, feet, meters. Area is a region and has units such as square inches, square feet, and square meters.

Up to this point in this section all problems we've worked with have involved only circles. Some applications, however, involve other geometric figures as well.

EXAMPLE 6: A circular window is located in the triangular peak of the front of a house as shown below. A piece of wallboard needs to be cut to fit this shape. (a) What quantity of wallboard will be cut out for this window? (b) What amount of wallboard remains to be painted?

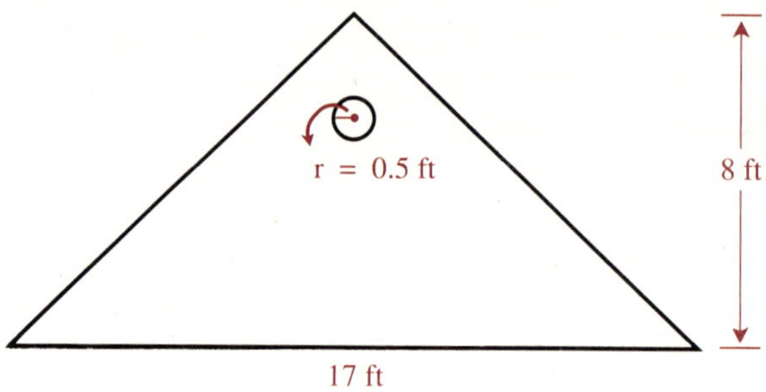

Solution:

a) The cut out quantity is the area of the circle with radius 0.5 ft.

$$A = \pi r^2 = (3.14)(0.5 \text{ ft})^2 = 0.785 \text{ sq ft}$$

There will be a 0.785 square foot cut out of wallboard.

b) The amount of wallboard remaining will be the area of the triangle minus the area of the cut out circle. First find the area of the triangle.

$$A = \frac{1}{2}bh = \frac{1}{2}(17 \text{ ft})(8 \text{ ft}) = 68 \text{ sq ft}$$

Now subtract the amount cut out for the window.

$$68 \text{ sq ft} - 0.785 \text{ sq ft} = 67.215 \text{ sq ft}.$$

Therefore, the amount of wallboard to be painted is 67.215 square feet.

PRACTICE PROBLEM 6: A circular tile commemorating the anniversary of a city's founding date is to be installed in a rectangular patio area in front of city hall. The remaining area of the patio is to be covered with cobblestones. (See the following diagram.) What area of patio will be covered with cobblestone?

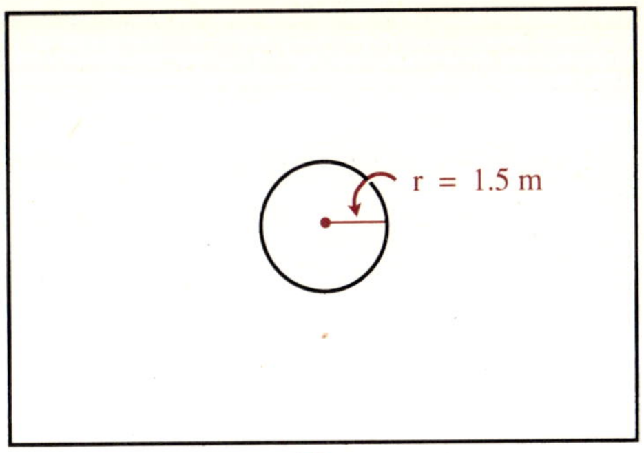

ANSWERS FOR PRACTICE PROBLEMS

1. $5\frac{1}{4}$ in **2.** 0.335 m **3.** 3.83 m **4.** 87.9 ft **5.** 50.24 sq yd

6. 142.935 sq m

EXERCISE SET 10.3A

Skill Development

Find each diameter.

1. $r = 3.5$ mm
2. $r = 120$ ft
3. $r = 2.406$ dm

Find each radius.

4. $d = 200$ mi
5. $d = 7.03$ cm
6. $d = 4\frac{3}{4}$ yd

Find the circumference of each circle.

7. $r = 3$ in
8. $r = 2.06$ m
9. $d = 36$ ft

Find the area of each circle.

10. $r = 6$ mm
11. $r = 10.4$ in
12. $d = 9$ km

Applications

13. A circular table has a diameter of 42 in. Find its circumference and its area.

14. The blade of a circular saw has a radius of 16 cm. Find its circumference and its area.

15. Find the area of the shaded portion in the following figure.

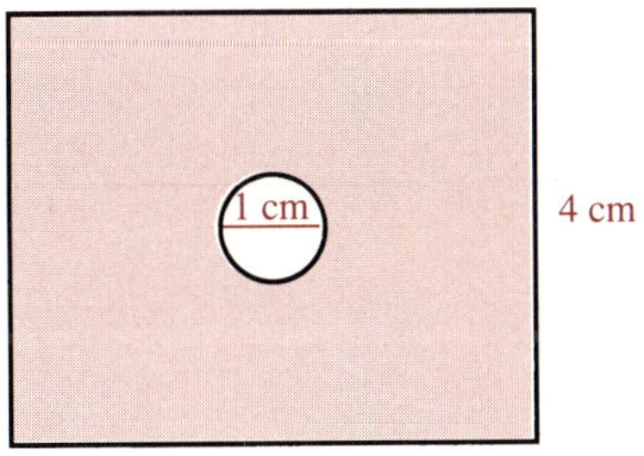

16. Find the area of the shaded portion in the following figure.

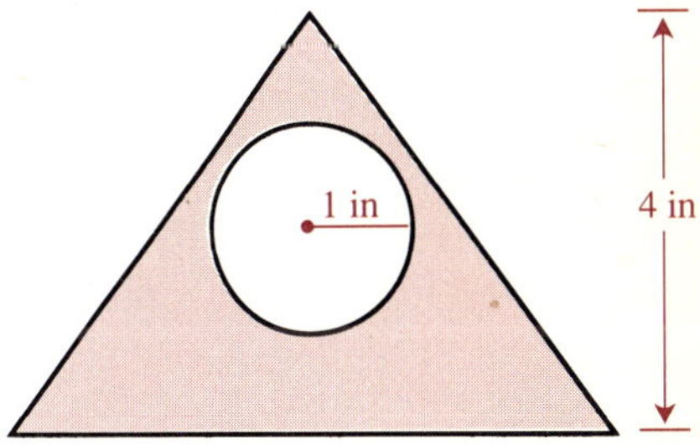

For Some Extra Thought

17. Find the area of a circle whose circumference is 10π cm.

18. The area of a circle is π square ft. What is the diameter of the circle?

Review Problems for Maintaining Your Skills

19. Round 44,444,444 to the nearest hundred thousand.

20. $\frac{11}{12} + \frac{3}{10} =$ **21.** $3\frac{1}{2} \times 2\frac{2}{5} =$

22. Divide 23.407 by 6.1 and round the quotient to the nearest hundredth.

23. 17 is what percent of 50?

24. Solve: $3x + 2x + x = -24$

25. $\frac{5^8}{5^6} =$

26. Multiply and express your result in scientific notation.
$(5.3 \times 10^{-2})(6.7 \times 10^{-1}) =$

27. Use a calculator to approximate $\sqrt{300}$ to the nearest tenth.

Determine the following conversions.

28. 4,500 lb = ________ t **29.** 45.6 cm = ________ dm

30. Find the perimeter of a square whose side measures 1.2 mm.

31. Find the perimeter of a rectangle whose length and width are 3.4 ft and 2.6 ft, respectively.

32. Find the area of a triangle whose base and height are 6 in and 4 in, respectively.

33. Find the area of a trapezoid whose bases measure 13 mm and 11 mm and whose height measures 3 mm.

EXERCISE SET 10.3B

Skill Development

Find each diameter.

1. $r = 6.75$ dm **2.** $r = 24$ mi **3.** $r = 0.04$ cm

Find each radius.

4. $d = 50$ yd **5.** $d = 1.25$ mm **6.** $d = \frac{1}{4}$ in

Find the circumference of each circle.

7. $r = 8$ cm **8.** $r = 6.04$ km **9.** $d = 0.7$ in

Find the area of each circle.

10. $r = 21$ m **11.** $r = 1.2$ mi **12.** $d = 30$ mm

Applications

13. A coin has a radius of 25 mm. Find its circumference and its area.

14. The circular entrance of a bird house has a diameter of 2.4 in. Find its circumference and its area.

15. Find the area of the shaded portion in the following figure.

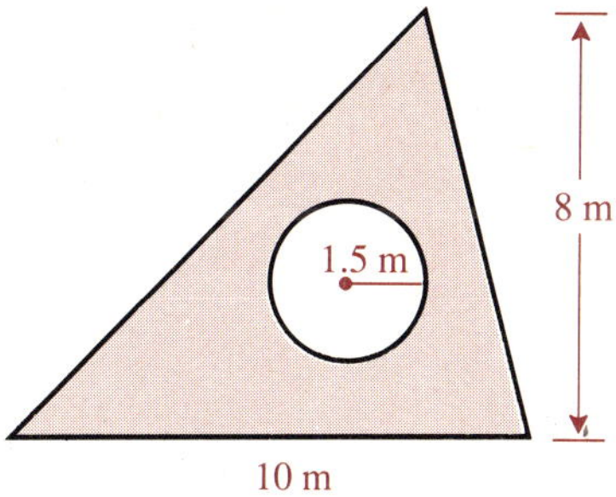

16. Find the area of the shaded portion in the following figure.

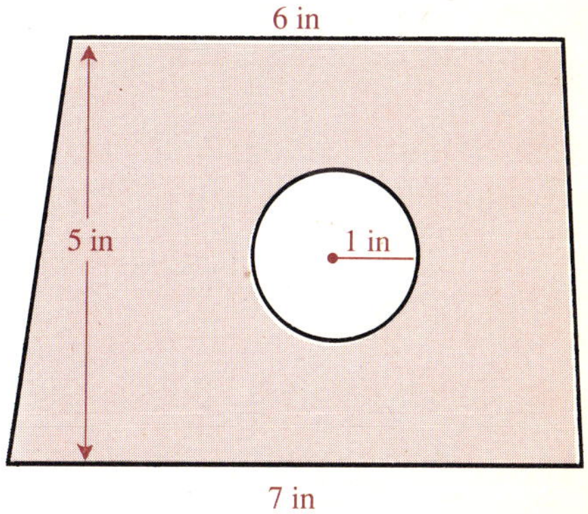

For Some Extra Thought

17. Find the circumference of a circle whose area is 16π square feet.

18. If the radius of a circle is doubled, how does this effect its area?

Review Problems for Maintaining Your Skills

19. Round 3,608,452 to the nearest ten thousand.

20. $\frac{2}{3} + \frac{1}{10} =$ **21.** $4\frac{3}{5} \times 1\frac{1}{3} =$

22. Divide 0.07 by 1.3 and round the quotient to the nearest tenth.

23. 126 is what percent of 200?

24. Solve: $7 - x = x + 9$

25. $\frac{10^5}{10^4} =$

26. Multiply and express your result in scientific notation.
$(4.6 \times 10^{-5})(3.8 \times 10^{8}) =$

27. Use a calculator to approximate $\sqrt{21}$ to the nearest hundredth.

Determine the following conversions.

28. 25 mi = _________ km **29.** 250 g = _________ kg

30. Find the perimeter of a square whose side measures 30 ft.

31. Find the perimeter of a rectangle whose length and width are 45.3 mm and 36.7 mm, respectively.

32. Find the area of a square whose side measures 0.005 cm.

33. Find the area of a parallelogram whose base and height measure 34.5 in and 24.8 in, respectively.

10.4 VOLUME

Preview Test

1. Find the volume of the cube shown below.

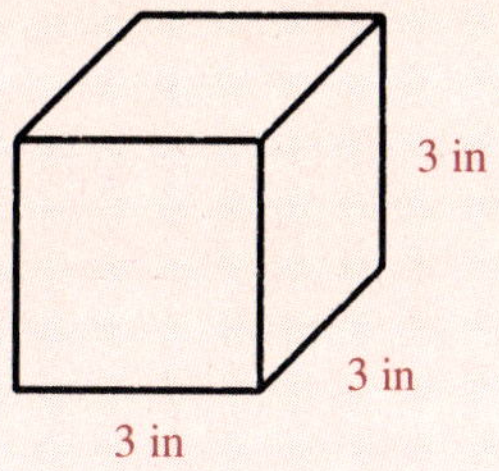

2. Find the volume of the rectangular solid shown below.

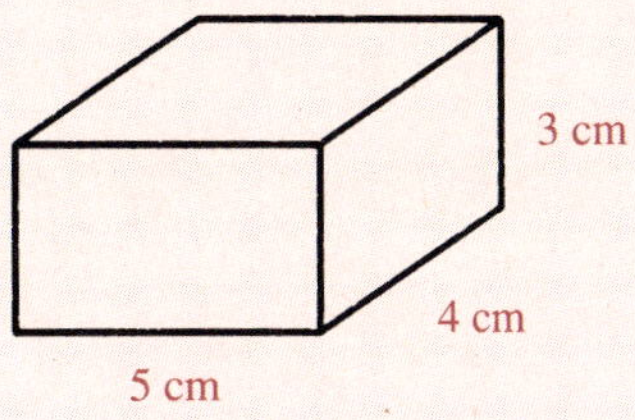

3. Find the volume of the cylinder shown below.

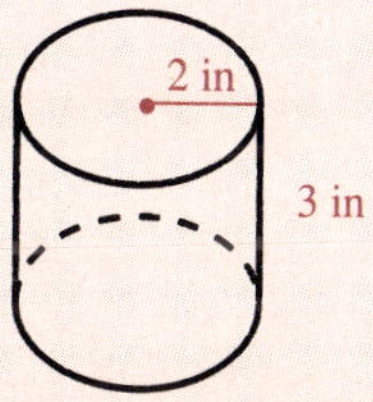

4. Find the volume of the sphere shown below.

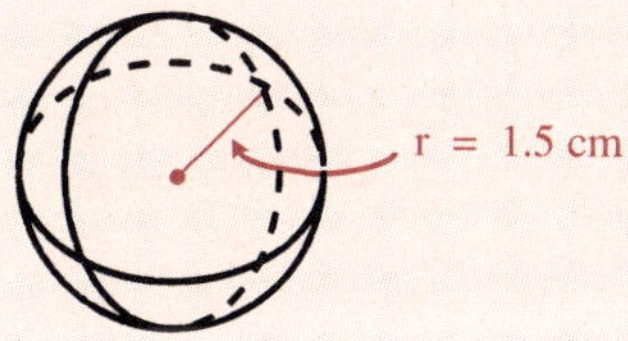

The size of a geometric solid can be measured by determining how many cubes are needed to fill the space. This measure is called the **volume** of the figure. For example, the cube shown on the left, whose edge is 1 cm, has a volume of 1 cubic centimeter.

ANSWERS FOR PREVIEW TEST

1. 27 cubic inches
2. 60 cubic centimeters
3. 37.68 cubic inches
4. 14.13 cubic millimeters

Notice that the rectangular solid on the right contains 24 of the 1 cubic centimeter cubes. Therefore, the volume of the rectangular solid is 24 cubic centimeters.

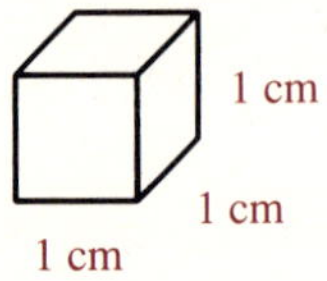

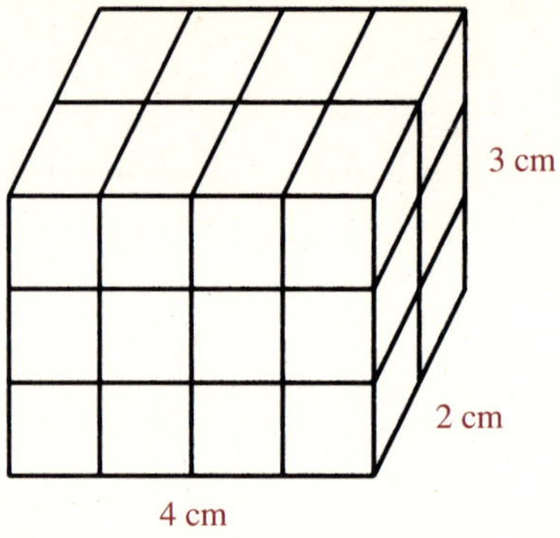

Instead of counting the number of cubes in a rectangular solid, we can multiply the area of the base (LW) by the height (H).

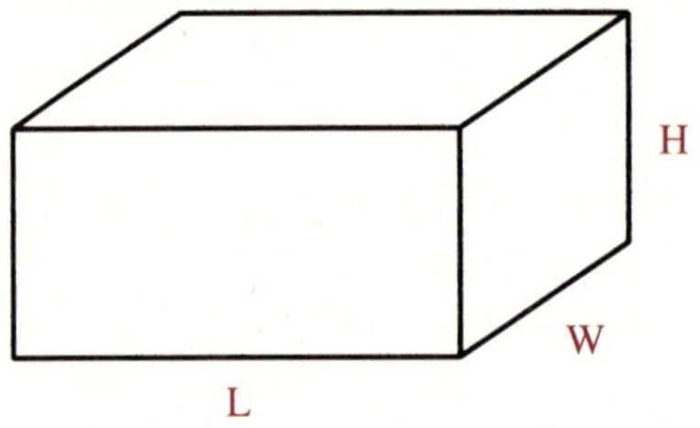

So, using the formula, the volume of the rectangular solid is

$$V = LWH$$
$$= (4 \text{ cm})(2 \text{ cm})(3 \text{ cm}) = 24 \text{ cubic centimeters}$$

A cube is a special kind of rectangular solid whose length, width, and height are equal. The length, width, or height of a cube is called its **edge**, and its volume V is given by the formula

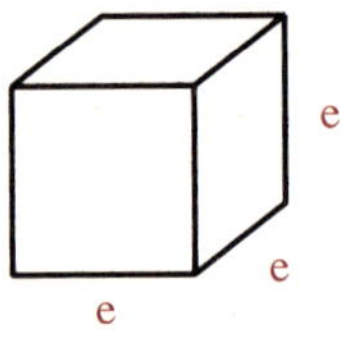

$V = e^3$, where e represents the edge.

EXAMPLE 1: Find the volume of each figure.

(a)

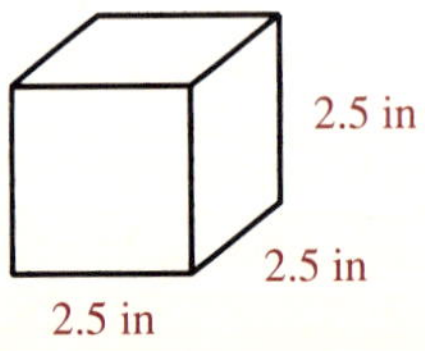

(b)

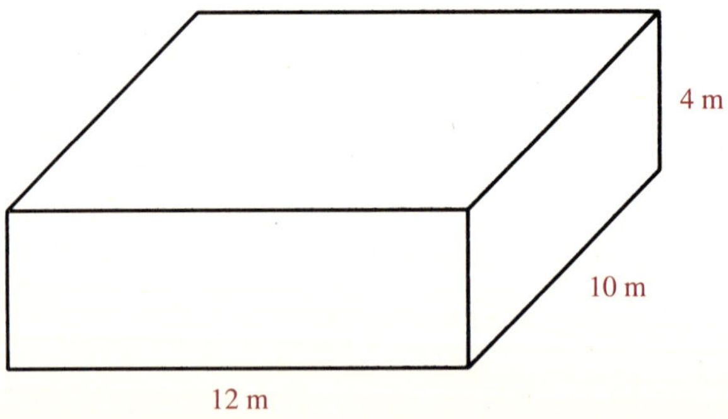

Solution:

a) Since the figure is a cube whose edge measures 2.5 in, its volume is

$$V = e^3 = (2.5)^3 = 15.625 \text{ cubic inches}$$

b) Since the figure is a rectangular solid whose length, width, and height are 12 m, 10 m, and 4 m, respectively, its volume is

$$\begin{aligned} V &= LWH \\ &= (12 \text{ m})(10 \text{ m})(4 \text{ m}) \\ &= 480 \text{ cubic meters.} \end{aligned}$$

PRACTICE PROBLEM 1: Find the volume of each figure.

(a)

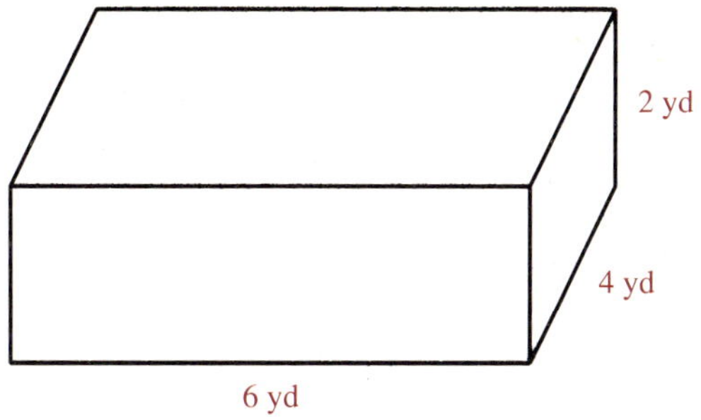

(b)

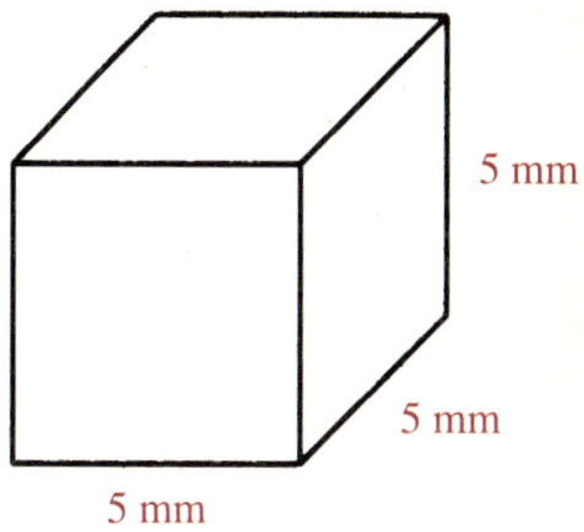

The can of soda shown below has the shape of a geometric solid called a **cylinder**. A cylinder has a circular base, and its volume is found by multiplying the area of its base (πr^2) by its height (h). So, the volume (V) of a cylinder with base radius (r) and height (h) is

$$V = \pi r^2 h$$

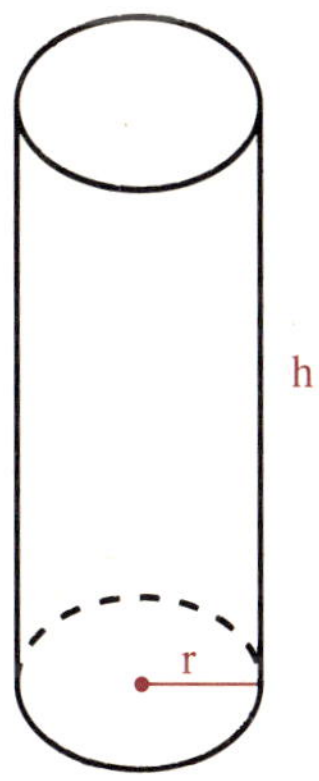

EXAMPLE 2: Find the volume of a cylinder whose base radius is 6 cm, and height 8 cm.

Solution:

$$v = \pi r^2 h$$

Now substitute 3.14 for π, 6 cm for r, and 8 cm for h.

$$V = (3.14)(6 \text{ cm})^2(8 \text{ cm}) = 904.32 \text{ cubic centimeters.}$$

PRACTICE PROBLEM 2: Find the volume of a cylinder whose base radius is 1.2 in and height 5 in.

A basketball has a geometric shape called a **sphere**. A sphere is a set of points in space that lie at a fixed distance, called its radius, from a fixed point, called its center.

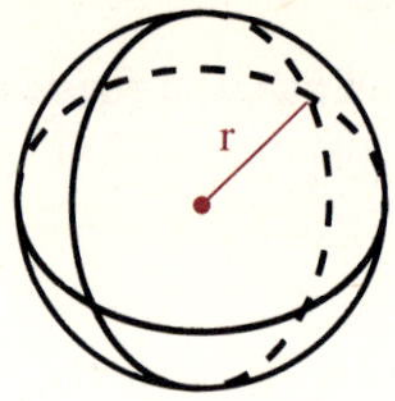

The volume (V) of a sphere whose radius is (r) is given by the formula

$$V = \frac{4}{3}\pi r^3$$

EXAMPLE 3: Find the volume of a sphere whose radius is 6 mm.

Solution:

$$V = \frac{4}{3}\pi r^3$$

Substitute 3.14 for π and 6 mm for r.

$$V = \frac{4}{3}\pi r^3 = \frac{4}{3}(3.14)(6\text{ mm})^3 = 904.32 \text{ cubic millimeters}$$

PRACTICE PROBLEM 3: Find the volume of a sphere whose radius is 0.3 ft.

ANSWERS FOR PRACTICE PROBLEMS

1. **a)** 48 cubic yds **b)** 125 cubic millimeters

2. 22.608 cubic inches **3.** 0.11304 cubic feet

EXERCISE SET 10.4A

Skill Development

Find the volume of each cube.

1. $e = 5$ mm **2.** $e = 7.2$ in **3.** $e = 4.6$ mi

Find the volume of each rectangular solid.

4. $L = 2$ cm, $W = 1.3$ cm, $H = 0.9$ cm

5. $L = 6.7$ in, $W = 2.4$ in, $H = 1.6$ in

6. $L = 3\frac{1}{4}$ ft, $W = 2\frac{1}{2}$ ft, $H = 1\frac{3}{4}$ ft

Find the volume of each cylinder.

7. $r = 3$ cm, $h = 10$ cm **8.** $r = 1.2$ in, $h = 2$ in

9. $r = 0.6$ dm, $h = 0.8$ dm

Find the volume of each sphere.

10. $r = 10$ in **11.** $r = 1.8$ cm **12.** $r = 2.1$ ft

Applications

13. A room measures 12 ft by 10 ft by 8 ft. Find its volume.

14. The edge of a cubical die measures 2 cm. Find its volume.

15. A juice can has a height of 6 in and a radius of 2 in. Find its volume.

16. A softball has a diameter of 9.6 cm. Find its volume.

For Some Extra Thought

17. If the edge of a cube is doubled, what effect does this have on its volume?

18. Find the volume of the largest sphere that can be placed inside a cube whose edge measures 2 m.

Review Problems for Maintaining Your Skills

19. Round 6,542 to the nearest ten

20. $3\frac{2}{5} - 1\frac{3}{4} =$ **21.** $\frac{17}{18} \div \frac{1}{12} =$

22. $2.56 \times 3.7 =$

23. To the nearest hundredth of a percent, 23 is what percent of 43?

24. Solve: $4(5x - 7) = x - 9$

25. Convert 0.0000000324 to scientific notation.

Determine the following conversions.

26. 5 in = __________ cm **27.** 52 oz = __________ lb

28. Find the perimeter of a rectangle whose length and width are 16 mm and 14 mm, respectively.

29. Find the area of a triangle whose base and height are 6 in and 4 in, respectively.

30. Find the circumference of a circle whose radius measures 36 mm.

31. Find the area of a circle whose radius measures 1 in.

EXERCISE SET 10.4B

Skill Development

Find the volume of each cube.

1. $e = 12$ cm **2.** $e = 3.6$ mm **3.** $e = 4.2$ yd

Find the volume of each rectangular solid.

4. $L = 4$ in, $W = 3$ in, $H = 2.5$ in

5. $L = 0.5$ cm, $W = 0.5$ cm, $H = 0.4$ cm

6. $L = \frac{2}{5}$ ft, $W = \frac{3}{5}$ ft, $H = \frac{4}{5}$ ft

Find the volume of each cylinder.

7. $r = 2.4$ mm, $h = 1$ mm **8.** $r = 4$ cm, $h = 2$ cm

9. $r = 3.6$ ft, $h = 3.5$ ft

Find the volume of each sphere.

10. $r = 3.6$ in **11.** $r = 0.6$ dm **12.** $r = 1.5$ m

Applications

13. A box measures 40 cm by 35 cm by 30 cm. Find its volume.

14. A small storage room is cubical. If its edge measures 9 ft, find its volume

15. A mailing cylinder has a radius of 6 cm and a length of 50 cm. Find its volume.

16. A balloon is inflated so its radius is 6 in. Find its volume.

For Some Extra Thought

17. The volume of a cube is 125 cubic centimeters. Find the measure of its edge.

18. The radius of a sphere is increased from 3 cm to 4 cm. By how much is its volume increased, to the nearest cubic centimeter?

Review Problems for Maintaining Your Skills

19. Round 42,608 to the nearest thousand.

20. $10\frac{1}{3} - 7\frac{2}{3} =$ **21.** $\frac{13}{14} \div \frac{3}{7} =$

22. $1.067 \times 100{,}000 =$

23. To the nearest tenth of a percent, 89 is what percent of 108?

24. Solve: $-3(4 - x) + 2x = 18$

25. Convert 2,400,000,000 to scientific notation.

26. 60 mi = ________ km **27.** 2 kg = ________ g

28. Find the perimeter of a square whose side measures 6 mm.

29. Find the area of a trapezoid whose bases measure 12 cm and 9 cm and whose height measures 4 cm.

30. Find the circumference of a circle whose diameter measures 10 yd.

31. Find the area of a circle whose radius measures 3 cm.

PRACTICE TEST 10A

1. Label these figures using all the appropriate terms—triangle, rectangle, square, parallelogram, trapezoid, circle.

(a) 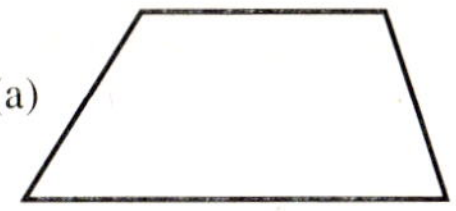 (b)  (c) 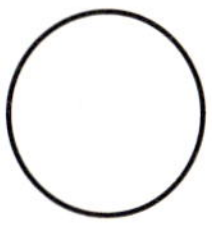 (d) 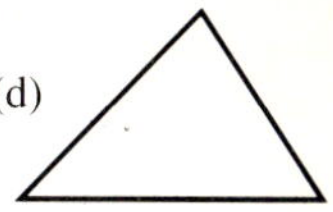

2. Find the perimeter of a square whose side measures 5.2 mm.

3. Find the perimeter of a rectangle whose length and width are 24.6 in and 18.4 in, respectively.

4. Find the perimeter of the figure shown below.

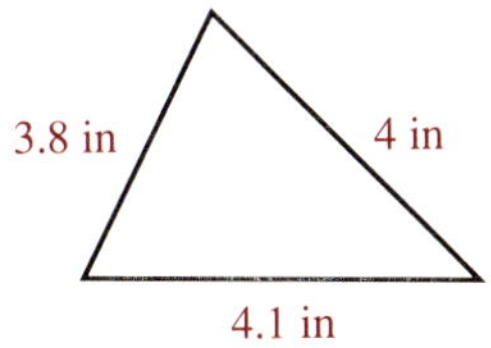

5. Find the perimeter of the figure shown below.

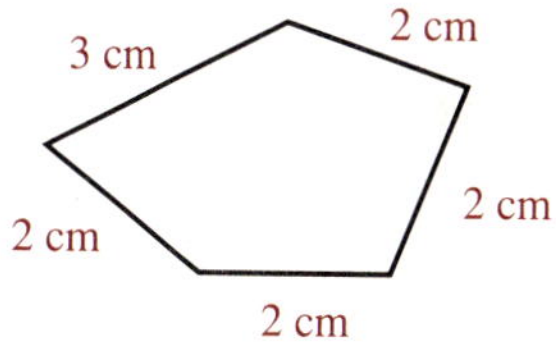

6. Find the area of the square whose side measures 6.75 mi.

7. Find the area of the parallelogram whose base and height are 6.4 cm and 2.1 cm, respectively.

8. Find the area of the figure shown below.

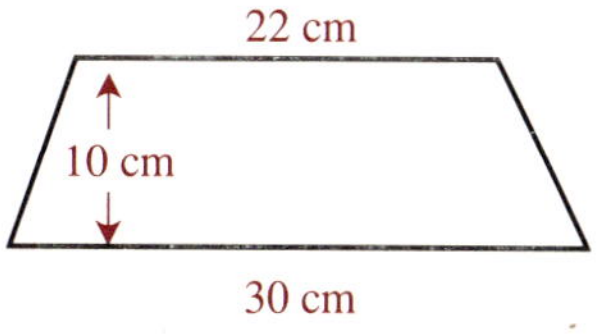

9. Find the area of the figure shown below.

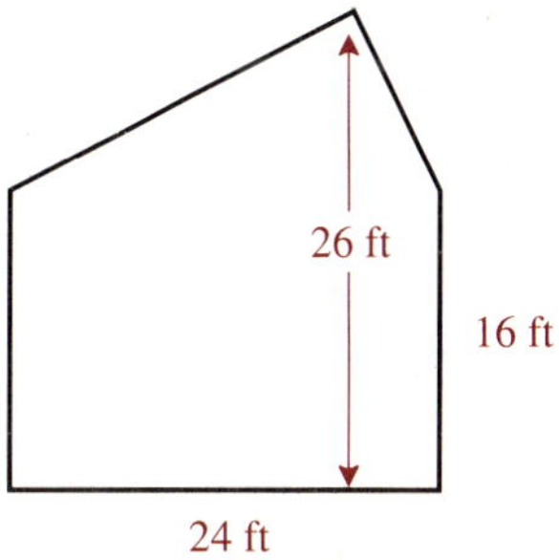

10. Find the circumference of a circle whose diameter measures 6 in.

11. Find the area of a circle whose radius measures 2.6 cm.

12. Find the volume of the rectangular solid whose length, width, and height are 3 ft, 2 ft and 0.5 ft, respectively.

13. Find the volume of a cylinder whose radius and height are 3 cm and 14 cm, respectively.

14. Find the volume of a sphere whose radius measures 9 mm.

PRACTICE TEST 10B

1. Label these figures using all the appropriate terms—triangle, rectangle, square, parallelogram, trapezoid, circle.

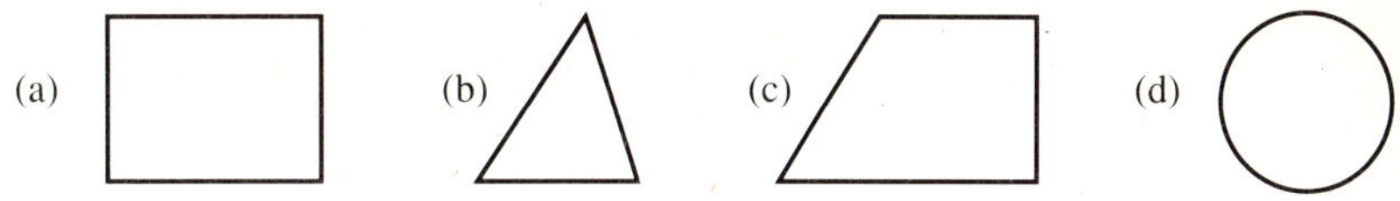

2. Find the perimeter of a square whose side measures 3.6 in.

3. Find the perimeter of a rectangle whose length and width are 15.6 cm and 12.7 cm, respectively.

4. Find the perimeter of the figure shown below.

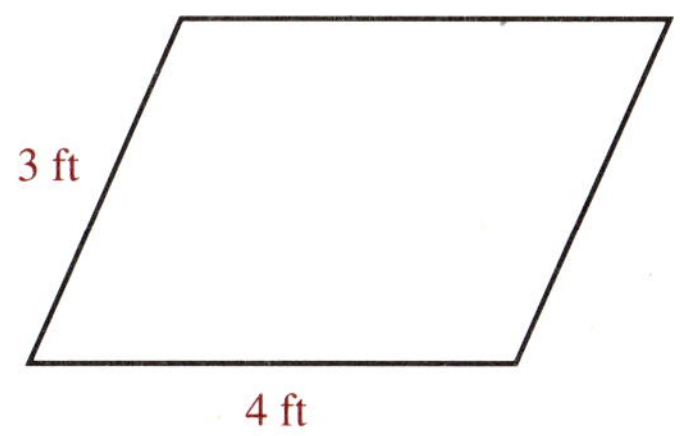

5. Find the perimeter of the figure shown below.

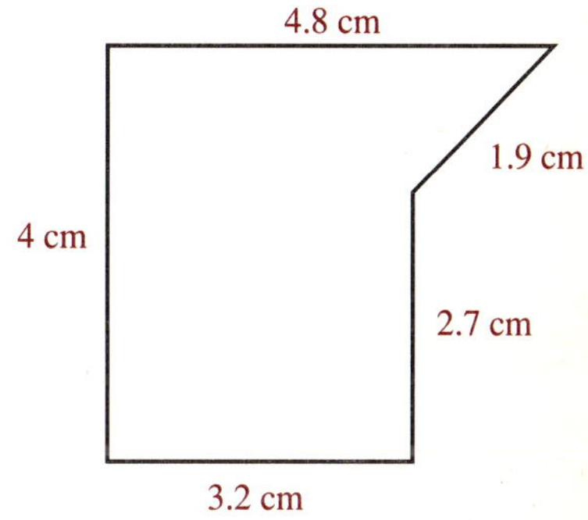

6. Find the area of the square whose side measures 4.8 cm.

7. Find the area of the trapezoid whose bases measure 8 ft and 6 ft and whose height measures 4 ft.

8. Find the area of the figure shown below.

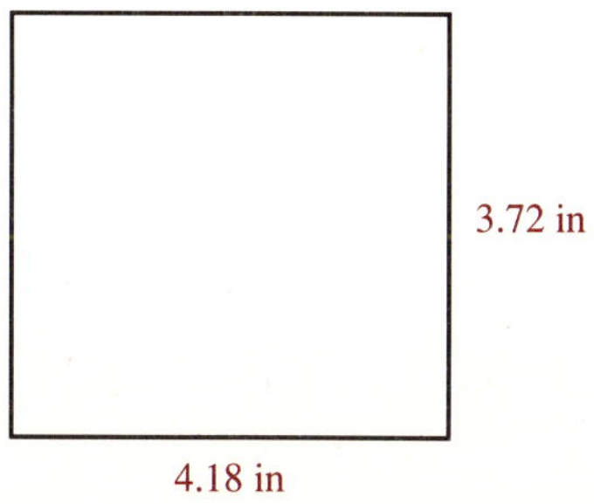

9. Find the area of the figure shown below.

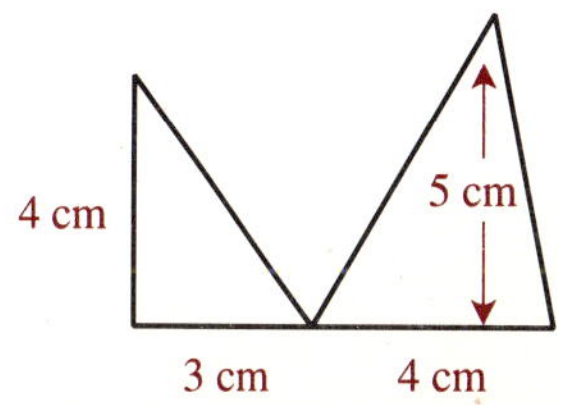

10. Find the circumference of a circle whose diameter measures 8.4 cm.

11. Find the area of a circle whose radius measures 4.2 mm.

12. Find the volume of the rectangular solid whose length, width, and height are 20 in, 14 in and 10 in, respectively.

13. Find the volume of a cylinder whose radius and height are 1.2 in and 5.6 in, respectively.

14. Find the volume of a sphere whose radius measures 6 cm.

CUMULATIVE REVIEW EXERCISES FOR CHAPTERS 1 THROUGH 10

1. Write 2,504,970 in words.

2. Round 4,506,800 to the nearest ten thousand.

In 3 through 6, reduce each fraction to its lowest terms and express any result that is an improper fraction as a mixed number.

3. $\frac{10}{13} \times \frac{4}{5} =$

4. $7 \div 1\frac{1}{3} =$

5. $6\frac{1}{4} + 7\frac{2}{5} =$

6. $\frac{11}{12} - \frac{2}{3} =$

7. Write 4.0145 in words.

8. Round 12.0406 to the nearest thousandth.

9. Round $1.62701 to the nearest cent.

10. Divide 413.6 by 73.25 and round the quotient to the nearest hundredth.

11. $30.8 - 2.5 \times 6.7 =$

12. Convert 4.56% to a decimal.

13. Convert 1.25 to a percent.

14. Convert 14% to a fraction reduced to its lowest terms.

15. To the nearest tenth of a percent, 17 is what percent of 19?

16. 2.5% of what amount equals $50?

17. Find 64% of $25,000.

18. If a person's weight drops from 250 pounds to 175 pounds, find the percent decrease.

19. Solve the proportion: $\frac{M}{5} = \frac{2}{3}$

20. $-20 + 34 - 17 - 2 =$

21. $(-4.23) - (+6.78) =$

22. $(1.23) \times (-600) =$

23. $(-5.6) \div (+0.14) =$

24. Write in words: $8w - 6x$

25. Solve: $2x - 3 = 17$

26. Solve: $3(4 - x) = x - 8$

27. Write $5 \times 5 \times 5 \times 5 \times 5 \times 5 \times 5$ in exponential form.

28. Write $(-2)^3$ as a product of factors.

29. Evaluate.

a) 10^4 **b)** $2^5 - 3^3$

30. Evaluate

a) $5^2 \cdot 3^2$ **b)** $\left(\frac{4}{5}\right)^3$

31. Write each number in scientific notation.

a) 0.00023 **b)** 3,800,000,000

32. Write as a whole number or decimal.

a) 5.4×10^4 **b)** 2.3×10^{-3}

33. Express the result in scientific notation.
$(2.4 \times 10^{12}) \times (3.1 \times 10^{-15})$

34. Find the square roots of 121.

35. Find the value of **a)** $-\sqrt{64}$ **b)** $\sqrt{225}$

36. Find two consecutive integers that bound $\sqrt{87}$.

37. Use a calculator to approximate each to the nearest hundredth.

a) $\sqrt{23}$ **b)** $\sqrt{145}$ **c)** $\sqrt{7}$

In 38 through 43, determine each conversion.

38. 45 ft = ________ yd **39.** 600 km = ________ m

40. 7 mi = ________ km **41.** 480 oz = ________ gal

42. 9,000 lb = ________ t **43.** 14 g = ________ mg

44. Label these figures using all the appropriate terms—triangle, rectangle, square, parallelogram, trapezoid, circle.

(a) 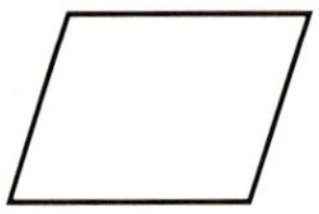 (b) 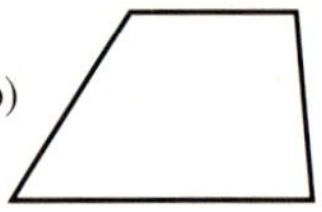 (c)  (d) 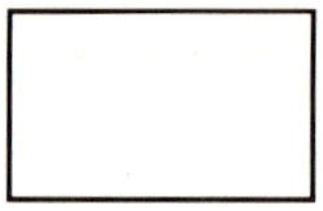

45. Find the perimeter of a rectangle whose length and width are 45.6 cm and 23.7 cm, respectively.

46. Find the perimeter of the figure shown below.

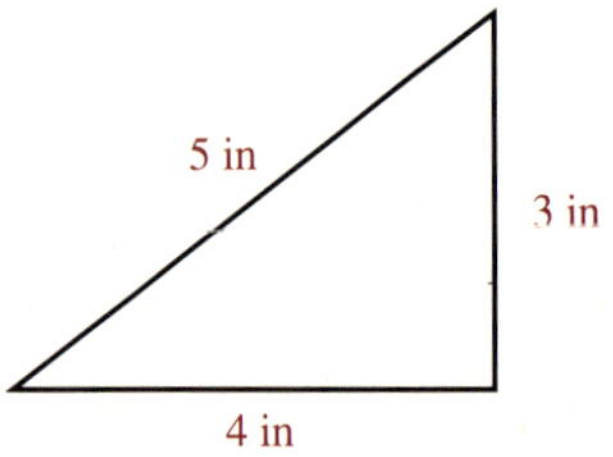

47. Find the area of the figure shown below.

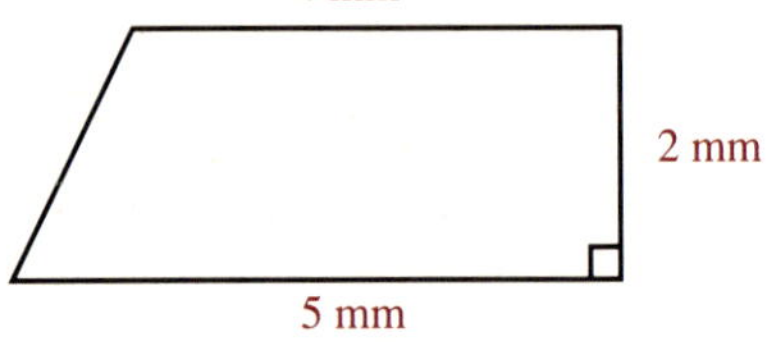

48. Find the area of a parallelogram whose base and height are 6.4 ft and 2.7 ft, respectively.

49. Find the circumference of a circle whose diameter measures 28 mm.

50. Find the area of a circle whose radius measures 5 in.

51. Find the volume of a cube whose edge measures 2.3 cm.

52. Find the volume of a cylinder whose radius and height are 1 cm and 8 cm, respectively.

APPENDIX

USING A CALCULATOR

There are many different calculators you can buy. Some can only be used for basic arithmetic. Others are used by engineers and financial managers to solve complex problems in their profession. The calculator illustrated here has the keys that are found on those that are designed for basic arithmetic.

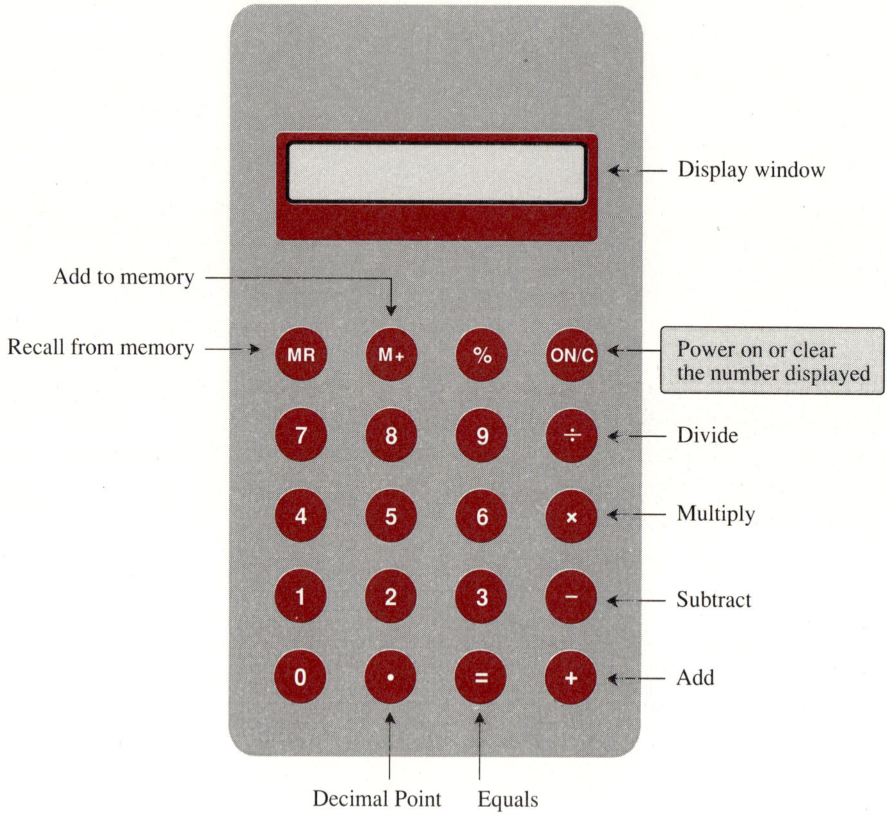

When the calculator is turned on, the display window will show a zero with a decimal point on its right.

To enter the number 237 the digit keys 2 3 7 are pressed and the display shows

In the examples in this section you will be told what number to enter and what function keys to press (+ , − , × , ÷ , M+ , MR , or =). The number in the window display will be shown after each step is completed.

EXAMPLE 1: Use a calculator to find the sum for 3.26 + 8.29

Solution:

	Window Display	
Press ON/C	0.	
Enter 3.26	3.26	
Press +	3.26	
Enter 8.29	8.29	
Press =	11.55	This is the sum for 3.26 + 8.29

If you divide 37 by 8 on your calculator, the display window will read

4.625

The number 4.625 has a whole number part (4) and a decimal part (.625). The whole number part represents the quotient for 37 divided by 8. To find the remainder you have to subtract the product of the quotient and divisor from the dividend. In fact, for any division problem

$$\text{REMAINDER} = \text{DIVIDEND} - \text{QUOTIENT} \times \text{DIVISOR}$$

EXAMPLE 2: Find the quotient and remainder for 37 ÷ 8.

Solution:

	Window Display	
Press ON/C	0.	
Enter 37	37.	
Press ÷	37.	
Enter 8	8.	
Press =	4.625	
Press ON/C	0.	Clear the display.
Enter 4	4.	Enter the dividend.
Press ×	4.	
Enter 8	8.	
Press =	32.	Find QUOTIENT × DIVISOR
Press M+	32.M	Add 32 to memory.
Enter 37	37.M	Enter the dividend.
Press −	37.M	Subtract QUOTIENT × DIVISOR
Press MR	32.M	from the DIVIDEND.
Press =	5.M	The remainder is 5.

So, for 37 ÷ 8 the result is $4\frac{5}{8}$.

In the next example the memory key is used to carry out operations in the correct order.

EXAMPLE 3: Use a calculator to find the value for $50 - 5 \times 3$.

Solution:

	Window Display	
Press ON/C	0.	
Enter 5	5.	
Press ×	5.	Multiply before
Enter 3	3.	subtracting.
Press =	15.	
Press M+	15.M	Add 15 to memory.
Enter 50	50.M	
Press −	50.M	
Press MR	15.M	Subtract 15 from 50.
Press =	35.M	

Therefore $50 - 5 \times 3 = 35$.

ANSWER KEY

CHAPTER 1

SET 1.1A

1. (a) hundred thousands (b) hundreds (c) tens
2. (a) millions (b) ten billions (c) ten millions
3. (a) ten billions (b) hundreds (c) ten thousands
4. sixty-seven thousand, thirty-nine
5. eight hundred forty-nine thousand
6. one thousand, three hundred twenty-eight
7. thirty-one million, six hundred thousand, five hundred twenty-one
8. twenty-six billion, four hundred twenty-six
9. eight hundred ninety-seven billion, two hundred seventy-six million, four hundred thousand
10. nine million
11. forty-three thousand, six hundred fifty-one
12. four hundred sixty-three billion, six hundred thousand, five
13. ninety-five billion, six hundred fifty million
14. 2,083 **15.** 5,743,009 **16.** 18,500
17. 73,600,400 **18.** 23,000,050,328
19. 692,117 **20.** 37,000,000,000 **21.** 9,000,008
22. 400,006,023,862 **23.** 842,070,000
24. 227,000,000
25. two hundred sixty-two thousand, seventeen square miles
26. hundred thousands
27. four hundred nine trillion, seven hundred thirty-six billion
28. 72,006,500,000,000 **29.** two billion

SET 1.2A

1. 5,700 **2.** 750,000 **3.** 498,700
4. 13,000,000 **5.** 24,000,000 **6.** 56,500
7. 5,720,000 **8.** 45,000 **9.** 1,200,000
10. 60 **11.** 7,000,000 **12.** 24,000,000,000
13. 1,600,000 **14.** 5,700 **15.** 400
16. 4,000,000,000,000
17. Yes. 200 rounded to the nearest hundred equals 200
18. (a) thousands (b) hundred thousands (c) tens
19. (a) four thousand, five hundred eighty-seven
(b) sixty-two million, eight thousand, four hundred
(c) three billion, eight hundred forty-five million, two hundred
20. (a) 52,719 (b) 9,060,508 (c) 230,000,100,000

SET 1.3A

1. 28 **2.** 25 **3.** 31 **4.** 257 **5.** 197
6. 1,545 **7.** 2,082 **8.** 12,376 **9.** 25,863
10. 122,864 **11.** 146,297 **12.** 2,383,852
13. 45 **14.** 57 **15.** 227 **16.** 315
17. 1,895 **18.** 2,516 **19.** 86,581
20. 1,294,826 **21.** 152,190 **22.** 21,380
23. \$612 **24.** 90 **25.** 249 **26.** 12
27. 20 **28.** 9 **29.** 1,300 **30.** 2,400
31. 3,900 **32.** 9,600 **33.** \$1,970 **34.** \$210
35. \$80 **36.** \$1,850 **37.** ten thousands
38. seven million, four hundred eighty-nine thousand, six hundred three
39. 9,000,670,000 **40.** 47,000 **41.** 1,400,000
42. 680

SET 1.4A

1. 33 **2.** 83 **3.** 52 **4.** 51 **5.** 130
6. 777 **7.** 1,413 **8.** 617 **9.** 1,068
10. 3,393 **11.** 11,349 **12.** 84,019 **13.** 44
14. 44 **15.** 524 **16.** 325 **17.** 355
18. 5,308 **19.** 17,459 **20.** 67,679
21. 15 days **22.** \$318 **23.** \$428
24. 43 miles **25.** 5,825 males **26.** 47

27. 1 **28.** 10 millions **29.** \$510 **30.** \$250
31. \$390 **32.** \$420 **33.** 2,400
34. fifty-six million, nine hundred seventy thousand, two hundred
35. 67,000,009,040 **36.** 2,600 **37.** 6,190,000
38. 45,000,000 **39.** 1,905 **40.** 40,111

SET 1.5A

1. 448 **2.** 378 **3.** 726 **4.** 10,396
5. 2,408 **6.** 1,780 **7.** 9,672 **8.** 40,716
9. 9,600 **10.** 270,000 **11.** 418,068
12. 8,563,270 **13.** 702 **14.** 2,394
15. 51,012 **16.** 43,100 **17.** 400,000
18. 47,112 **19.** 1,000,000 **20.** 360,962
21. 6,750 square feet **22.** \$350,000
23. \$1,820 **24.** 168 hours **25.** \$7,400
26. 13 and 15 **27.** 8
28. No, because 99 × 99 = 9,801 is the largest product of a pair of two-digit whole numbers.
29. 2,400 **30.** 3,200 **31.** 21,000
32. 42,000 **33.** 20,000 **34.** 120,000
35. \$1,000,000
36. three billion, eight hundred seventy-three thousand
37. 78,000 **38.** 890 **39.** 352 **40.** 3,895
41. 86 **42.** 4,157

SET 1.6A

1. 24 **2.** 95 R 4 **3.** 69 **4.** 1,357
5. 301 R 1 **6.** 23 R 8 **7.** 3 R 11
8. 44 R 11 **9.** 13 R 9 **10.** 28
11. 308 R 17 **12.** 43 R 279 **13.** 9 R 4
14. 28 R 1 **15.** 90 **16.** 20 R 1 **17.** 108
18. 138 R 3 **19.** 6 **20.** 7 R 4 **21.** 10 R 37
22. 10 R 113 **23.** 13 R 87 **24.** 306 R 100
25. 9,189 R 1 **26.** 9,482 R 29 **27.** 4,702 R 26
28. 34,042 R 26 **29.** 2,280 R 80
30. 10,969 R 13 **31.** \$300,000 **32.** 27 miles
33. \$4,239 **34.** 7 cases **35.** 4
36. Yes, if the divisor is 1. For example, 901/1 = 901.
37. 16 **38.** 8 **39.** 5 **40.** 8 **41.** 8
42. 10 **43.** 5 **44.** 7 **45.** \$13
46. eighty-seven million, six hundred fifty-two thousand, one hundred eighty-nine
47. 13,000 **48.** 4,600,000 **49.** 1,544
50. 293 **51.** 122 **52.** 247 **53.** 10,166
54. 240,000

SET 1.7A

1. 12 **2.** 21 **3.** 219 **4.** 23 **5.** 172
6. 159 **7.** 28 **8.** 31 **9.** 9 **10.** 364
11. \$6,600 **12.** \$1,120 **13.** 90 **14.** 10
15. 4 **16.** 4,600,000 **17.** 2,350 **18.** 3,825
19. 426 **20.** 1,440,000 **21.** 44,496
22. 12 R 9 **23.** 70 R 10

PRACTICE TEST 1A

1. hundred thousands
2. two hundred four thousand, eight hundred seventy-three
3. 7,063,904 **4.** 54,300 **5.** 6,790,000
6. 5,146 **7.** 25,097 **8.** 3,977 **9.** \$1,086
10. 42,441 **11.** 3,030,000 **12.** 18 R 12
13. 2 R 307 **14.** \$547 **15.** 69 **16.** 5,220

PRACTICE TEST 1B

1. tens **2.** two million, five hundred thirty
3. 823,017 **4.** 46,000 **5.** 4,600 **6.** 2,131
7. 2,664 **8.** 8,668 **9.** \$1,700 **10.** 13,872
11. 4,100,000 **12.** \$581 **13.** 3 R 27
14. 3 R 194 **15.** 32 **16.** 18

CHAPTER 2

SET 2.1A

1. $\frac{21}{22}$ **2.** $\frac{4}{5}$ **3.** $\frac{4}{5}$ **4.** $\frac{31}{60}$ **5.** $\frac{41}{45}$
6. $\frac{87}{100}$ **7.** $\frac{27}{36}$ **8.** $\frac{35}{63}$ **9.** $\frac{5}{35}$ **10.** $\frac{30}{32}$
11. $\frac{18}{39}$ **12.** $\frac{15}{55}$ **13.** $\frac{5}{8}$ **14.** $\frac{2}{5}$ **15.** $\frac{11}{24}$
16. $\frac{32}{80}$ **17.** Yes **18.** 1,000 **19.** 867,000
20. 1,751 **21.** 547 **22.** 13,500,000
23. 7,254 **24.** 33 R 6 **25.** 26

SET 2.2A

1. $1\frac{7}{10}$ 2. $1\frac{4}{15}$ 3. $17\frac{1}{7}$ 4. $1\frac{28}{43}$ 5. $1\frac{9}{31}$
6. $8\frac{9}{10}$ 7. $\frac{21}{5}$ 8. $\frac{15}{8}$ 9. $\frac{73}{7}$ 10. $\frac{68}{11}$
11. $\frac{32}{9}$ 12. $\frac{36}{13}$ 13. $4\frac{5}{8}$ 14. $5\frac{2}{3}$ 15. $\frac{31}{2}$
16. $\frac{5}{4}$ 17. the 21-ounce block 18. 95 quarters
19. 41,690,000 20. 524 21. 50,792
22. 4 R 51 23. 66 24. $\frac{4}{5}$ 25. $\frac{24}{28}$

SET 2.3A

1. $\frac{1}{6}$ 2. $\frac{21}{32}$ 3. $\frac{1}{12}$ 4. $\frac{5}{13}$ 5. $\frac{25}{28}$
6. $\frac{15}{77}$ 7. $\frac{1}{10}$ 8. $\frac{5}{28}$ 9. $\frac{120}{169}$ 10. $4\frac{1}{5}$
11. $48\frac{2}{5}$ 12. $32\frac{1}{2}$ 13. $1\frac{9}{40}$ 14. $16\frac{8}{25}$
15. $2\frac{7}{9}$ 16. $\frac{27}{50}$ 17. $\frac{5}{14}$ 18. $\frac{21}{32}$ acres
19. $\frac{18}{25}$
20. No, because the product of two numbers, each larger than 1, must be larger than 1.
21. 3 22. 2,000,000 23. 277 24. 4,183
25. 5 R 4 26. 2 27. $\frac{15}{17}$ 28. $\frac{44}{60}$
29. $1\frac{6}{13}$ 30. $\frac{33}{7}$

SET 2.4A

1. $\frac{5}{24}$ 2. $\frac{36}{65}$ 3. $\frac{15}{16}$ 4. $2\frac{13}{36}$ 5. $\frac{10}{33}$
6. $\frac{5}{7}$ 7. $\frac{3}{7}$ 8. $\frac{19}{240}$ 9. 28 10. $3\frac{29}{49}$
11. $2\frac{44}{51}$ 12. $\frac{21}{188}$ 13. $2\frac{7}{15}$ 14. $3\frac{3}{5}$
15. $6\frac{8}{15}$ 16. 20 17. 4 18. $\frac{1}{20}$ 19. 28
20. $\frac{1}{3}$
21. Yes, since if N is a whole number then $N \div \frac{1}{7} = N \times 7$ which must be a whole number.
22. 1,700 23. 2,335 24. 2 R 23 25. 7
26. $\frac{7}{9}$ 27. $\frac{36}{48}$ 28. $1\frac{4}{5}$ 29. $\frac{41}{7}$ 30. $\frac{13}{21}$
31. 33 32. $8\frac{15}{16}$

SET 2.5A

1. 120 2. 84 3. 120 4. $\frac{5}{7}$ 5. $1\frac{1}{5}$
6. $\frac{24}{25}$ 7. $1\frac{1}{4}$ 8. $\frac{44}{45}$ 9. $\frac{61}{200}$ 10. $\frac{127}{180}$
11. $\frac{1}{3}$ 12. $1\frac{13}{54}$ 13. $1\frac{16}{21}$ 14. $\frac{29}{60}$
15. $1\frac{37}{105}$ 16. $2\frac{5}{6}$ 17. $5\frac{31}{40}$ 18. $3\frac{29}{56}$
19. $6\frac{1}{15}$ 20. $13\frac{1}{16}$ 21. $18\frac{5}{8}$ 22. $9\frac{1}{10}$
23. $7\frac{1}{9}$ 24. $7\frac{7}{18}$ 25. $\frac{7}{15}$ 26. $\frac{3}{4}$
27. $\frac{7}{18}$ gallons 28. $\frac{24}{35}$ 29. $\frac{11}{17}$
30. All are $\frac{1}{20}$ 31. 56,000 32. 1,891
33. 7,872 34. 18 R 8 35. $\frac{2}{3}$ 36. $\frac{140}{200}$
37. $3\frac{4}{5}$ 38. $\frac{38}{7}$ 39. $\frac{1}{3}$ 40. $5\frac{1}{4}$
41. $3\frac{1}{2}$ 42. $\frac{15}{17}$

SET 2.6A

1. $\frac{13}{40}$ 2. $\frac{6}{13}$ 3. $\frac{11}{100}$ 4. $\frac{3}{8}$ 5. $\frac{37}{60}$
6. $\frac{5}{12}$ 7. $\frac{5}{12}$ 8. $\frac{13}{56}$ 9. $\frac{1}{400}$ 10. $5\frac{1}{2}$
11. $1\frac{7}{40}$ 12. $4\frac{1}{12}$ 13. $7\frac{3}{10}$ 14. $5\frac{5}{36}$
15. $9\frac{9}{35}$ 16. $2\frac{10}{13}$ 17. $1\frac{1}{7}$ 18. $\frac{1}{16}$
19. $\frac{7}{16}$ 20. $\$\frac{1}{16}$ 21. $1\frac{7}{10}$ tons 22. $6\frac{1}{16}$ feet
23. $\frac{6}{5}$ and $\frac{2}{5}$ 24. $\frac{8}{13}$ 25. 6,810,000
26. 11 R 6

27. $\frac{12}{19}$ **28.** $\frac{35}{84}$ **29.** $7\frac{4}{5}$ **30.** $\frac{51}{8}$ **31.** $\frac{5}{26}$
32. $\frac{2}{3}$ **33.** $3\frac{9}{11}$ **34.** $1\frac{1}{28}$ **35.** $1\frac{5}{8}$ **36.** $5\frac{5}{7}$

SET 2.7A

1. $1\frac{5}{32}$ **2.** $\frac{7}{12}$ **3.** $\frac{2}{3}$ **4.** $3\frac{1}{5}$ **5.** $\frac{23}{30}$
6. $8\frac{1}{4}$ **7.** $\frac{1}{10}$ **8.** $\frac{9}{16}$ **9.** $2\frac{1}{2}$ **10.** $\frac{17}{48}$
11. $\frac{2}{9}$ **12.** 0 **13.** 2,100,000 **14.** 41
15. 21,060 **16.** $\frac{4}{13}$ **17.** $\frac{66}{78}$ **18.** $1\frac{1}{73}$
19. $\frac{11}{9}$ **20.** $\frac{4}{7}$ **21.** $18\frac{5}{12}$ **22.** $\frac{1}{4}$ **23.** $1\frac{11}{20}$
24. $3\frac{1}{2}$ **25.** $5\frac{2}{9}$ **26.** $\frac{17}{22}$ **27.** $8\frac{3}{8}$ **28.** $4\frac{9}{10}$

PRACTICE TEST 2A

1. (a) $\frac{5}{14}$ (b) $\frac{37}{39}$ (c) $\frac{1}{3}$ **2.** (a) $\frac{42}{54}$ (b) $\frac{28}{64}$
3. (a) $2\frac{5}{7}$ (b) $6\frac{1}{4}$ (c) $2\frac{11}{20}$
4. (a) $\frac{49}{9}$ (b) $\frac{182}{15}$ (c) $\frac{12}{7}$ **5.** (a) $\frac{2}{15}$ (b) $20\frac{9}{10}$
6. (a) $\frac{2}{3}$ (b) 20 (c) $2\frac{13}{16}$ **7.** 800 shares
8. 120 **9.** (a) $\frac{17}{30}$ (b) $5\frac{7}{10}$ (c) $13\frac{23}{28}$
10. (a) $\frac{9}{20}$ (b) $2\frac{2}{3}$ (c) $9\frac{1}{10}$ **11.** $\frac{1}{20}$ **12.** $\frac{1}{2}$
13. $\frac{5}{36}$

PRACTICE TEST 2B

1. (a) $\frac{11}{12}$ (b) $\frac{6}{17}$ (c) $\frac{3}{50}$ **2.** (a) $\frac{27}{72}$ (b) $\frac{130}{1{,}000}$
3. (a) $2\frac{1}{2}$ (b) $1\frac{12}{31}$ (c) $1\frac{4}{11}$
4. (a) $\frac{73}{15}$ (b) $\frac{131}{10}$ (c) $\frac{23}{8}$ **5.** (a) $\frac{11}{20}$ (b) $28\frac{1}{2}$
6. (a) 4 (b) $2\frac{11}{16}$ (c) $2\frac{29}{38}$ **7.** $\frac{3}{20}$ tons
8. 60 **9.** (a) $1\frac{1}{30}$ (b) $6\frac{19}{30}$ (c) $11\frac{11}{24}$
10. (a) $\frac{11}{20}$ (b) $\frac{3}{4}$ (c) $7\frac{13}{20}$ **11.** $\frac{13}{20}$ **12.** $\frac{39}{40}$
13. $\frac{19}{21}$

CUMULATIVE REVIEW EXERCISES FOR CHAPTERS 1 AND 2

1. ten thousands
2. twenty-one million, three hundred forty-one thousand
3. 1,080,200 **4.** 23,500 **5.** 3,783 **6.** 3,273
7. \$952 **8.** 25 R 9 **9.** 184 **10.** $\frac{24}{29}$
11. $\frac{18}{48}$ **12.** $3\frac{3}{25}$ **13.** $\frac{33}{7}$ **14.** $\frac{4}{5}$ **15.** $1\frac{2}{3}$
16. $7\frac{9}{10}$ **17.** $\frac{11}{21}$ **18.** $\frac{1}{3}$

CHAPTER 3

SET 3.1A

1. thousandths, 8/1,000
2. hundred thousandths, 2/100,000
3. tenths, 9/10 **4.** ten-thousandths, 8/10,000
5. ten-thousandths, 7/10,000
6. thousandths, 6/1,000
7. hundred-thousandths, 3/100,000
8. hundredths, 7/100
9. seven and six thousand twenty-three ten-thousandths
10. seven thousand twenty-three millionths
11. seventy-five and nine hundred eighty-one ten-thousandths
12. five hundred thirty-one millionths
13. five and nine tenths **14.** six millionths
15. three hundred forty-five and two thousand two hundred seventy-three ten-thousandths
16. three and four thousand sixty-three ten-thousandths
17. 4.7 **18.** 0.0035 **19.** 0.000004

20. 0.0000006 **21.** 2.00023 **22.** 0.000016
23. 42.1 **24.** 6.0032 **25.** hundredths, 7/100
26. 0.345 **27.** two and sixty-three hundredths
28. twenty-three thousand, four hundred sixty-two and seven thousandths
29. 6,0040 **30.** 22,000 **31.** 10 R 15 **32.** 26
33. $\frac{2}{3}$ **34.** $2\frac{3}{7}$ **35.** $\frac{1}{5}$ **36.** $9\frac{4}{9}$ **37.** $1\frac{3}{20}$
38. $10\frac{7}{12}$ **39.** $\frac{91}{108}$ **40.** $\frac{1}{2}$

SET 3.2A

1. 45.2 **2.** 6.90 **3.** 12.2505 **4.** 1.25763
5. 14.56 **6.** 0.071 **7.** 43.642846
8. 0.1437895 **9.** 5.4 **10.** \$4.56
11. \$33.26 **12.** \$3.87 **13.** 2.64 inches
14. 3.1 **15.** 84.371 pounds **16.** 2,143.7 pesos
17. \$4.053 **18.** 1.02 **19.** 43,600,000
20. 101 **21.** 56,000 **22.** $\frac{21}{56}$ **23.** $\frac{33}{5}$
24. $\frac{1}{10}$ **25.** $\frac{7}{20}$ **26.** $\frac{31}{40}$ **27.** $6\frac{5}{9}$ **28.** $1\frac{5}{12}$
29. $3\frac{1}{4}$ **30.** thousandths, 6/1,000
31. one and four hundred seventy-eight thousandths
32. 17.000062

SET 3.3A

1. 29.5 **2.** 278.2 **3.** 5.63 **4.** 38.8
5. 4.23 **6.** 11.541 **7.** 3.168 **8.** 8.65
9. 7.16 **10.** 8.00346 **11.** 68.25 **12.** 45.67
13. 19.488 **14.** 950.3967 **15.** 6.8
16. 11.64 **17.** 3.333 **18.** 0.0214
19. \$1,675.54 **20.** \$1,554.78 **21.** 7.9 degrees
22. 0.00658 centimeters **23.** 0.0005 and 0.0005
24. 2 months **25.** $0.3 + 0.03 + \cdots + 0.00000003$
26. 98 **27.** 53 **28.** \$37 **29.** \$37.20
30. two million, eight thousand, three hundred forty-five
31. 5 R 9 **32.** $\frac{4}{9}$ **33.** $5\frac{2}{11}$ **34.** $1\frac{11}{40}$
35. $3\frac{4}{7}$ **36.** $1\frac{1}{10}$
37. ten-thousandths, 5/10,000
38. five and four hundred ninety-six ten-thousandths
39. 6.0486 **40.** \$2.46

SET 3.4A

1. 76.8 **2.** 5,695.0 **3.** 1,398.60 **4.** 173.88
5. 5.046 **6.** 3.42000 **7.** 1.38212
8. 4.78314 **9.** 0.0115482 **10.** 156.256
11. 8,338.09 **12.** 175.86 **13.** 0.54
14. 0.0483 **15.** 238.74 **16.** 1,200
17. 0.035 **18.** 0.0567 **19.** 0.00023
20. \$87.60 **21.** 39.6 **22.** 21.8091 sq. ft.
23. \$11.84 **24.** \$11.29 **25.** 0.0000085
26. 0.5 **27.** 0.001 **28.** 32 **29.** 9
30. 0.014 **31.** 180,000 **32.** 56,020,000,000
33. 922 **34.** 5,100 **35.** $\frac{5}{7}$ **36.** $\frac{35}{8}$
37. $7\frac{9}{10}$ **38.** $\frac{17}{30}$ **39.** $\frac{5}{11}$
40. twenty-two and five hundred sixty-eight thousandths
41. 4.51 **42.** \$5.13 **43.** 21.35 **44.** 1.876

SET 3.5A

1. 8.9 **2.** 8.3 **3.** 5.23 **4.** 19 **5.** 9
6. 0.0462 **7.** 0.094 **8.** 3.5 **9.** 60
10. 2 **11.** 5 **12.** 19.0 **13.** 0.2 **14.** 2.68
15. 21.48 **16.** 1.816 **17.** 44.262 **18.** \$8.25
19. \$4.85 **20.** \$6.78 **21.** \$2.63 **22.** 0.75
23. 0.063 is larger than 0.057 **24.** 0.2 **25.** 8
26. 45 **27.** 20 **28.** 1 **29.** 5,100,000
30. 9 R 10 **31.** $\frac{4}{11}$ **32.** $2\frac{5}{17}$ **33.** $1\frac{29}{130}$
34. $1\frac{1}{8}$ **35.** $1\frac{9}{16}$ **36.** 2.41 **37.** \$3.67
38. 14.22 **39.** 3.447 **40.** 245 **41.** 32.487

SET 3.6A

1. 51.075 **2.** 63.9 **3.** 8.3 **4.** 36 **5.** 1
6. 5.9015 **7.** 13.81 **8.** 10,000 **9.** 0.355
10. 28 **11.** \$2,700 **12.** 3.54 **13.** \$500.50
14. 1 **15.** 0
16. one million, twenty-six thousand, eight hundred
17. 529 **18.** 376,808 **19.** $\frac{3}{7}$ **20.** $\frac{31}{4}$
21. $5\frac{1}{10}$ **22.** $\frac{47}{60}$ **23.** $1\frac{1}{8}$ **24.** 2.0474
25. \$5.87 **26.** 15.73 **27.** 66.892 **28.** 542
29. 1,560.09 **30.** 0.03267 **31.** 15,800
32. 35.94 **33.** 2.4

PRACTICE TEST 3A

1. four and five hundred three thousandths
2. 35.0017 3. 4.5 4. 0.07829 5. $41.24
6. 16.92 7. 2.847 8. 0.58 9. 3.952
10. 4.68 11. $10.83 12. 84.65 13. 2.63
14. $0.36 15. 12.46

PRACTICE TEST 3B

1. twenty-three and six hundred seventy-two ten-thousandths
2. 0.000076 3. 2.41 4. 364.3 5. $1.34
6. 15.23 7. 4.358 8. 2.89 degrees
9. 2.768 10. 6.25 11. $11,242 12. 80
13. 5.931 14. $0.02 15. 8.82

CUMULATIVE REVIEW EXERCISES FOR CHAPTERS 1 THROUGH 3

1. three million, sixty-seven thousand, one hundred twenty-one
2. 12,500 3. 877 4. 4 R 9 5. $\frac{6}{7}$
6. $\frac{55}{8}$ 7. $14\frac{2}{7}$ 8. $\frac{11}{21}$ 9. $4\frac{5}{16}$
10. $\frac{19}{24}$ 11. $10\frac{2}{7}$
12. four and one hundred five thousandths
13. 0.000007 14. 12.0574 15. $1.24
16. 89.66 17. 1.023 18. 6.48 19. 3.211
20. 3.8 21. 130.4 22. 63.22

CHAPTER 4

SET 4.1A

1. 0.29 2. 0.36 3. 0.053 4. 0.7814
5. 0.029 6. 0.007 7. 0.00019 8. 0.0425
9. 0.156 10. 0.99 11. 47% 12. 9%
13. 25% 14. 7.6% 15. 6.12% 16. 0.7%
17. 15.9% 18. 320% 19. 500%
20. 4,100% 21. $\frac{31}{100}$ 22. $\frac{37}{50}$ 23. $\frac{9}{100}$
24. $\frac{93}{1,000}$ 25. $\frac{41}{1,000}$ 26. $\frac{13}{400}$ 27. $\frac{7}{1,000}$
28. $\frac{19}{400}$ 29. $\frac{21}{200}$ 30. $\frac{3}{800}$ 31. 82%
32. 85% 33. 80% 34. 93% 35. 25%
36. 51.4% 37. 23.1% 38. 26.4%
39. 216.7% 40. 755.6% 41. 0.015
42. .8% 43. $\frac{43}{50}$ 44. 12.5% 45. 0.3%
46. $\frac{7}{800}$ 47. 4,560 48. 5 R 7 49. $5\frac{17}{20}$
50. $\frac{11}{20}$ 51. $\frac{15}{16}$ 52. 4.202 53. $0.20
54. 17.42 55. 55.63 56. 28.35 57. 200
58. 2.68 59. 11.38 60. 4.56

SET 4.2A

1. $2,028 2. $29.60 3. 11,200 4. 161.2
5. $292.50 6. $1,608 7. $12.40
8. $3,600 9. 39.2 10. 100 11. $364
12. $1,020 13. 10 14. 20 15. $506.25
16. $153 17. $4.35 18. $675 19. 4.8
20. 0.48 21. $24 22. $1,312.50
23. $1,674 24. 221 25. 125 26. 100%
27. 92 28. $140 29. 52 30. 160
31. four million, sixty thousand, seventeen
32. 293,748 33. $1\frac{1}{3}$ 34. $14\frac{1}{4}$ 35. $1\frac{1}{6}$
36. 2.41 37. $3.94 38. 5.461 39. 21,467
40. 27.586 41. 44.12 42. 0.073 43. 7.8%
44. $\frac{9}{125}$ 45. 95%

SET 4.3A

1. 70% 2. 39% 3. 46% 4. 5%
5. 75% 6. 68% 7. 500% 8. 100%
9. 25% 10. 54% 11. 87.5% 12. 45%
13. 69.2% 14. 69.7% 15. 61.3%
16. 73.4% 17. 21.3% 18. 16.7%
19. 26.3% 20. 67.8% 21. 20% 22. 5.2%
23. 73.9% 24. 1.5% 25. 150% 26. 2
27. 62.5% 28. 46% 29. 95% 30. 10%
31. 3,600,000 32. 4 R 1 33. $5\frac{1}{3}$ 34. $\frac{31}{50}$
35. $5\frac{13}{15}$ 36. 0.005 37. $12.40 38. 8.24
39. 2,300 40. 10.81 41. 0.0562

42. 14.38% 43. $\frac{2}{125}$ 44. 62.5%
45. \$33.28 46. \$60 47. 24 48. \$150

SET 4.4A

1. 400 2. 300 3. \$1,150 4. \$4,000
5. 3,000 6. \$256 7. 620 8. \$605
9. 700 10. 6,000 11. \$100 12. \$16
13. 500 14. 1,440 15. \$12,000 16. \$840
17. \$384.62 18. \$13.33 19. \$485.71
20. \$37.14 21. \$18,500 22. \$160,000
23. 2,625,000 24. \$3,250,000 25. \$12
26. If the rate is more than 100% 27. \$120
28. \$73 29. 207 30. 29
31. seventeen thousand, five hundred six
32. 14 R 25 33. $4\frac{13}{20}$ 34. $\frac{91}{100}$ 35. $\frac{63}{152}$
36. 2.61 37. \$1.48 38. 30.82 39. 0.397
40. 500% 41. 70 42. \$279.50 43. 24%
44. 400% 45. 56.1%

SET 4.5A

1. 468 2. 280 3. 495.8 4. 1,600
5. 85 6. 79.2% 7. 350.4 8. 1,800
9. 350% 10. 400 11. 1,417,500
12. 2.63% 13. 40 14. 45% 15. 3,400
16. 18% 17. \$243,750 18. \$82,000
19. No 20. No 21. 50,000 22. 453,264
23. $\frac{31}{40}$ 24. $6\frac{1}{4}$ 25. $5\frac{2}{3}$ 26. 0.057
27. \$2.14 28. 121.62 29. 0.0605
30. 17.5% 31. 51 32. 8,000 33. 65%
34. 5.9% 35. 400

SET 4.6A

1. 12% 2. 16% 3. 5% 4. 45%
5. 15% 6. 10% 7. 13.5% 8. 13.8%
9. 4% 10. 24% 11. 7% 12. 10%
13. 4% 14. 1% 15. 8.97% 16. 17.78%
17. 207.1% 18. 39.2% 19. 173.3%
20. 7.4% 21. 200% 22. No
23. one million, two hundred five thousand, seven hundred
24. 19 R 18 25. $4\frac{3}{20}$ 26. $\frac{31}{60}$ 27. $\frac{1}{4}$
28. 2.305 29. \$14.56 30. 13.5 31. 0.1756
32. $\frac{3}{8}$ 33. 0.7% 34. 38% 35. 7.5
36. \$1,562.50

PRACTICE TEST 4A

1. 0.478 2. 4.25% 3. $\frac{6}{25}$ 4. 73.7%
5. 700 6. \$100 7. 36% 8. 22%
9. \$702 10. 168 inches 11. 15%
12. 43.1%

PRACTICE TEST 4B

1. 4.5 2. 17.3% 3. $\frac{1}{40}$ 4. 31.3%
5. 34% 6. \$31 7. \$405 8. \$25,000
9. 53.1% 10. \$9,660 11. 17.14%
12. 26%

CUMULATIVE REVIEW EXERCISES FOR CHAPTERS 1 THROUGH 4

1. forty-two million, six hundred fifty-seven thousand, twenty-three
2. 1,270,000 3. 3 R 25 4. $\frac{2}{3}$ 5. $2\frac{3}{10}$
6. $\frac{34}{9}$ 7. $\frac{4}{13}$ 8. 23 9. $7\frac{7}{18}$ 10. $\frac{19}{36}$
11. two and nine thousandths 12. 5.00023
13. 0.005455 14. \$45.80 15. 11.059
16. 123,000 17. 11.98 18. 162 19. 0.1734
20. 0.3% 21. $\frac{31}{50}$ 22. 77.36% 23. \$5,000
24. 380 25. 11.5% 26. 387 27. \$80,000
28. 62.5% 29. 51.5%

CHAPTER 5

SET 5.1A

1. 5 to 4 2. 8:5 3. 7:9 4. $\frac{2}{3}$ 5. $\frac{3}{2}$
6. $\frac{12}{25}$ 7. 5:2, $\frac{5}{2}$ 8. 1:6, $\frac{1}{6}$ 9. 23:3, $\frac{23}{3}$

10. 2:7, $\frac{2}{7}$ **11.** 1:1, $\frac{1}{1}$ **12.** 13:19, $\frac{13}{19}$
13. 3 to 5, $\frac{3}{5}$ **14.** 17 to 50, $\frac{17}{50}$ **15.** 6 to 5, $\frac{6}{5}$
16. 4 to 1, $\frac{4}{1}$ **17.** 5 to 83, $\frac{5}{83}$
18. 100 to 1, $\frac{100}{1}$ **19.** (a) 4:5 (b) $\frac{1}{4}$
20. (a) 7 to 12 (b) 12:7 **21.** 1 to 2
22. 43:180 **23.** 30 **24.** Yes **25.** 4,700
26. 19,110 **27.** $\frac{23}{40}$ **28.** $2\frac{1}{3}$ **29.** $4.26
30. 38.83 **31.** 0.0345 **32.** 200% **33.** 35%
34. $2,000 **35.** $660 **36.** 12% **37.** 3.1%

SET 5.2A

1. Yes **2.** No **3.** Yes **4.** No **5.** No
6. Yes **7.** Yes **8.** Yes **9.** Yes
10. No **11.** No **12.** Yes **13.** No
14. Yes **15.** Yes **16.** Yes **17.** $5
18. $75
19. one million, nine hundred fifty-six thousand, four hundred twenty-two
20. 11 R 13 **21.** $12\frac{9}{10}$ **22.** $\frac{34}{39}$ **23.** 0.0453
24. 6.51 **25.** 2.50 **26.** 46% **27.** $671.25
28. 96% **29.** $65 **30.** 15% **31.** 1:4
32. 4 to 5

SET 5.3A

1. $N = 3$ **2.** $B = 15$ **3.** $A = 2$ **4.** $Z = \frac{1}{4}$
5. $M = 1\frac{1}{4}$ **6.** $X = 0.5$ **7.** $K = \frac{1}{3}$
8. $P = 1$ **9.** $S = 6$ **10.** 9.6 inches
11. 40 cubic centimeters **12.** $640 **13.** $2,800
14. 3,200 grams **15.** 18 **16.** $17,600
17. $825 **18.** $160 **19.** 8,976
20. 560,000,000 **21.** 2,130,000 **22.** $\frac{58}{11}$
23. $4\frac{8}{23}$ **24.** $12.06 **25.** 11.46 **26.** 0.2375
27. 2.4% **28.** $182.25 **29.** 56.7%
30. $1,680 **31.** 3:2 **32.** $\frac{29}{37}$ **33.** No
34. Yes

PRACTICE TEST 5A

1. (a) 11 to 7 (b) 1:1 (c) $\frac{1}{2}$ **2.** 2:3
3. (a) No (b) Yes **4.** Yes
5. (a) $N = 10$ (b) $W = 5.4$ **6.** 156 miles

PRACTICE TEST 5B

1. (a) 4 to 1 (b) 2:5 (c) $\frac{3}{4}$ **2.** $\frac{4}{7}$
3. (a) No (b) Yes **4.** No
5. (a) $R = 12$ (b) $N = 125$ **6.** $360

CUMULATIVE REVIEW EXERCISES FOR CHAPTERS 1 THROUGH 5

1. one billion, three million, five hundred thousand
2. 23,900 **3.** 15 R 9 **4.** $\frac{6}{11}$ **5.** $11\frac{1}{9}$
6. $\frac{33}{7}$ **7.** $24\frac{5}{9}$ **8.** $9\frac{3}{4}$ **9.** $1\frac{8}{15}$ **10.** $7\frac{3}{4}$
11. one and five hundred eighty-seven ten-thousandths
12. 0.0093 **13.** 12.484 **14.** $3.07
15. 5.536 **16.** 21.253 **17.** 2.1 **18.** 9.24
19. 0.0225 **20.** 350% **21.** $\frac{9}{20}$ **22.** 56.7%
23. 13% **24.** 970 **25.** $1,260 **26.** $1,192
27. $4,000 **28.** 40% **29.** 21.2%
30. (a) 4:5 (b) 5 to 6 **31.** (a) Yes (b) No
32. (a) $R = 6.5$ (b) $T = 40$ **33.** 65

CHAPTER 6

SET 6.1A

1. $>$ **2.** $>$ **3.** $<$ **4.** $<$ **5.** $>$
6. $<$ **7.** $>$ **8.** $>$ **9.** $<$ **10.** $>$
11. $>$ **12.** $<$ **13.** 5 **14.** 17 **15.** 12
16. 0 **17.** 2.5 **18.** 3.5 **19.** 1.6 **20.** 22.5
21. 8 degrees **22.** $422.17 **23.** Rising
24. $0.75 **25.** $<$ **26.** $<$ **27.** Left
28. 29,000 **29.** 30 R 14 **30.** $\frac{11}{30}$ **31.** $6\frac{1}{2}$

32. 13.76 **33.** \$231 **34.** 56.7% **35.** \$125
36. 16% **37.** 6 to 7 **38.** No **39.** Yes
40. $R = 2$ **41.** $N = 1$

SET 6.2A

1. −14 **2.** 4 **3.** 13 **4.** 0 **5.** −8
6. −12 **7.** −24 **8.** −6 **9.** −9.9
10. −2.54 **11.** 2 **12.** −6 **13.** −1.28
14. −9.6 **15.** $-\frac{1}{2}$ **16.** $\frac{1}{6}$ **17.** $\frac{1}{2}$
18. $-\frac{11}{12}$
19. Yes, since $-3 - 7 - 2 - 6 - 1 = -19$
20. \$25.75 **21.** 9 degrees **22.** −2 degrees
23. −4.2
24. No, since no matter how large the sum of the first two numbers is there will always exist a negative number to make the sum −15. For example, $50 + 1{,}000{,}000 - 1{,}000{,}065 = -15$
25. two million, three hundred fifty-eight thousand, six hundred seventy-four
26. $\frac{7}{18}$ **27.** $\frac{79}{8}$ **28.** 17.06 **29.** 67,400
30. \$408.50 **31.** \$170 **32.** 75%
33. $M = 1.5$ **34.** $R = 3$ **35.** $<$ **36.** $>$
37. 1.65 **38.** 75

SET 6.3A

1. −16 **2.** 27 **3.** 1 **4.** 21 **5.** −25
6. −105 **7.** 0 **8.** −39 **9.** 5.4
10. 6.90 **11.** 5.08 **12.** 2.82 **13.** $-1\frac{2}{5}$
14. $-\frac{3}{8}$ **15.** $\frac{6}{7}$ **16.** $1\frac{1}{6}$ **17.** −17 degrees
18. −10 degrees **19.** 68 years **20.** 12.5 games
21. −9 **22.** Yes. For example $(-5) - (-7) = 2$
23. 12,700 **24.** 8 R 9 **25.** $2\frac{1}{3}$ **26.** $\frac{4}{5}$
27. 0.27 **28.** \$125 **29.** 42% **30.** \$1,170
31. No **32.** Yes **33.** −15 **34.** −17.25

SET 6.4A

1. −48 **2.** 14 **3.** 36 **4.** −180
5. 105 **6.** −162 **7.** −72 **8.** 1,000
9. 11.7 **10.** −18.788 **11.** 33.48
12. −81.6 **13.** $\frac{1}{12}$ **14.** $-\frac{1}{2}$ **15.** $-7\frac{1}{12}$
16. $8\frac{9}{14}$
17. The product of each of the 40 pairs of negative factors is positive.
18. $-\frac{1}{7}$ **19.** one billion, six hundred five million
20. $\frac{37}{47}$ **21.** 18.35 **22.** \$1.29 **23.** \$70
24. 14.29% **25.** 7% **26.** $W = 210$ **27.** $>$
28. $>$ **29.** 13 **30.** 15.9 **31.** 2
32. 11.55

SET 6.5A

1. 6 **2.** −15 **3.** 3 **4.** −1 **5.** 4
6. −5 **7.** −8 **8.** −8 **9.** −20
10. 0.024 **11.** 120 **12.** −1,230 **13.** $-1\frac{1}{2}$
14. $\frac{35}{52}$ **15.** $-\frac{5}{8}$ **16.** $-3\frac{1}{2}$ **17.** Yes
18. Yes **19.** 22,000 **20.** 16 R 7 **21.** $14\frac{7}{12}$
22. 53.58 **23.** \$406.25 **24.** 14%
25. $N = 25$ **26.** $<$ **27.** 3 **28.** −2
29. −180 **30.** 3.75

SET 6.6A

1. 19 **2.** −53 **3.** −68 **4.** −38
5. −50 **6.** 7 **7.** 20 **8.** −32 **9.** 4
10. 44 **11.** 58 **12.** 20
13. two hundred thirty-seven thousand, nine hundred eight
14. $\frac{29}{3}$ **15.** $1\frac{11}{12}$ **16.** 89.088 **17.** 22.9%
18. \$475 **19.** $H = 3.6$ **20.** $>$ **21.** −17
22. −4.96 **23.** −105 **24.** −0.0002
25. −800 **26.** 6

PRACTICE TEST 6A

1. (a) $>$ (b) $>$ **2.** (a) 6.7 (b) 4.29
3. (a) −30 (b) 6.9 (c) −10 (d) $-\frac{9}{10}$
4. (a) −23 (b) 9.26 **5.** 2.9 degrees

6. (a) -63 (b) 27.6 (c) $-22\frac{2}{3}$ (d) $-\frac{7}{12}$

7. (a) -3 (b) -122 (c) 2 (d) $3\frac{1}{5}$ 8. -16

PRACTICE TEST 6B

1. (a) > (b) > 2. (a) 6.89 (b) 5.44
3. (a) -16 (b) -11 (c) $-\frac{5}{14}$ (d) $\frac{13}{20}$
4. (a) 7 (b) -20.2 5. 87 years
6. (a) -216 (b) -8.04 (c) $-2\frac{1}{10}$ (d) $\frac{2}{3}$
7. (a) -5 (b) 30 (c) $-1\frac{1}{6}$ (d) -60 8. -27

CUMULATIVE REVIEW EXERCISES FOR CHAPTERS 1 THROUGH 6

1. twenty-three million, five hundred eighty-four thousand, one hundred
2. 40,000 3. $\frac{11}{14}$ 4. $3\frac{11}{15}$ 5. $10\frac{1}{8}$ 6. $\frac{2}{3}$
7. twenty-three ten-thousandths 8. 4.361
9. \$12.61 10. 5.226 11. 4.91 12. 20.92
13. 0.042 14. 17.5% 15. $\frac{6}{25}$ 16. \$3,675
17. \$157.50 18. 52% 19. \$1,057.50
20. \$25,000 21. 7% 22. Yes 23. $N = 2.5$
24. (a) < (b) > 25. (a) 34 (b) 4.09
26. (a) -22 (b) -3 27. (a) 2.98 (b) 800
28. (a) 35 (b) -65 29. (a) -40 (b) 2,000
30. 42

CHAPTER 7

SET 7.1A

1. 3 2. 2 3. 2 4. 1 5. -8 6. 1
7. 0.25 8. -1 9. $-2x$ 10. $8w - 3x$
11. $2z - 9y$ 12. $9M + 11K$ 13. $14x - 35$
14. $8x + 9$ 15. $63w - 27$ 16. $28y - 9$
17. six times x minus five
18. three times w plus five times y minus two times z
19. A minus three times B minus ten
20. three times x minus four times y plus two times z minus one
21. (a) $2.5A$ (b) $A + 20{,}000$
22. (a) $x + y$ (b) $|x - y|$ (c) $y + 2.6$
23. (a) $0.04S$ (b) $B + 0.15S$
24. (a) $F + S$ (b) $G - (F + S)$ 25. $8w$
26. -2 27. 7,930,000,000 28. 2 R 43
29. $14\frac{29}{30}$ 30. \$21 31. 6.5% 32. \$703
33. Yes 34. > 35. 10 36. -7.8
37. -560 38. -200 39. -26 40. -134

SET 7.2A

1. $x = 19$ 2. $w = -12$ 3. $4x = 13$
4. $-5z = -13$ 5. $0.2S = 400$
6. $0.05S = 550$ 7. $z = -2$ 8. $x = 2\frac{1}{8}$
9. $x + 2 = 3$ 10. $y - 2 = 2$ 11. $w = 12$
12. $x = -6$
13. (a) equivalent (b) not equivalent (c) equivalent
14. (a) not equivalent (b) equivalent (c) equivalent
15. (a) not equivalent (b) equivalent (c) equivalent
16. $x + 5 = 8$ 17. $2w + 10 = 30$
18. $3y + 1 = 13$
19. five million, six hundred four thousand, eight hundred
20. 5 R 63 21. $1\frac{7}{24}$ 22. $4\frac{1}{10}$ 23. 4.0369
24. 387.9 25. 73.6% 26. \$228 27. $N = 2$
28. -16 29. -2.78 30. 1.75 31. $-2{,}000$
32. 112 33. 3 34. -7 35. $8w - 2x$
36. $-19r + 35$
37. seven times N plus five times P minus two and five-tenths

SET 7.3A

1. $x = 10$ 2. $y = 20$ 3. $z = -11$
4. $y = -2$ 5. $w = 0$ 6. $x = -11$
7. $z = -8$ 8. $x = 3$ 9. $w = 0.3$
10. $y = 4.8$ 11. $z = -16.5$ 12. $x = 0.16$
13. $x = -\frac{1}{8}$ 14. $y = 1\frac{3}{20}$ 15. $z = -\frac{6}{7}$
16. 28 miles per gallon 17. \$590 18. 541
19. $x = 0.63$ 20. $w = 0.02$ 21. 5,600
22. $2\frac{3}{4}$ 23. $\frac{7}{12}$

24. five and thirty-four thousandths **25.** 1.9
26. \$56.55 **27.** 10.4% **28.** -15.1
29. $-3{,}000$ **30.** 9,440 **31.** $-14.2x + 12.7y$
32. $81N - 75$ **33.** $x = -6$
34. (a) equivalent (b) not equivalent

SET 7.4A

1. $w = 14$ **2.** $x = -4$ **3.** $z = 1$
4. $y = -6$ **5.** $x = -20$ **6.** $x = 13\frac{1}{3}$
7. $x = 400$ **8.** $z = 20$ **9.** $S = 60{,}000$
10. $x = 12$ **11.** $y = 30$ **12.** $z = 9$
13. $w = 8$ **14.** $x = -28$ **15.** $z = \frac{16}{21}$
16. 7,000 **17.** \$1.17 **18.** 63
19. $S = 25{,}000$ **20.** $z = 0$ **21.** 5,400,000
22. 5 R 2 **23.** $\frac{11}{40}$ **24.** $8\frac{1}{10}$ **25.** 5.03789
26. 27.5% **27.** $B = 12$ **28.** 3.62 **29.** -63
30. $-14.4w + 9y$ **31.** $58H - 24$ **32.** $x = -8$
33. $w = -1.5$

SET 7.5A

1. $x = 6$ **2.** $x = -1$ **3.** $x = -3\frac{1}{2}$
4. $x = 12\frac{6}{7}$ **5.** $x = 1\frac{1}{6}$ **6.** $-\frac{5}{11}$
7. $x = -2$ **8.** $x = 1\frac{1}{7}$ **9.** $x = 4\frac{1}{2}$
10. $x = 1\frac{5}{6}$ **11.** $x = -2\frac{1}{6}$ **12.** $x = -1\frac{5}{6}$
13. $x = \frac{40}{47}$ **14.** $x = 2\frac{9}{11}$ **15.** $x = 6\frac{6}{7}$
16. $x = -2\frac{4}{5}$ **17.** $x = -2\frac{14}{17}$ **18.** $x = 2\frac{10}{19}$
19. \$10 **20.** 68 degrees **21.** 14 **22.** $x = \frac{4}{7}$
23. No solution **24.** 5,000 **25.** $6\frac{3}{20}$
26. $6\frac{2}{3}$ **27.** 14.17 **28.** \$720 **29.** \$1,500
30. -20 **31.** -24.62 **32.** $46x - 62$
33. $x = -28.1$ **34.** $z = -19$ **35.** $x = -0.8$
36. $w = -48$

PRACTICE TEST 7A

1. 2 **2.** $57x - 34$
3. four times A minus five times B plus two
4. (a) equivalent (b) not equivalent
5. $w = 1.57$ **6.** $x = -9$ **7.** $x = 32$
8. $x = 13$ **9.** $x = -19$ **10.** \$12.50

PRACTICE TEST 7B

1. 2 **2.** $44x - 38$
3. seven times x minus three times w plus five
4. (a) not equivalent (b) equivalent **5.** $A = 0$
6. $w = -4{,}600$ **7.** $z = \frac{15}{16}$ **8.** $x = 1$
9. $x = 0$ **10.** 35

CUMULATIVE REVIEW EXERCISES FOR CHAPTERS 1 THROUGH 7

1. one million, two hundred three thousand, five hundred
2. 1,507,000 **3.** $22\frac{3}{4}$ **4.** $2\frac{4}{13}$ **5.** $5\frac{19}{20}$
6. $\frac{97}{240}$ **7.** 1.2038 **8.** \$1.24 **9.** 30,400
10. 25.06 **11.** 21.2 **12.** 0.0545 **13.** 2.5%
14. $\frac{9}{25}$ **15.** 5% **16.** \$675 **17.** \$130
18. 6.1% **19.** $R = 17.5$ **20.** (a) $<$ (b) $>$
21. (a) 2 (b) -10.75 **22.** (a) 4 (b) -1.082
23. (a) 28 (b) -76.8 **24.** (a) -8 (b) $\frac{3}{5}$
25. 13 **26.** 3
27. two times x minus seven times w
28. (a) not equivalent (b) not equivalent
29. $x = 26$ **30.** $x = -6$ **31.** $w = 6\frac{2}{3}$
32. $x = 3$ **33.** \$56

CHAPTER 8

SET 8.1A

1. 5^4 **2.** 8^7 **3.** $(-4)^3$ **4.** $(1.01)^2$
5. $6 \cdot 6 \cdot 6 \cdot 6$ **6.** $2 \cdot 2 \cdot 2 \cdot 2 \cdot 2 \cdot 2 \cdot 2$

7. $(-3)(-3)(-3)(-3(-3)$ 8. $(1.02)(1.02)(1.02)$
9. 64 10. 49 11. 1 12. 1,296
13. -32 14. -1 15. 0.008 16. -0.343
17. 387 18. $-2{,}081$ 19. 27 20. 0
21. $551.27 22. 88 23. 192 feet 24. 1
25. 7 seconds
26. two million, three hundred five thousand, nine hundred
27. 2 R 41 28. $\frac{31}{50}$ 29. $10\frac{9}{20}$
30. five and two hundred sixty-seven ten thousandths
31. 68.42% 32. $640 33. -4.69
34. -136 35. $x = -26$ 36. $w = -1.8$
37. $x = 8$ 38. $z = 18$ 39. $x = -10$
40. $x = -2$

SET 8.2A

1. 4,096 2. 1,000,000 3. $-7{,}776$ 4. 729
5. 625 6. 1 7. 81 8. 10 9. 64
10. 1 11. 1 12. 1 13. $\frac{1}{49}$ 14. $-\frac{1}{125}$
15. $\frac{1}{3}$ 16. 25 17. $-\frac{1}{2}$ 18. 1 19. 144
20. 8,000 21. $\frac{1}{225}$ 22. $\frac{8}{125}$ 23. $\frac{1}{243}$
24. $\frac{81}{256}$ 25. 243 26. 256 27. 1,240
28. $\frac{71}{75}$ 29. $1\frac{2}{7}$ 30. 16.23 31. 609
32. 15.1% 33. -16.64 34. 710
35. $X = -8$ 36. $x = 1$ 37. 5^6
38. $(-7)(-7)(-7)(-7)$ 39. 64 40. -243

SET 8.3A

1. 8.64×10^6 2. 6.0×10^3 3. 2.4×10^{11}
4. 1.0×10^4 5. 3.21×10^{-4} 6. 8.9×10^{-6}
7. 5.03×10^{-3} 8. 1.0×10^{-3} 9. 620
10. 380,000 11. 30,000 12. 96 13. 0.0021
14. 0.00000078 15. 0.5 16. 0.000081
17. 1.47×10^4 18. 2.4×10^{11}
19. 1.274×10^{-5} 20. 1.36×10^{-1}
21. 3.0×10^9 22. 9.0×10^7 23. 2.0×10^{-11}
24. 2.5×10^4 25. 6.378×10^6
26. 2.54×10^{-4} 27. 1.36×10^9
28. 2.0×10^0 29. 5.8×10^{-9} 30. 12,400,000
31. $\frac{6}{7}$ 32. $\frac{34}{9}$ 33. 22.4% 34. $N = 12\frac{1}{2}$
35. 4.35 36. -448 37. $x = -4$
38. $x = 7$ 39. 216 40. -1 41. 9
42. 81

SET 8.4A

1. 6 and -6 2. 10 and -10 3. 7 and -7
4. 1 and -1 5. 5 6. -2 7. -30
8. 8 9. -1 10. 14 11. 9 12. -3
13. 4 and 5 14. 9 and 10 15. 5 and 6
16. 11 and 12 17. 4.6 18. 8.2 19. 12.4
20. 6.4 21. 14.4 22. 8.9 23. 11.6
24. 2.6 25. 5.2 26. 8.6 27. 3.6
28. 7.1 29. 1.4 30. 8.4
31. -12 and -11 32. 0
33. two billion, one hundred eighty-seven million, twenty-three
34. $\frac{1}{6}$ 35. $36\frac{3}{4}$ 36. 14.6233 37. $618.24
38. 11.1% 39. 14.9 40. -52.5
41. $x = -3$ 42. $x = -\frac{1}{7}$ 43. 512
44. -32 45. 9 46. 216
47. (a) 6.87×10^3 (b) 7.2×10^{-6}
48. (a) 610,000 (b) 0.00000038 49. 1.785×10^4
50. 2.0×10^5

PRACTICE TEST 8A

1. 7^5 2. $(-6)(-6)(-6)(-6)(-6)(-6)(-6)$
3. (a) 625 (b) 64 (c) 25
4. (a) 128 (b) 27 (c) $\frac{1}{32}$
5. (a) 1.23×10^7 (b) 8.7×10^{-6}
6. (a) 523,000 (b) 0.0021
7. (a) 7.68×10^7 (b) 3.0×10^8 8. 11 and -11
9. (a) -8 (b) 10 10. 5 and 6
11. (a) 6.9 (b) 12.0 (c) 14.1

PRACTICE TEST 8B

1. $(-3)^4$ 2. $5 \cdot 5 \cdot 5 \cdot 5 \cdot 5 \cdot 5 \cdot 5 \cdot 5$
3. (a) 81 (b) 625 (c) 61
4. (a) 729 (b) 4 (c) $\frac{8}{125}$
5. (a) 1.16×10^{10} (b) 2.13×10^{-8}
6. (a) 378,000,000 (b) 0.000067
7. (a) 2.25×10^{16} (b) 5.0×10 8. 9 and -9

9. (a) 6 (b) -7 **10.** 8 and 9
11. (a) 7.81 (b) 12.37 (c) 4.12

CUMULATIVE REVIEW EXERCISES FOR CHAPTERS 1 THROUGH 8

1. twenty million, three hundred four thousand
2. 21,000 **3.** $\frac{1}{10}$ **4.** $2\frac{3}{5}$ **5.** $1\frac{17}{24}$ **6.** $5\frac{6}{7}$
7. one and five ten-thousandths **8.** 21.580792
9. \$41.21 **10.** 2.4518 **11.** 13.72
12. 15.84 **13.** 0.1275 **14.** 4.8% **15.** $\frac{1}{25}$
16. \$3,640 **17.** 3% **18.** \$2,100 **19.** 18%
20. $M = 250$ **21.** -38 **22.** -31
23. 15.41 **24.** -160 **25.** 2
26. four times x plus two times w minus seven
27. $x = 12$ **28.** $x = -15$ **29.** $x = -21$
30. $(-8)^3$ **31.** $7 \cdot 7 \cdot 7 \cdot 7 \cdot 7$
32. (a) 625 (b) -243 (c) 200
33. (a) 4,500 (b) 648 (c) $\frac{8}{125}$
34. (a) 2.3×10^5 (b) 4.1×10^{-4}
35. (a) 64,100 (b) 0.073 **36.** 3.15×10^3
37. 15 and -15 **38.** (a) 20 (b) -8
39. 8 and 9 **40.** (a) 4.69 (b) 11.58 (c) 9.64

CHAPTER 9

SET 9.1A

1. 7 **2.** 10,560 **3.** $8\frac{1}{2}$ **4.** 1.2 **5.** 1,207
6. 67 **7.** 30,000 **8.** 1.2 **9.** 0.008
10. 5,280 **11.** $13\frac{1}{3}$ **12.** $8\frac{1}{3}$ yards
13. 5,280 feet **14.** $6\frac{2}{3}$ yards **15.** 760 meters
16. 83 millimeters **17.** \$4.50 **18.** \$9,600
19. 1,145 meters **20.** 22,850,000 **21.** $\frac{43}{56}$
22. $1\frac{1}{4}$ **23.** 18.48 **24.** \$1,645 **25.** \$182
26. -9.3 **27.** -42 **28.** $x = -23$
29. $x = 1$ **30.** 625 **31.** -243 **32.** 49
33. 9 **34.** 4.53×10^{10}
35. (a) 0.0375 (b) 4,900,000 **36.** 4.08×10^9
37. 3.0×10^9 **38.** 7 and -7 **39.** -20
40. 6 and 7 **41.** 4.4

SET 9.2A

1. 10.16 **2.** 2.13 **3.** 12.43 **4.** 1.10
5. 7.28 **6.** 204.10 **7.** 5.72 **8.** 10.26
9. 0.15 **10.** 6.89 **11.** (c) **12.** (b)
13. (a) **14.** (b) **15.** 26,103 miles
16. 366 meters **17.** 51 centimeters
18. 8.89 centimeters **19.** 60 cm for \$3.10
20. twenty-one million, six hundred eight thousand, three
21. $\frac{23}{60}$ **22.** $3\frac{11}{14}$ **23.** 8.8434 **24.** 29.4%
25. 281.25 **26.** $T = 4$ **27.** 6 **28.** $x = -10$
29. 1,024 **30.** -5 **31.** 64 **32.** 5.62×10^7
33. (a) 57,000 (b) 0.000081 **34.** 2.312×10^4
35. 3.0×10^{12} **36.** 8 and -8 **37.** 12
38. 8 and 9 **39.** 11.5 **40.** (a) 54 (b) 6,000

SET 9.3A

1. 5 **2.** 14 **3.** 4.5 **4.** $\frac{1}{2}$ **5.** $\frac{5}{8}$
6. $11\frac{1}{3}$ **7.** $6\frac{5}{6}$ **8.** 24 **9.** 7.5 **10.** $6\frac{1}{4}$
11. 1,584 **12.** 0.563 **13.** 5,000 **14.** 0.018
15. 7.02 **16.** 90,000 **17.** 0.2 **18.** 88.2
19. 2.5 **20.** 20 **21.** 236.8 **22.** 640 qt
23. 20 pt **24.** 0.24L **25.** 50 daL
26. 38,400 bottles **27.** 4 ounces for \$1.15
28. 11 bottles **29.** 13,000,000 **30.** $4\frac{13}{20}$
31. $\frac{1}{6}$ **32.** 26.51 **33.** \$51 **34.** \$1,145
35. -21.6 **36.** $w = -1$ **37.** 625 **38.** 81
39. 4.12×10^7 **40.** 160,000 **41.** 2.0×10^2
42. -12 **43.** 11.8 **44.** (a) 1.5 (b) 2.05
45. (a) 30,000 (b) 6.4008

SET 9.4A

1. 192 **2.** 3 **3.** 0.8 **4.** 6,400 **5.** 140
6. 1.5 **7.** 1 **8.** 67,200 **9.** 12 **10.** 0.6
11. 0.012 **12.** 500 **13.** 426 **14.** 45
15. 25 **16.** 50 **17.** 5 lb **18.** 0.75 t
19. 0.9 kg **20.** 21,250 mg **21.** 44 lb or 19.8 kg

22. 1 ton for \$1,600 **23.** 425 lb
24. two million, seventy-five thousand, three hundred
25. $\frac{7}{15}$ **26.** $\frac{1}{2}$ **27.** 31.248 **28.** 67.4%
29. \$740 **30.** $x = -1$ **31.** 64 **32.** 3,125
33. 4.03×10^{-6} **34.** 480,000 **35.** 4.8×10^{-8}
36. 12.1 **37.** 2.05 **38.** 0.006 **39.** 24
40. 0.0074 **41.** 10 **42.** 0.672

PRACTICE TEST 9A

1. 15,840 **2.** 15 **3.** 5.64 **4.** 0.004
5. 2.7432 **6.** 24.135 **7.** 2 **8.** 60.96
9. 2.25 **10.** 1,113.6 **11.** 0.00045 **12.** 5.4
13. 240 **14.** 24 **15.** 108,800 **16.** 90

PRACTICE TEST 9B

1. 1.3 **2.** 17,600 **3.** 40,000 **4.** 2.4
5. 2.0574 **6.** 0.125 **7.** 2.2526 **8.** 365.76
9. 624 **10.** 20.8 **11.** 4.375 **12.** 0.012
13. 80 **14.** 600 **15.** 0.9125 **16.** 1.8

CUMULATIVE REVIEW EXERCISES FOR CHAPTERS 1 THROUGH 9

1. twelve million, six hundred forty-three thousand
2. 45,600 **3.** 8 **4.** $5\frac{1}{2}$ **5.** $1\frac{1}{3}$ **6.** $3\frac{5}{12}$
7. five and twelve thousandths **8.** 1.451
9. \$3.12 **10.** 12.40 **11.** 28.7 **12.** 2.25
13. 0.487% **14.** $\frac{13}{50}$ **15.** 28% **16.** \$208
17. \$1,280 **18.** 13% **19.** $R = 2\frac{1}{2}$ **20.** 7
21. -51 **22.** -19.76 **23.** 160
24. five times x minus seven times w plus two
25. $x = -12$ **26.** $x = 1$ **27.** $(-1)^3$
28. $5 \cdot 5 \cdot 5 \cdot 5 \cdot 5 \cdot 5$ **29.** (a) 36 (b) -32 (c) 15
30. (a) 2,025 (b) $\frac{8}{125}$
31. (a) 4.5×10^7 (b) 8.23×10^{-4}
32. (a) 0.00429 (b) 344,000 **33.** 1.386×10^6
34. 14 and -14 **35.** (a) -11 (b) 15
36. 6 and 7 **37.** (a) 6.6 (b) 12.2 (c) 8.8
38. 15 **39.** 4,600 **40.** 11.7457 **41.** 3.46875
42. 160 **43.** 0.740

CHAPTER 10

SET 10.1A

1. (a) straight angle (b) right angle (c) right angle
(d) straight angle
2. (a) parallel lines
(b) perpendicular lines, intersecting lines, and right angles
(c) parallel lines (d) intersecting lines
3. (a) trapezoid (b) circle
(c) square, rectangle, and parallelogram
(d) triangle (e) parallelogram
(f) rectangle and parallelogram
4. 10 cm **5.** 12.2 m **6.** 10.6 ft **7.** 25 cm
8. 420 ft **9.** 59 cm **10.** 37 ft **11.** 6.6 cm
12. No. The sum of the lengths of the other two sides must be more than 12 inches
13. 5 cm **14.** 1,610,000 **15.** $8\frac{13}{20}$ **16.** $\frac{1}{4}$
17. 1.65 **18.** \$4,312.50 **19.** $x = 5$ **20.** 125
21. 4.0×10^9 **22.** 9.64 **23.** 216 **24.** 482.7
25. 0.75 **26.** 2.7 **27.** 10,000 **28.** 2,000

SET 10.2A

1. 25 sq cm **2.** 51.84 sq m **3.** 21.8089 sq ft
4. 10,000 sq in **5.** $\frac{9}{16}$ sq mi **6.** $6\frac{1}{4}$ sq cm
7. 25.9 sq in **8.** 37.72 sq ft **9.** 30 sq mi
10. 62.05 sq m **11.** 7.5 sq cm
12. 10.005 sq km **13.** 699.36 sq mm
14. 6.48225 sq m **15.** 11.5 sq yd
16. 27.3 sq in **17.** $\frac{7}{16}$ sq ft **18.** $12\frac{15}{16}$ sq mi
19. 9 sq cm **20.** 1.95 sq ft **21.** 50.975 sq cm
22. 6,150 sq mm **23.** 240,000 sq m
24. 20.4 sq ft **25.** 10 sq ft **26.** 15 sq cm
27. 45.5 sq m **28.** 48 sq yd **29.** 6.15 cm
30. 96 sq ft
31. two million, one hundred five thousand, seven
32. $1\frac{9}{10}$ **33.** 7 **34.** 20,041 **35.** 58.6%
36. $x = -6$ **37.** -216 **38.** 5.6×10^{-7}
39. 13 **40.** 158.4 **41.** 127 **42.** 12,000
43. 170.2 m **44.** 4 mi

SET 10.3A

1. 7 mm **2.** 240 ft **3.** 4.812 dm
4. 100 mi **5.** 3.515 cm **6.** $2\frac{3}{8}$ yd
7. 18.84 in **8.** 12.9368 m **9.** 113.04 ft
10. 113.04 sq mm **11.** 339.6224 sq in
12. 63.585 sq km
13. Circumference = 131.88 in, Area = 1,384.74 sq in
14. Circumference = 100.48 cm, Area = 803.84 sq cm
15. 19.215 sq cm **16.** 8.86 sq in
17. 25π sq cm **18.** 2 ft **19.** 44,400,000
20. $1\frac{13}{60}$ **21.** $8\frac{2}{5}$ **22.** 3.84 **23.** 34%
24. $x = -4$ **25.** 25 **26.** 3.551×10^{-2}
27. 17.3 **28.** 2.25 **29.** 4.56 **30.** 4.8 mm
31. 12 ft **32.** 12 sq in **33.** 36 sq mm

SET 10.4A

1. 125 cubic mm **2.** 373.248 cubic in
3. 97.336 cubic mi **4.** 2.34 cubic cm
5. 25.728 cubic in **6.** $14\frac{7}{32}$ cubic ft
7. 282.6 cubic cm **8.** 9.0432 cubic in
9. 0.90432 cubic dm **10.** 4,186.6667 cubic in
11. 24.41664 cubic cm **12.** 38.77272 cubic ft
13. 960 cubic ft **14.** 8 cubic cm
15. 75.36 cubic in **16.** 463.01184 cubic cm
17. The volume is multiplied by 8
18. 4.1866667 cubic m **19.** 6,540 **20.** $1\frac{13}{20}$
21. $11\frac{1}{3}$ **22.** 9.472 **23.** 53.49% **24.** $x = 1$
25. 3.24×10^{-8} **26.** 12.7 **27.** 3.25
28. 60 mm **29.** 12 sq in **30.** 226.08 mm
31. 3.14 sq in

PRACTICE TEST 10A

1. (a) trapezoid
(b) square, rectangle, and parallelogram
(c) circle (d) triangle
2. 20.8 mm **3.** 86 in **4.** 11.9 in **5.** 11 cm
6. 45.5625 sq mi **7.** 13.44 sq cm
8. 260 sq cm **9.** 504 sq ft **10.** 18.84 sq in
11. 21.2264 sq cm **12.** 3 cubic ft
13. 395.64 cubic cm **14.** 3,052.08 cubic mm

PRACTICE TEST 10B

1. (a) rectangle and parallelogram (b) triangle
(c) trapezoid (d) circle
2. 14.4 in **3.** 56.6 cm **4.** 14 ft **5.** 16.6 cm
6. 23.04 sq cm **7.** 28 sq ft **8.** 15.5496 sq in
9. 16 sq cm **10.** 26.376 cm
11. 55.3896 sq mm **12.** 2,800 cubic in
13. 25.32096 cubic in **14.** 904.32 cubic cm

CUMULATIVE REVIEW EXERCISES FOR CHAPTERS 1 THROUGH 10

1. two million, five hundred four thousand, nine hundred seventy
2. 4,510,000 **3.** $\frac{8}{13}$ **4.** $5\frac{1}{4}$ **5.** $13\frac{13}{20}$
6. $\frac{1}{4}$
7. four and one hundred forty-five ten-thousandths
8. 12.041 **9.** \$1.63 **10.** 5.65 **11.** 14.05
12. 0.0456 **13.** 125% **14.** $\frac{7}{50}$ **15.** 89.5%
16. \$2,000 **17.** \$16,000 **18.** 30%
19. $M = 3\frac{1}{3}$ **20.** -5 **21.** -11.01
22. -738 **23.** -40
24. eight times w minus six times x **25.** $x = 10$
26. $x = 5$ **27.** 5^7 **28.** $(-2)(-2)(-2)$
29. (a) 10,000 (b) 5 **30.** (a) 225 (b) $\frac{64}{125}$
31. (a) 2.3×10^{-4} (b) 3.8×10^{9}
32. (a) 54,000 (b) 0.0023 **33.** 7.44×10^{-3}
34. 11 and -11 **35.** (a) -8 (b) 15
36. 9 and 10 **37.** (a) 4.80 (b) 12.04 (c) 2.65
38. 15 **39.** 600,000 **40.** 11.263 **41.** 3.75
42. 4.5 **43.** 14,000
44. (a) parallelogram (b) trapezoid (c) circle
(d) rectangle and parallelogram
45. 138.6 cm **46.** 12 in **47.** 9 sq mm
48. 17.28 sq ft **49.** 87.92 mm **50.** 78.5 sq in
51. 12.167 cubic cm **52.** 25.12 cubic cm

INDEX

Absolute value, 296
Addition
  carrying, 24–25
  of decimals, 157–58
  of fractions, 117–19
  mental estimation, 26, 159–60
  of mixed numbers, 118–19
  property of exponents, 403
  of signed numbers, 301–05
  of whole numbers, 23–26
Algebraic expressions, 344–48
  definition, 345
  like terms, 347
  numerical coefficient, 346
  removing parentheses, 347–48
  term, 346
  variables, 344
Amount
  mental calculation of, 220–21
  in the percent equation, 219–21, 241–43
Angle(s), 483–86
  degrees, 484
  right, 484
  straight, 484
  vertex, 483
Area, 501–08
  of a circle, 519
  of a parallelogram, 504
  of a rectangle, 503
  of a square, 503
  of a trapezoid, 504
  of a triangle, 504

Base
  mental calculation of, 236
  in the percent equation, 235–36, 241–43
Base(s)
  of a cylinder, 529
  of a parallelogram, 504
  of a trapezoid, 504
  of a triangle, 504
Borrowing in subtraction
  mixed numbers, 126
  whole numbers, 36–37

Calculator, 544–46
Center
  of a circle, 487, 517
  of a sphere, 530
Circle, 487, 517–21
  area of, 519
  center of, 487
  circumference, 518
  definition of, 487, 517
  diameter of, 517
  radius of, 487, 517
Circumference
  of a circle, 518
Common divisor, 87
Converting
  improper fractions to mixed numbers, 95
  mixed numbers to improper fractions, 96
Cross multiplication, 269
Cube
  edge of, 528
  volume of, 528
Cylinder
  base of, 529
  base radius of, 529
  height of, 529
  volume of, 529

Decimal point, 144
Decimals, 143–207
  addition of, 157–58
  common conversions from fractions, 228
  division of, 183–88
  mixed operations, 197–98
  multiplication of, 169–73
  names for, 144–46
  on the number line, 151
  and percents, 211–12
  rounding, 151–52
  subtraction of, 158–59
Decrease, percent, 249–50
Degrees, 484
Denominator, 87
Diameter
  of a circle, 517
Difference, 36, 159
Dividend, 62
Division
  checking, 62
  of decimals, 183–88
  of fractions, 109–11
  mental estimation, 65–66, 187–88
  of mixed numbers, 111
  by powers of ten, 186–87
  of signed numbers, 325–26
  of whole numbers, 61–66
Divisor, 62

Edge
  of a cube, 528
Equations, 355–83
  equivalent, 355–57
  solution, 355
  solving by addition, 363–65
  solving by multiple operations, 379–83
  solving by multiplication or division, 371–74
Equivalent equations, 355–57
  definition of, 356
Equivalent fractions, 86–89
Expanded form, 4
Exponent(s)
  addition property of, 403
  definition, 396
  finding powers, 396–98
  multiplication property of, 404
  negative property of, 406
  order of operations, 397
  scientific notation, 413–16
  subtraction property of, 405
  zero property of, 406
Exponential form, 396

Factor(s)
  in multiplication, 50
Fraction(s), 85–141
  addition of, 117–19
  conversions to decimals, 212, 228
  division of, 109–11
  equivalent, 86–89
  improper, 95
  mixed numbers, 95
  mixed operations, 131–32
  multiplication of, 101–03
  and percents, 212–13
  proper, 95
  raising to higher terms, 89
  reducing, 87–88
  subtraction of, 125–26

Geometric plane figures
  circle, 487, 517–21
  parallelogram, 486, 504
  rectangle, 486, 490, 503
  square, 486, 490, 503
  trapezoid, 487, 504
  triangle, 487, 504
Geometric solids
  cube, 528
  cylinder, 529
  rectangular, 528
  sphere, 530
Geometry, 481–541
  angles, 483–86
  area, 501–08
  figures, 486–87
  lines, 485–86
  perimeter, 488–91
  volume, 527–30

Height
of a cylinder, 529
of a parallelogram, 504
of a trapezoid, 504
of a triangle, 504
Hindu-Arabic System of Numeration, 3

Improper fraction(s), 95
and mixed numbers, 95–96
Increase, percent, 249–50
Intersecting lines, 485–86
Least common denominator, 117
Least common multiple, 117
Length measurements, 436–41, 447–50
charts for converting between metric and U.S. Customary, 447, 449
metric, 439
U.S. Customary, 437
Like terms, 347
Lines, 485–86
intersecting, 485–86
parallel, 485–86
perpendicular, 485–86
Liquid volume measurements, 455–60
charts for converting between metric and U.S. Customary, 458, 459
metric, 457
U.S. Customary, 455

Measurements, 435–80
area, 501–08, 519
length, 436–41
liquid volume, 455–60
perimeter, 488–91
metric system. *See* Metric system of measurement.
U.S. Customary System. *See* U.S. Customary System of Measurement.
volume, 527–30
weight, 465–69
Mental estimation or calculation
of the amount in the percent equation, 220–21
of the base in the percent equation, 236
of differences, 37–38, 159–60
of products, 52, 172–73
of quotients, 65–66, 187–88
of the rate in the percent equation, 228–29
of sums, 26, 159–60
Metric system of measurement, 435–80
charts for converting to U.S. Customary, 447, 449, 458, 459, 468
length, 439–41, 447–50
liquid volume, 457–60
weight, 466–69
Minuend, 36, 159
Mixed number(s), 95
addition of, 118–19
division of, 111
exponential terms, 397
and improper fractions, 95–96
multiplication of, 103
subtraction of, 125–26
Mixed operations
decimals, 197–98
exponential terms, 397
fractions, 131–32
signed numbers, 331–32
whole numbers, 75–76
Multiplication
of decimals. 169–73
of fractions, 101–03
mental estimation, 52, 172–73
of mixed numbers, 103
by powers of ten, 51–52, 170–72
property of exponents, 404
as repeated addition, 47–48
of signed numbers, 317–20
of whole numbers, 47–52
by zero, 51
Multiplicand, 50
Multiplier, 50

Negative
numbers, 294
property of exponents, 406
Number(s)
decimals. *See* Decimals.
fractions. *See* Fractions.
mixed. *See* Mixed Numbers.
negative, 294
positive, 294
prime, 117
signed. *See* Signed Numbers.
whole. *See* Whole Numbers.
Number line
adding signed numbers, 301–02
decimals, 151
signed numbers, 295
whole numbers, 15–16
Numerator, 87
Numerical coefficient, 346

Order of operations
decimals, 197–98
exponential terms, 397
fractions, 131–32
signed numbers, 331–32
whole numbers, 75–76

Parallel lines, 485–86
Parallelogram, 486, 504
area of, 504
base of, 504
height of, 504
Partial product, 50
Percents, 209–60
and decimals, 211–12
definition, 210
finding the amount, 219–21, 241–43
finding the base, 235–36, 241–43
finding the rate, 227–29, 241–43
and fractions, 212–13
percent increase and decrease, 249–50
Perimeter, 488–91
of a rectangle, 490
of a square, 490
Perpendicular lines, 485–86
Pi, (π), 518
Place value chart
decimals, 145
whole numbers, 3–4, 6
Positive numbers, 294
Powers
finding, 396–98
Powers of ten
division by, 186–87
multiplication by, 51–52, 170–72
scientific notation, 413–16
Prime number, 117
Product, 50
Proper fraction, 95
Proportions, 269–78
checking by cross-multiplication, 269
definition, 269
solving, 275–78

Quotient, 62

Radius
of a circle, 487, 517
of a cylinder, 529
of a sphere, 530
Rate
mental calculation of, 228–29
in the percent equation, 227–29, 241–43
Ratio, 262–64
colon form, 263
fraction form, 263
verbal form, 263
Reciprocal, 373
Rectangle, 486, 490, 503
area of, 503
perimeter of, 490
Rectangular solid
volume of, 528
Reducing fractions, 87–88
Remainder, 62
Right angle, 484
Rounding
decimals, 151–52
whole numbers, 15–17

Scientific notation, 413–16
Signed numbers, 293–342
absolute value, 296
addition of, 301–05
comparison of, 295
division of, 325–26
mixed operations, 331–32
multiplication of, 317–20
on the number line, 295, 301–02
subtraction of, 311–12
Solving equations, 363–83
by addition, 363–65
by multiple operations, 379–83
by multiplication or division, 371–74
Solving proportions, 275–78
Sphere
center of, 530
definition, 530
radius of, 530
volume of, 530
Square, 486, 490, 503
area of, 503
perimeter of, 490

Square root(s), 396, 421–24
Straight angle, 484
Subtraction
  borrowing, 36–37
  checking, 36
  of decimals, 158–59
  of fractions, 125–26
  mental estimation, 37–38, 159–60
  of mixed numbers, 125–26
  property of exponents, 405
  of signed numbers, 311–12
  of whole numbers, 35–38
Subtrahend, 36, 159

Term, 346
  like, 347
  numerical coefficient of, 346
Trapezoid, 487, 504
  area of, 504
  bases of, 504
  height of, 504
Triangle, 487, 504
  area of, 504
  base of, 504
  height of, 504

U.S. Customary System of Measurement, 435–80
  charts for converting to metric, 447, 449, 458, 459, 468
  length, 437–38, 447–50
  liquid volume, 455–60
  weight, 465–69

Variables, 344
Vertex, 483
Volume, 527–30
  of a cube, 528
  of a cylinder, 529
  liquid, 455–60
  of a rectangular solid, 528
  of a sphere, 530

Weight measurements, 465–69
  chart for converting between metric and U.S. Customary, 468
  metric, 466–69
  U.S. Customary, 465–69
Whole numbers, 1–84
  addition of, 23–26
  division of, 61–66
  expanded form, 4
  mixed operations, 75–76
  multiplication of, 47–52
  names for, 5–7
  on the number line, 15–16
  rounding, 15–17
  subtraction of, 35–38

Zero
  multiplication by, 51
  property of exponents, 406